The Minerals, Metals & Materials Series

The Minerals, Metals & Materials Series publications connect the global minerals, metals, and materials communities. They provide an opportunity to learn about the latest developments in the field and engage researchers, professionals, and students in discussions leading to further discovery. The series covers a full range of topics from metals to photonics and from material properties and structures to potential applications.

More information about this series at https://link.springer.com/bookseries/15240

Michael L. Free

Hydrometallurgy

Fundamentals and Applications

Second Edition

Michael L. Free
Department of Materials Science
and Engineering
University of Utah
Salt Lake City, UT, USA

ISSN 2367-1181   ISSN 2367-1696   (electronic)
The Minerals, Metals & Materials Series
ISBN 978-3-030-88089-7   ISBN 978-3-030-88087-3   (eBook)
https://doi.org/10.1007/978-3-030-88087-3

This Springer imprint is published by the registered company Springer Nature Switzerland AG
The registered company address is: Gewerbestrasse 11, 6330 Cham, Switzerland

Preface

This book provides a college-level overview of chemical processing of metals in water-based solutions. It is an expanded version of a previous textbook, "Chemical Processing and Utilization of Metals in Aqueous Media," with two editions written by the author. The information in this book is relevant to engineers using, producing, or removing metals in water. The metals can take the form of dissolved ions, mineral particles, or metal. The material in each chapter in this textbook could be expanded into individual textbooks. It is clearly not comprehensive in its coverage of relevant information. Other resources, such as the 4 volume series "Principles of Extractive Metallurgy" by Fathi Habashi, provide more details for specific metal processing methods. Thus, this text presents a condensed collection of information and analytical tools. These tools can be used to improve the efficiency and effectiveness with which metals are extracted, recovered, manufactured, and utilized in aqueous media in technically viable, reliable, environmentally responsible, and economically feasible ways.

The author expresses gratitude to family, colleagues, teachers, and students who have contributed in various ways to the completion of this text.

The author has used his best efforts to prepare this text. However, the author and publisher make no warranty of any kind, expressed or implied, with regard to the material in this book. The author and publisher shall not be liable in any event for incidental or consequential damages in connection with or arising out of the use of the material in this book.

Salt Lake City, USA   Michael L. Free

v

Contents

Chapter 1
Introduction

> *Metals form a foundation for our modern standard of living. We would live with Stone Age technology if we did not have metals.*

Key Chapter Learning Objectives and Outcomes:
Understand the importance of metals
Know some important uses of metals
Know the value of metals
Understand how metal-bearing ore bodies can form
Understand how metals are naturally cycled in different forms on earth
Understand the importance of water in metal processing
Identify basic methods needed to produce metal from metal-bearing ore

1.1   The Importance of Metals

Metals are needed for survival. Our bodies rely on metal to perform many vital functions. Iron is used to transport oxygen to cells. Calcium is needed for bones. Sodium and potassium are needed to regulate many necessary biological functions. Many other trace metals are needed for a variety of critical activities within our bodies.

Metals form a foundation for our modern standard of living. We use large quantities of metals to build transportation vehicles that range from bicycles and cars to ships and airplanes. We rely on metals for structural support for buildings, bridges, and highways. We need metals to build computers and electronic devices. Metals are also necessary to generate electricity. Metals are the foundation for the myriad of motors that mechanize our factories and homes. Metals are critical to our way of life. We would live with Stone Age technology if we did not have metals.

Countless things are made of metal or have metal in them that is critical to their function. Metals items are found in nearly every part of the world, and metals are produced in a variety of locations around the world. A summary of metals and uses is presented in Table 1.1. Worldwide production of metals is shown in Table 1.2.

© The Minerals, Metals & Materials Society 2022
M. L. Free, *Hydrometallurgy*, The Minerals, Metals & Materials Series,
https://doi.org/10.1007/978-3-030-88087-3_1

Table 1.1 Comparison of selected metals and common uses [1–3]

Metal	Symbol	Common Uses
Actinium	Ac	Thermoelectric power, source of neutrons
Aluminum	Al	Alloys, containers, aerospace structures, outdoor structures, also in ceramics as Al_2O_3
Antimony	Sb	Semiconductors, alloys, flame proofing compounds
Barium	Ba	Used mostly in nonmetallic forms such as barite and barium sulfate, specialty alloys
Beryllium	Be	Specialty alloys, electrodes, X-ray windows
Bismuth	Bi	Specialty alloys, thermocouples, fire detection, cosmetics
Cadmium	Cd	Specialty alloys, coatings, batteries, solar cells
Calcium	Ca	Used mostly in nonmetallic forms, reducing agent, deoxidizer, specialty alloys
Cerium	Ce	Used mostly in nonmetallic forms, catalysts, specialty alloys
Cesium	Cs	Catalyst, oxygen getter, atomic clocks
Chromium	Cr	Stainless steel, coatings, catalyst, nonmetallic forms in colorants, corrosion inhibitors
Cobalt	Co	High strength, high temperature alloys, magnets, nonmetallic forms as colorants
Copper	Cu	Electrical wiring, tubing, heat transfer applications, alloys such as bronze and brass
Dysprosium	Dy	Specialty magnets, specialty alloys, nuclear control applications
Erbium	Er	Nonmetallic forms as colorants
Europium	Eu	Specialty products in nonmetallic forms and alloys
Gadolinium	Gd	Specialty products in nonmetallic forms and alloys
Gallium	Ga	Solar cells, semiconductors, neutrino detectors, specialty alloys, coatings
Gold	Au	Jewelry, bullion/investment tool, decoration, specialty coatings, alloys, coins
Hafnium	Ha	Specialty alloys, nuclear control rods
Holmium	Ho	Specialty alloys, filaments, neutron absorption
Indium	In	Specialty alloys, solar cells, thermistors, solder
Iridium	Ir	Specialty alloys and coatings, jewelry, electrodes
Iron	Fe	Steels, cast iron, many alloys – most widely used, lowest cost metal

(continued)

Table 1.1 (continued)

Metal	Symbol	Common Uses
Lanthanum	La	Specialty products in nonmetallic forms and alloys
Lead	Pb	Batteries, radiation shielding, cable covering, ammunition, specialty alloys
Lithium	Li	Heat transfer and specialty alloys, batteries, nonmetallic forms in glasses, medicine
Lutetium	Lu	Catalysts, specialty alloys
Magnesium	Mg	Specialty alloys, reducing agent, pyrotechnics, many nonmetallic form uses
Manganese	Mn	Steel and specialty alloys, nonmetallic form uses in batteries, colorants, chemistry
Mercury	Hg	Chlor-alkali production, amalgams, specialty uses
Molybdenum	Mo	Primarily used for steel alloys, catalysts, heating elements

(Part II) Comparison of metals and common uses [1–3]

Metal	Symbol	Common Uses
Neodymium	Nd	Specialty magnets, lasers; also nonmetallic forms such as glass colorants.
Neptunium	Np	Neutron detection
Nickel	Ni	Many specialty alloys, batteries, tubing, coatings, magnets, catalysts
Niobium	Nb	Specialty alloys, magnets
Osmium	Os	Specialty, high cost, hard alloys,
Palladium	Pd	Alloys (decolorizes gold), catalysts, used for hydrogen gas purification
Platinum	Pt	Catalyst, jewelry, thermocouples, glass making equipment, electrochemistry
Plutonium	Pu	Nuclear weapons and fuel
Polonium	Po	Neutron source, satellite thermoelectric power
Potassium	K	Reducing agent, most uses in nonmetallic forms such as fertilizer
Praseodymium	Pr	Specialty alloys, nonmetallic forms such as glass colorant
Radium	Ra	Neutron source
Rhenium	Re	Specialty alloys, thermocouples, catalysts
Rhodium	Rh	Specialty alloys, thermocouples, catalysts, jewelry
Rubidium	Rb	Specialty alloys, catalysts, specialty glasses (nonmetallic)
Ruthenium	Ru	Specialty alloys, catalysts

(continued)

Table 1.1 (continued)

Metal	Symbol	Common Uses
Samarium	Sm	Magnets, specialty alloys, catalysts
Scandium	Sc	Specialty alloys
Silver	Ag	Jewelry, silverware, solder, batteries, anti-microbial applications, coins
Sodium	Na	Reagent in chemical reactions, batteries, used mostly in nonmetallic forms
Strontium	Sr	Specialty alloys, zinc refining, fireworks, glass colorant
Tantalum	Ta	Specialty alloys, capacitors, surgical appliances
Technetium	Tc	Radioactive tracers,
Thallium	Tl	Specialty alloys, photovoltaic devices
Thorium	Th	Specialty alloys, portable gas light mantles, catalysts
Thullium	Tm	Specialty alloys, radiation source if previously exposed
Tin	Sn	Specialty alloys, solder, coatings, semiconductors
Titanium	Ti	Aerospace alloys, corrosion resistant alloys, implants, paint pigment (TiO_2)
Tungsten	W	Filaments, alloys, tool steels and hard materials
Turbium	Tb	Dopant in solid-state devices, specialty alloys
Uranium	U	Nuclear fuel, nuclear weapons
Vanadium	V	Specialty steels, catalysts, nuclear applications
Ytterbium	Yb	Specialty alloys, radiation source
Yttrium	Y	Alloys, catalysts, nonmetallic colorant applications
Zinc	Zn	Galvanized metal coatings, alloys such as brass, solder, batteries, coins
Zirconium	Zr	Nuclear fuel canisters, corrosion resistant tubing, explosive primers

Table 1.2 World metal production, price and value estimates based on data from USGS mineral commodity summaries 2020 [1]

Metal	Estimated worldwide production 2020 (metric tons)	Estimated metal price (2020) (US $/kg)	Estimated total value (price times production in US$)
Aluminum	65,200,000	$1.96	$127,930,509,932
Antimony	153,000	$8.77	$1,342,489,914
Beryllium	240	$620.00	$148,800,000
Bismuth	17,000	$5.97	$101,567,495
Cadmium	23,000	$2.30	$52,900,000
Chromium	40,000	$7.90	$316,000,000
Cobalt	140,000	$30.86	$4,321,082,916
Copper	25,000,000	$5.95	$148,812,804,515
Gallium	300	$570.00	$171,000,000
Gold	3200	$57,033.89	$182,508,432,726
Indium	900	$400.00	$360,000,000
Iron/steel	1,800,000,000	$1.26	$2,268,000,000,000
Lead	4,400,000	$1.98	$8,710,950,418
Lithium	82,000	$8.00	$656,000,000
Magnesium	1,000,000	$5.51	$5,511,585,352
Mercury	1930	$100.00	$193,000,000
Molybdenum	300,000	$20.00	$6,000,000,000
Nickel	2,500,000	$14.11	$35,274,146,255
Niobium	78,000	$24.00	$1,872,000,000
Palladium	210	$67,674.47	$14,211,638,781
Platinum	170	$27,392.05	$4,656,648,081
Rhenium	53	$1000.00	$53,000,000
Silver	25,000	$644.52	$16,112,969,140
Tin	270,000	$17.42	$4,702,484,623
Titanium	210,000	$6.90	$1,449,000,000

The total annual value of raw metal produced in 2020 was estimated to be more than 2.8 trillion US dollars. This is only the value of the raw metal. The value of finished products is much higher. The value to the world economy is several times that amount when the value of the final products that use metal is considered.

Metals undergo continual cyclical processing on our planet as shown in Fig. 1.1. Metal comes from the earth. Metals are continuously transported and transformed by geological processes.

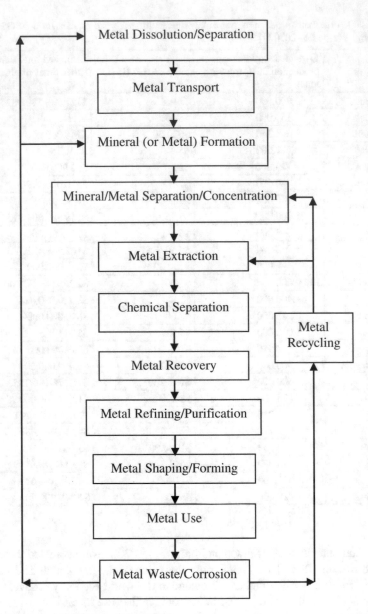

Fig. 1.1 Overview of the general metal cycle

Metals usually originate as minerals that are discovered, mined, and transformed into metal. All metal in use today is undergoing a very gradual or rapid process of corrosion or degradation. The process of corrosion is essentially a chemical transformation from metallic form to mineral or ion form. Atmospheric corrosion of metals leads generally to the formation of metal oxide minerals. The oxygen in the atmosphere and water causes the oxidation. Oxide mineral products of corrosion are often the same as the minerals in ores. Thus, metals are found in various forms depending on their position in the metal cycle. However, the time between metal cycle events can be enormous.

> *Metals usually originate as minerals that are discovered, mined, and transformed into metal.*

The processing paths used to produce individual metals are often very different. Many processing methods involve water or water-containing (aqueous) media. Thus, the term hydrometallurgy is often used to describe this topic. This textbook covers fundamental chemical metallurgy principles. It also presents many important methods of processing, utilizing and evaluating metals and metal processes in aqueous media. This book includes many associated topics such as metal extraction, electrodeposition, energy storage and generation, electroforming, environmental issues, economics, and statistics.

1.2 Mineral Deposition

Metals generally originate in the earth's crust as metal oxides, sulfides, and other minerals. Many metals such as aluminum, iron, calcium, sodium, magnesium, potassium, and titanium are abundant as shown in Table 1.3. In some locations mineralization processes have concentrated specific metals. Metals such as manganese, barium, strontium, zirconium, vanadium, chromium, nickel, copper, cobalt, lead, uranium, tin, tungsten, mercury, silver, gold, and platinum are generally scarce. Economic extraction of scarce metals requires a suitable ore deposit.

Ores are natural materials containing a concentrated resource. An ore deposit or ore body contains a large volume of ore. Some rare metals and minerals must be concentrated by a factor of 1000 or more above their average natural abundance to form an economically viable ore body as indicated in Table 1.4 (see Appendix A for atomic weights to convert to an atomic basis).

The method of concentration within the earth's crust varies widely. It is dependent on the metal and associated mineral(s) as shown in Table 1.5. Some metals such as chromium are concentrated by precipitation or crystallization in magma. Precipitation is dependent on solubility, which is dependent on temperature. As the magma cools different local temperature zones are created. Each temperature zone may correspond to the formation of a specific mineral (or set of similar minerals). Thus, temperature zones can create mineral zones or veins.

Table 1.3 Abundance of elements in the earth's crust [4]

Element	Abundance (%)	Element	Abundance (%)
Oxygen	46.4	Vanadium	0.014
Silicon	28.2	Chromium	0.010
Aluminum	8.2	Nickel	0.0075
Iron	5.6	Zinc	0.0070
Calcium	4.1	Copper	0.0055
Sodium	2.4	Cobalt	0.0025
Magnesium	2.3	Lead	0.0013
Potassium	2.1	Uranium	0.00027
Titanium	0.57	Tin	0.00020
Manganese	0.095	Tungsten	0.00015
Barium	0.043	Mercury	0.000008
Strontium	0.038	Silver	0.000007
Rare earths	0.023	Gold	<0.000005
Zirconium	0.017	Platinum metals	<0.000005

Table 1.4 Enrichment factors necessary for ore-body formation [5]

Approximate enrichment factor from natural occurrence level to form economical ore body	
Aluminum	4
Chromium	3000
Cobalt	2000
Copper	140
Gold	2000
Iron	5
Lead	2000
Manganese	380
Molybdenum	1700
Nickel	175
Silver	1500
Tin	1000
Titanium	7
Tungsten	6500
Uranium	500
Vanadium	160
Zinc	350

Ores are natural materials containing a concentrated resource.
An ore body is a local region in the earth's crust containing desired minerals in concentrations sufficient for commercial extraction.

Table 1.5 Methods of metal/metal-bearing mineral deposition [5]

	Mag.	Sub V.	Hydro.	Weath.	Sed.	Plac.	Metas.
Aluminum				X			
Chromium	X			X			
Cobalt			X				
Copper	X	X		X	X		X
Gold			X			X	X
Iron	X		X	X	X		X
Lead		X	X		X		X
Manganese		X		X	X		X
Molybdenum			X				X
Nickel	X		X	X			
Silver		X	X				X
Tin		X			X		X
Titanium	X					X	
Tungsten			X			X	X
Uranium			X		X		
Zinc		X	X		X		X

Mag. = magma crystallization
Sub V. = submarine volcanic exhalative processes
Hydro. = hydrothermal solution deposition
Weath. = weathering
Sed. = sedimentation
Plac. = placer deposition
Metas. = contact metasomatism

Mineral precipitation or crystallization is similar to the creation of frost. Frost forms because the solubility of water in air decreases significantly at 0 °C. As water is forced out of the gas phase below its freezing point it forms frost. In frost formation a pure component leaves a mixture (air and water vapor) to form a pure deposit. The pure deposit of frozen water is often localized. Frost often forms locally on windshields because they are colder than surroundings. Similarly, minerals form on locally colder surfaces. Figure 1.2 illustrates crystallization from magma and air. Such crystallization on local surfaces concentrates products.

Ore-grade aluminum deposits are formed by the dissolution and removal of contaminants such as silicates through extensive weathering. Weathering is most rapid in tropical environments. Correspondingly, most commercial aluminum deposits are in tropical regions.

Other metals such as silver and tin are concentrated hydrothermally. They are first dissolved in hydrothermal solutions within the earth's crust. Next, they are transported to regions where solution conditions change. Solution conditions can change due to changes in geological formations. Disruptions in rock layers and related aquifers can force water upward to cooler zones. Changes such as temperature cause changes in solubilities. If a dissolved metal is no longer soluble, it

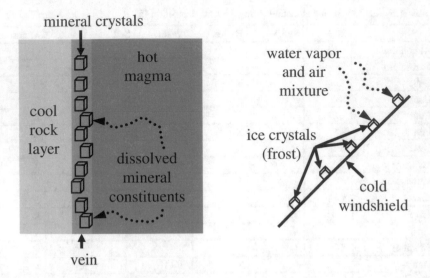

Fig. 1.2 Illustrations of (left-hand side) mineral crystal formation near a hot magma source from dissolved mineral constituents and (right-hand side) ice crystal formation from a water vapor and air mixture on a cold windshield. The ice is pure H_2O in a more locally concentrated form than the air–water vapor mixture from which it came. Similarly, mineral crystals contain concentrated metal

precipitates. These precipitates are generally minerals. Precipitation on a large scale can create an ore body. An ore body is a local region in the earth's crust containing desired minerals in concentrations sufficient for commercial extraction.

Metals such as gold can be transported as suspended particles. Particles settle to form "placer" deposits where water flows slowly. Thus, some gold ores are placer deposits.

Metals have some solubility in water as ions. Metal ions can be transported in water. Temperature changes in subsurface water alters solubilities of metal ions. Consequently, hydrothermal ore formation processes are very important. In fact, four deposition methods in Table 1.5 involve water. Hydrothermal processing is the most common ore body formation method in Table 1.5. Correspondingly, metals can

> *Metal ore body formation is the result of large-scale concentrated mineral deposits in specific areas.*

often be deposited from solution. Conversely, they can also be dissolved and processed using appropriate solutions. However, extreme processing conditions are often required for metal extraction.

Metals such as platinum and gold require highly oxidizing environments as well as complexing agents to make their dissolution in water possible. Other metals such as magnesium tend to dissolve easily.

Aqueous processing of metals has been performed for several centuries. Copper was recovered from acidic mine waters using metallic iron as early as the sixteenth century in Europe [6]. Evidence of metal dissolution due to bacterial activity in Rio Tinto, Spain was reported as early as 1670 [7]. The dissolution of precious metals by cyanide has been known since 1783 [8]. However, widespread application of aqueous metal processing is relatively recent.

Hydrometallurgical processing of metals is expanding. Worldwide hydrometallurgical copper production has generally been increasing [9]. However, because copper oxide deposits, which are generally processed by hydrometallurgy, are diminishing. Thus, further increases are likely to require more widespread hydrometallurgical processing of sulfides. New hydrometallurgical processes for copper sulfides may allow the increase to continue.

In addition to hydrometallurgical extraction, there is considerable hydrometallurgical refining. Most of the world's zinc, uranium, silver, gold, and copper is purified by aqueous means.

Historical events associated with aqueous metal processing and utilization are listed in Table 1.6. There are many new hydrometallurgical technologies that have been developed. Some of the new technologies are likely to be more widely used in the future. However, many of the core commercial processes were developed in past decades.

Table 1.6 Selected historical events involving chemical processing of metals in aqueous media

Event	Ref	Year	Entity or place
Separation of silver from gold leaf	[10]	1556	Europe
Electroless deposition of Cu	[11]	1670	RioTinto, Spain
Discovery of precious metal cyanidation	[8]	1783	
Invention of electrochemical battery	[12]	1800	Pavia, Italy
Precipitation of copper using iron	[11]	1820	Strafford, VT
Invention of electroforming	[13]	1838	St. Pet., Russia
Invention of aqueous-based fuel cell	[14]	1839	Swansea, G.B
Electrolytic reduction of copper	[12]	1869	Swansea
Cation exchange on zeolites	[10]	1876	(Lemberg)
Cyanidation of precious metals patented	[10]	1887	(MacArthur)
Bayer caustic leach process for aluminum	[10]	1892	(Bayer)
Alkaline rechargeable battery patented	[15]	1916	USA
Large scale electroforming of reflectors	[16]	1924	Newark, NJ
Large scale electrowinning	[10]	1926	Arizona
Synthesis of anion exchange resins	[10]	1935	(Adams, Holmes)
Recovery of uranium by ion exchange	[10]	1948	(Bross)
Ammoniacal leaching/gaseous red. of Ni	[11]	1954	(Sherrit-Gordon)
Large-scale uranium solvent extraction	[10]	1955	Shiprock, NM
Pressure sulfuric acid leach of oxide Ni ore	[10]	1958	Mao Bay, Cuba
Large-Scale bacterial dump leaching	[10]	1960	Bingham, UT

(continued)

Table 1.6 (continued)

Event	Ref	Year	Entity or place
Aqueous fuel cells U. S. Space Program	[17]	1965	USA
Large-scale Cu solvent extract,/electrow,	[11]	1968	Bluebird, AZ
Use of activated carbon for gold processing	a	1970	Cortez Gold
Pachuca leach of Cu ore with sulfuric acid	[10]	1974	(Nchanga LTD)
Commercializ. of press, oxid. of zinc sulf.	[10]	1981	(Cominco)
Commercialization biooxid. gold-ore conc	[7]	1984	(Gencor)
Demonstration of 11 MW aqueous fuel cell	[15]	1991	Japan
Pressure oxidation of copper concentrate	[19]	2002	Bagdad, AZ

a (https://p2infohouse.org/ref/33/32771.htm)

1.3   Importance of Water

Water makes aqueous processing of metals possible. Water is an unusual substance from a chemical perspective. Most substances with similar molecular weights, such as methane and ammonia, are gases at room temperature. However, water is a liquid at room temperature. Water's unusual properties are due primarily to hydrogen bonding effects. These effects are related to the tendency of hydrogen to donate electrons and the tendency of oxygen to accept electrons.

> *Water can play an important role in ore body formation and metal processing.*

Consequently, water molecules tend to b accordingly. The oxygen in water molecules has more electron abundance and maintains a slightly negative charge. Hydrogen maintains a slightly positive charge. Electrostatic attraction helps the hydrogen atoms associate with the oxygen atoms of adjacent molecules. The networked structure of associated molecules is nearly tetrahedral (4.4 nearest neighbors [20]). The resulting structure has open pores large enough for water molecules. The open pores accommodate ion diffusion. In fact, the structure is so porous, that water is compressible. High pressures can create a solid (ice VII) with a density 65% greater than that of liquid water [21].

Another important property of water is its role in electrochemical reactions. Molecules of water can break apart into oxygen and hydrogen. This molecular breakage is sometimes referred to as hydrolysis. However, the term hydrolysis is more commonly used to describe the decomposition of water by association with another compound. Water splitting occurs at either very high or very low oxidation potentials. At high potentials oxygen gas and hydrogen ions form. At low potentials the products are hydrogen gas and hydroxide ions. The associated chemistry is presented in Eqs. 1.1 and 1.2. Electrons are liberated by electrolysis at high oxidation potentials. Electrolysis is the removal or acquisition of electrons caused by the application of an electrochemical potential or voltage. In contrast, electrons are consumed at low oxidation potentials.

$$2H_2O \leftrightarrow 4H^+ + 4e^- + O_2 \tag{1.1}$$

$$2H_2O + 2e^- \leftrightarrow H_2 + 2OH^- \tag{1.2}$$

Equation 1.1 or its chemical equivalent involving hydroxide rather than hydrogen, $2H_2O + O_2 + 4e^- \leftrightarrow 4OH^-$, is crucial in many metal oxidation/reduction reactions. It is fundamental to atmospheric corrosion. The reverse reactions are also critical. The reaction $4H^+ + 4e^- + O_2 \leftrightarrow 2H_2O\ H^+ + 4e^- + O_2 \leftrightarrow 2H_2O\ H^+ + 4e^- + O_2 \leftrightarrow 2H_2O\ H^+ + 4e^- + O_2 \leftrightarrow 2H_2O$ is necessary for metabolism in oxygen-breathing organisms.

Water molecule structure is important to chemical reaction events. Cement and plaster harden in response to events such as hydration. Hydration is the process of aquiring water. Hydration increases effective ion size and decreases mobility. All ions hydrate to varying degrees. In other words, multiple water molecules partially bond with each ion. In addition, many interfacial phenomena are altered by hydration.

1.4   Aqueous Processing and Utilization of Metals

Metal atoms can be found in a wide variety of forms. They can exist as dissolved ions and ion complexes. They are also found as pure metals and metal-bearing minerals. Minerals can be transformed from minerals to metallic metals by hydrometallurgical processing. These transformations occur through dissolution, concentration, and recovery processes.

Metals must first be obtained from the earth. Consequently, metal extraction is discussed in the first sections of this text. After metals are extracted, purified, and recovered, they can be utilized. Thus, metal utilization is discussed after extraction, purification, and recovery.

Metal atoms are metallically bonded in pure metals. They have no net charge in the metal lattice. Metal with no net charge is sometimes written with a superscript "o" (M^o) (often, no "o" is written). Metal atom removal from a solid piece of metal nearly always requires the loss of at least one electron ($M^o \leftrightarrow M^+ + e^-$). Electron removal converts a metal atom to an oxidized state. An oxidized metal is deficient in electrons. Oxidized metal can be found as ions or in compounds. Examples of oxidized metal ions are Fe^{2+} and Fe^{3+}. Metal ions, as well as other types of molecues with a net charge, are generally soluble in water. Metal-bearing compounds that are ions include $Fe(OH)^+$, $CuCl_3^-$, and WO_4^{3-}. Examples of oxidized metal in compounds are FeO, Fe_3O_4, and Fe_2O_3. Ions can combine with other ions to form ionic or nonionic compounds. Examples include $Fe^{3+} + SO_4^{2-} \leftrightarrow FeSO_4^+$ and $Fe^{2+} + SO_4^{2-} \leftrightarrow FeSO_4$. Oxidized metal may also remain as metal ions.

> *Minerals can be transformed from minerals to metallic metals through hydrometallurgical dissolution, concentration, and recovery processes.*

In contrast, metal can be removed from some compounds without electron loss. An example of this is $2H^+ + CuO \leftrightarrow Cu^{2+} + H_2O$. In this example copper is oxidized to its divalent (+2) state in CuO and in Cu^{2+}. No electron transfer is needed for this exchange reaction. Metals found as metal-bearing minerals are already oxidized. Consequently, metal removal from a mineral may not require electron removal. Some metals in minerals require additional oxidation to be extracted.

> *Chemical reactions are needed to extract metals from minerals.*

However, in order to remove or extract a metal from a mineral a chemical bond must be severed. Thus, chemical reactions are needed to separate metals from minerals. In order to break a chemical bond, a more favorable alternative must be provided. In the case of the copper oxide reaction with hydrogen shown previously, the oxygen is more satisfied in the water molecule product than in the copper oxide reactant. Thus, the copper-oxygen bond was severed to form a more favorable set of products.

Reactivity of metals is based on atomic structure. Atomic structure is a function of electron orbitals and the number of protons and neutrons. The basic structure is most commonly described using a periodic table. An example of a periodic table is

> *Reactivity of metals is based on atomic structure.*

shown in Fig. 1.3. Metal elements on the far left, such as sodium and potassium, are very reactive. Their high reactivity limits their use in metallic form to minor, specialty applications. Metals in the second column from the left, which includes magnesium, are also very reactive, but they can be used in metallic form. Metals in other columns have a variety of properties based on their atomic structure. Metals in the lower sections are generally more dense.

One key to hydrometallurgy is contact between the metal and water. This can occur in a variety of ways. Metals and metal-bearing minerals can contact water in chemical reactors, lakes, rivers, oceans, ponds, etc. They can also contact water through condensation, rain, or water vapor. Contact can occur above or below ground level.

After contact, a reaction is needed to solubilize the metal. Reactions involving electron transfer require an electron donor and an acceptor. Other chemical reactions require a reactive species. The reactive species must exchange with or transform the nonmetal part of the mineral. The reaction converts the metal to an ion form to make it soluble. After the metal is solubilized it can be processed in solution. Dissolved metals are then concentrated, purified, or removed. These aqueous processing steps for metals are presented in Fig. 1.4.

Concentration processes include liquid–liquid or solvent extraction, ion exchange, carbon adsorption, reverse osmosis, and precipitation. In most industrial processes unwanted ions are dissolved along with the desired ions. Consequently, the desired ions must be separated from unwanted ions using a concentration

Group Period	1	2	3	4	5	6	7	8	9	10	11	12	13	14	15	16	17	18
1	1 H																	2 He
2	3 Li	4 Be											5 B	6 C	7 N	8 O	9 F	10 Ne
3	11 Na	12 Mg											13 Al	14 Si	15 P	16 S	17 Cl	18 Ar
4	19 K	20 Ca	21 Sc	22 Ti	23 V	24 Cr	25 Mn	26 Fe	27 Co	28 Ni	29 Cu	30 Zn	31 Ga	32 Ge	33 As	34 Se	35 Br	36 Kr
5	37 Rb	38 Sr	39 Y	40 Zr	41 Nb	42 Mo	43 Tc	44 Ru	45 Rh	46 Pd	47 Ag	48 Cd	49 In	50 Sn	51 Sb	52 Te	53 I	54 Xe
6	55 Cs	56 Ba	57- 71	72 Hf	73 Ta	74 W	75 Re	76 Os	77 Ir	78 Pt	79 Au	80 Hg	81 Tl	82 Pb	83 Bi	84 Po	85 At	86 Rn
7	87 Fr	88 Ra	89- 103	104 Rf	105 Db	106 Sg	107 Bh	108 Hs	109 Mt	110 Ds	111 Rg	112 Cn	113 Uut	114 Fl	115 Uup	116 Lv	117 Uus	118 Uuo

Lanthanoids	57 La	58 Ce	59 Pr	60 Nd	61 Pm	62 Sm	63 Eu	64 Gd	65 Tb	66 Dy	67 Ho	68 Er	69 Tm	70 Yb	71 Lu
Actinoids	89 Ac	90 Th	91 Pa	92 U	93 Np	94 Pu	95 Am	96 Cm	97 Bk	98 Cf	99 Es	100 Fm	101 Md	102 No	103 Lr

Fig. 1.3 Periodic table of the elements

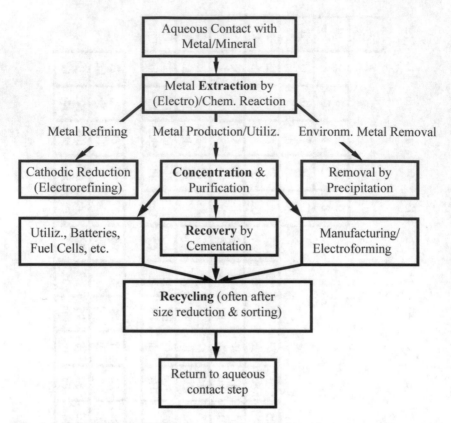

Fig. 1.4 Flow sheet possibilities for metal dissolution/extraction, concentration, recovery, production, utilization, and processing in aqueous media. (Items in bold font are the main processing steps.)

process. Concentration processes can also involve the separation of solvent from desired ions. Solvent or solution extraction is a common industrial concentration process. Concentration processes are also referred to as purification processes.

Solvent extraction consists of contacting the metal-bearing aqueous solution with an organic compound. The organic compound is dissolved in a nonaqueous medium. The nonaqueous medium is vigorously mixed with the aqueous medium. Vigorous mixing provides intimate contact between the media. Mixing allows the metal ions to be transferred readily from aqueous to organic media.

> *Concentration processes include liquid-liquid or solvent extraction, ion exchange, carbon adsorption, reverse osmosis, and precipitation.*

The solvent extraction process is designed to be selective. Consequently, unwanted entities remain in the aqueous phase. After extraction of the metal into the organic or nonaqueous phase, the nonaqueous medium is separated from the aqueous medium by settling. The extracted metal is then stripped from the nonaqueous phase. Stripping is performed in a smaller volume to concentrate the ions. Stripping solutions are maintained with low levels of impurities.

Ion exchange is essentially the same as solvent extraction, except that the active organic compound is anchored to a stationary medium such as resin beads, rather than dissolved in a nonaqueous medium. Carbon adsorption is often treated as a specialized form of ion exchange. Reverse osmosis is a process of molecular-level filtration.

Precipitation consists of adding chemical compounds that react with dissolved entities. The resulting reactions result in precipitate formation. Precipitates may consist of waste or valuable materials. Precipitates are separated from solution by solid–liquid separation processes.

After metal has been concentrated in aqueous media it can be recovered as a pure metal. Often, pure metal is the desired form of recovered metal. However, it can also be manufactured into products such as coatings and electroformed objects. Additionally, it can be utilized in devices such as batteries, electrodes, and sensors. In some applications such as batteries and fuel cells, metals and/or metal compounds are used as electrodes in aqueous media.

Recovery of metals from concentrated solutions is made by electrochemical reduction. In electrolytic reduction, the dissolved metal ions are plated on a cathode. Electrolytic reduction results from an impressed or applied voltage. The voltage is applied between anodic (anode) and cathodic (cathode) electrodes. In other forms of reduction, the electrons necessary to reduce the dissolved metal to its elemental or metallic state are provided by a reductant. The reductant can be supplied through solution species (noncontact reduction) or surface compounds (contact reduction).

> *Recovery of metals from concentrated solutions is made by electrochemical reduction.*

An example of extraction, concentration, and recovery for copper is presented in Fig. 1.5. The extraction step for most metals involves acid, oxidizing agents, and/or complexing agents. The concentration step is often solvent extraction or ion exchange, which releases acid or other ions back into solution. The recovery step provides electrons needed to reduce the metal to its metallic state.

Refining or additional purification of metals is often performed electrolytically. This technique, referred to as electrorefining, is utilized to purify metals. High purity metal can be obtained by electrorefining. Electrorefining is the electrolytic process of refining an impure metal. It occurs by electrically forcing metal dissolution and redeposition. Dissolution occurs at the anode and redeposition occurs on the cathode. Often the contaminants are not dissolved and remain as fine particles. Less reactive metals tend not to dissolve. More rective metals tend to remain solubilized. Thus, the final product is purer than the initial material.

Removal of impurity metals is accomplished by many techniques. Most contaminant metal removal is performed by precipitation. Metals can also be removed by ion exchange processes. Alternatively, carbon adorption is commonly used for metal removal. Reverse osmosis or membrane filtration is another important removal technique.

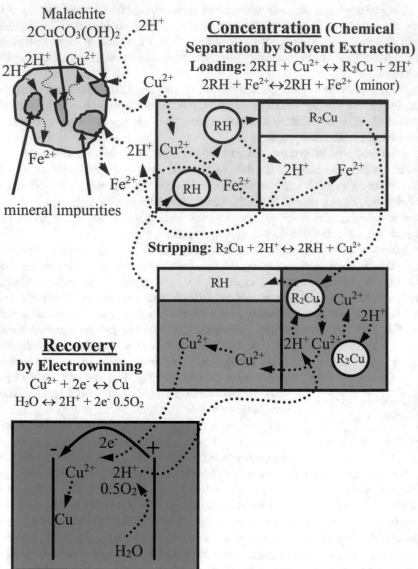

Extraction (Leaching from ore)
$$2CuCO_3(OH)_2 + 4H^+ \leftrightarrow 2Cu^{2+} + 3H_2O + CO_2$$

Malachite
$2CuCO_3(OH)_2$ $2H^+$

Concentration (Chemical **Separation by Solvent Extraction**)

Loading: $2RH + Cu^{2+} \leftrightarrow R_2Cu + 2H^+$
$2RH + Fe^{2+} \leftrightarrow 2RH + Fe^{2+}$ (minor)

$2H^+$ Cu^{2+}
$2H^+$

Cu^{2+}

R_2Cu

RH

$2H^+$ Cu^{2+}

$2H^+$ Fe^{2+}

RH Fe^{2+}

Fe^{2+}

Fe^{2+}

mineral impurities

Stripping: $R_2Cu + 2H^+ \leftrightarrow 2RH + Cu^{2+}$

RH

R_2Cu Cu^{2+}

$2H^+$

Recovery
by Electrowinning
$$Cu^{2+} + 2e^- \leftrightarrow Cu$$
$$H_2O \leftrightarrow 2H^+ + 2e^- \; 0.5O_2$$

Cu^{2+} Cu^{2+}

Cu^{2+}

$2H^+$ Cu^{2+}

R_2Cu

$2e^-$

$-$ $+$

Cu^{2+} $2H^+$

$0.5O_2$

Cu

H_2O

Fig. 1.5 Metal extraction, concentration, and recovery example for copper from mineral to metal by hydrometallurgical processing

After metals have been utilized in the form of consumer products such as wires, batteries, vehicle parts, electronic devices, etc., the parts can be recycled by the same means that allowed their creation, although recycling of consumer products generally requires sorting and size reduction prior to dissolution as indicated in Fig. 1.4.

1.5   Overview of Fundamentals and Applications

The main steps in hydrometallurgical processing are extraction, concentration, and recovery. A discussion of specific applications for these steps has been provided. Next, a discussion of underlying fundamentals will be provided. The fundamentals of hydrometallugy are related to the application.

A determination of whether or not reactions occur spontaneously is made using thermodynamics. Effectively, thermodynamics provides answers regarding chemical stability and reaction viability. Thermodynamic calculations can determine whether or not the addition of a specified amount of a compound will dissolve or precipitate. Such calculations can be used to determine whether or not leaching of a specific mineral is favorable under a specified condition. Thermodyanmics can be used to determine the voltage needed for recovery of a desired metal under specified conditions. The heat generated by a reaction can also be determined using thermodynamics. Thus, thermodynamics is a critical tool in hydrometallurgical processing. Thermodynamics is discussed in Chap. 2. Chapter 3 discusses speciation and equilibria, which are subsets of thermodynamics.

> *Thermodynamics provides quantitative answers regarding chemical stability and reaction viability.*

The rate of reactions is important to hydrometallurgy. If a reaction occurs rapidly it may lead to an explosion. If a reaction takes 10,000 years to complete it may be of no commercial interest. Reaction kinetics is an evaluation of the rates of reactions. Factors that affect rates such as temperature, concentration, and area are considered in reaction kinetics calculations. Overall reaction rates are often dominated by mass transport of reacting species. Species must be transported in solution by processes such as diffusion and convection in order to arrive at a reacting surface. Consequently, reaction kinetics is coupled with mass transport. Thus, rate processes, such as reaction kinetics and mass transport are discussed together in Chap. 4.

> *Rates of reactions are critical to safety and economics in hydrometallurgy.*

Metal extraction, which has already been discussed in the hydrometallurgical context, is presented in Chap. 5. Concentration and recovery, which have also been discussed, are presented in Chaps. 6 and 7, respectively.

> *Metals are used in a wide variety of products. Their response to their environment, such as corrosion, often determines their useful life.*

Utilization of metals in aqueous media, which is important to many areas of technology, is discussed in Chap. 8. Metals are commonly exposed to outdoor environments that include water. Exposure of metals to water often leads to corrosion. In batteries, energy is harvested from metal corrosion reactions. Correspondingly, the utilization chapter discusses these topics as well as some niche metal manufacturing in aqueous media.

Hydrometallurgical processing has important environmental ramifications. In order for hydrometallurgical processing to be performed in a sustainable way, it must be performed in an envi-

> *Appropriate responses to environmental issues are critical to hydrometallurgy.*

ronmentally sensitive way. Chapter 9 discusses environmental regulations and issues related to hydrometallurgy.

> *Hydrometallurgical processes must be technically and economically viable.*

The production of metal from minerals in large commercial settings requires appropriate equipment and design for organized treatment of material. Chapter 10 discusses principles associated with designing hydrometallurgical processes.

Commercial hydrometallurgical processing must be performed in an economically viable way. Engineers need to understand how their technical solutions impact the oveall process economics. Engineers are often required to do preliminary economic assessments for technical solutions they offer to corporate management. Solutions to engineering challenges must be economically viable as well as technically sound in order to have them applied on a large scale. Chapter 11 discusses important economic principles that can be applied to hydrometallurgical processes.

All hydrometallurgical engineers are faced with data collection and analysis. Statistics provides robust and trusted methods for data collection and analysis. Chapter 12 discusses statistical principles and methods that have application in hydrometallurgical data collection and analysis.

> *Statistical analyses are important tools for evaluating data and making decisions.*

References

1. USGS mineral commodity summaries (2012)
2. http://www.webelements.com/
3. http://www.chemicool.com/
4. B.A. Wills, *Mineral Processing Technology*, 3rd edn. (Pergamon Press, Oxford, 1985)

5. W. Dennen, M. Resources, *Geology Exploration, and Development* (Taylor & Francis, New York, 1989)
6. R.H. Lamborn, *The Metallurgy of Copper* (Lockwood and Company, London, 1875)
7. C.L. Brierly, Bacterial leaching. CRC Crit. Rev. Microbiol. **6**(3), 207 (1978)
8. F. Habashi, *Principles of Extractive Metallurgy*, vol. 2 (Gordon and Breach, New York, 1970)
9. Industry Newswatch, Copper production reaches 13.2 Mt during 2000. Min. Eng. **53**(5), 18 (2001)
10. J.O'M. Bockris, A.K.N. Reddy, *Modern Electrochemistry* (Plenum Press, New York, 1970)
11. L. Pauling, *General Chemistry* (Dover Publications Inc., New York, 1970)
12. M.E. Wadsworth, Hydrometallurgy course notes (1995)
13. N.A. Arbiter, A.W. Fletcher, Copper hydrometallurgy—evolution and milestones. Min. Eng. **46**(2), 118 (1994)
14. C.A Vincent, B. Scrosati, *Modern Batteries* (Wiley, London, 1997), p. 1
15. S.G. Bart, Historical reflections on electroforming, in *Electroforming—Applications, Uses, and Properties of Electroformed Metals* (ASTM Publication No. 318, ASTM, Philadelphia, 1962), p. 172
16. J.A.A. Ketelaar, History, Chapter 1 in *Fuel Cell Systems* ed. by L.J.M.J. Blomen, M.N. Mugerwa (Plenum Press, New York, 1993), p. 20
17. C.A Vincent, B. Scrosati, *Modern Batteries* (Wiley, London, 1997), p. 162
18. S.G. Bart, Historical reflections on electroforming, in *Electroforming—Applications, Uses, and Properties of Electroformed Metals* (ASTM Publication No. 318, ASTM, Philadelphia, 1962), p. 180
19. C.C. Morrill, *Proceedings of the Annual Power Sources Conference*, vol. 19, No. 32 (1965)
20. R. Anahara, Research, development and demonstration of phosphoric acid fuel cell systems, Chapter 8 in *Fuel Cell Systems*, ed. by L.J.M.J. Blomen, M.N. Mugerwa (Plenum Press, New York, 1993), p. 20
21. R.A. Carter, Pressure leach plant shows potential. Eng. Min. J. **204**(6), 26–28 (2003)

Problems

(1-1) What is the metal cycle and how is it related to hydrometallurgy?

(1-2) Name three methods of ore body formation that involve hydrometallurgy.

(1-3) Discuss briefly the importance of water in hydrometallurgy.

(1-4) List the main metal processing steps.

(1-5) List five metals that have large production value.

Chapter 2
Chemical Fundamentals
of Hydrometallurgy

> *Chemical reactions provide the foundation for all hydrometallurgical processing.*

Key Chapter Learning Objectives and Outcomes:

Identify types of hydrometallurgical reactions
Formulate relevant hydrometallurgical reactions
Understand free energy and its relationship with equilibria
Determine reaction viability
Calculate standard and nonstandard free energies of reactions
Understand the role of species activities in determining equilibria
Calculate species activity coefficients for relevant species
Calculate species equilibria for hydrometallurgical reactions
Determine electrochemical potentials of electrochemical reactions

2.1 General Reactions

Metals and minerals interact with other species in aqueous media. Reactions involving metals and other species can be classified. Reaction classifications include precipitation, hydrolysis, electrochemical, conversion, complexation, solvation, and ionic disassociation. Many reactions can be classified by more than one type of reaction. There are some interelationships amongst these groups. Examples of seven basic types of hydrometallurgical reactions are given as follows:

Precipitation

$$Ag^+ + Cl^- \leftrightarrow AgCl \qquad (2.1)$$

$$Fe^{2+} + 2OH^- \leftrightarrow Fe(OH)_2 \qquad (2.2)$$

© The Minerals, Metals & Materials Society 2022
M. L. Free, *Hydrometallurgy*, The Minerals, Metals & Materials Series,
https://doi.org/10.1007/978-3-030-88087-3_2

Hydrolysis

$$Fe^{3+} + H_2O \leftrightarrow Fe(OH)^{2+} + H^+ \tag{2.3}$$

$$PCl_3 + 3H_2O \leftrightarrow P(OH)_3 + 3HCl \tag{2.4}$$

Electrochemical

$$Fe^{3+} + e^- \leftrightarrow Fe^{2+} \tag{2.5}$$

$$O_2 + 4H^+ + 4e^- \leftrightarrow 2H_2O \tag{2.6}$$

Conversion

$$CaCO_3 + 2HCl \leftrightarrow CaCl_2 + CO_2 + H_2O \tag{2.7}$$

$$CuO + H_2SO_4 \leftrightarrow CuSO_4 + H_2O \tag{2.8}$$

Complexation

$$Au^+ + 2CN^- \leftrightarrow Au(CN)_2^- \tag{2.9}$$

$$Cu^{2+} + nNH_3 \leftrightarrow Cu(NH_3)_n^{2+} \tag{2.10}$$

Solvation

$$CO_2(g) \leftrightarrow CO_2(aq) \ (or \ H_2O + CO_2(g) \leftrightarrow H_2CO_3(aq)) \tag{2.11}$$

$$O_2(g) \leftrightarrow O_2(aq) \tag{2.12}$$

Ionic Disassociation

$$Na_2SO_4 \leftrightarrow Na^+ + SO_4^2 \tag{2.13}$$

$$MgCl_2 \leftrightarrow Mg^{2+} + Cl^- \tag{2.14}$$

Chemical reactions can be insightful if they are written correctly. Incorrect reactions can be very problematic. Determining a correct chemical expression for equilibrium in aqueous media can be challenging. There are often several possible expressions. Assuming the initial and final species are known, the expressions are often most effectively written in a six-step process:

Hydrometallurgical reaction writing steps:

(1) Write the basic, unbalanced expression.
(2) Balance all elements except for oxygen and hydrogen.
(3) Balance oxygen using water molecules. (In some unusual cases involving water and, hydrogen, hydroxide or hydrogen peroxide, molecular oxygen, O_2, or hydroxide, OH^-, is used.)
(4) Balance hydrogen using H^+ ions.
(5) Balance charge using electrons, each with a negative charge.
(6) Write the expression as a reduction reaction (electrons on left-hand side) *if electrons are explicitly present on one side of the equation to balance the charge* (For example a reaction such as $Fe^{2+} \leftrightarrow Fe^{3+} + e^-$ would need to be reversed, but a reaction such as $Fe^{2+} + SO_4^{2-} \leftrightarrow FeSO_4$ does not involve an electron and isn't affected by this step). This step is not necessary if electrochemical reactions are written as reduction reactions.

In cases where the final species are not known, it is often useful to perform free energy calculations. More commonly, free energy minimization software is used to determine the most probable final species.

> *The systematic 6 -step hydrometallurgical reaction writing method can help individuals write useful hydrometallurgical reactions.*

Example 2.1 Determine the balanced equilibrium expression for the reaction of ferric sulfate in water with calcium dihydroxide (slaked lime) to form ferric hydroxide and calcium sulfate.

Initial expression:	$Fe_2(SO_4)_3 + Ca(OH)_2 \leftrightarrow Fe(OH)_3 + CaSO_4$
Iron balance:	$Fe_2(SO_4)_3 + Ca(OH)_2 \leftrightarrow 2Fe(OH)_3 + CaSO_4$
Calcium balance:	$Fe_2(SO_4)_3 + 3Ca(OH)_2 \leftrightarrow 2Fe(OH)_3 + 3CaSO_4$
Other balances:	Not necessary because no electrons are exchanged.

Example 2.2 Determine the balanced equilibrium expression for the reaction of iron in water to form hematite, a ferric iron oxide:

Initial expression:	$Fe \leftrightarrow Fe_2O_3$
Iron balance:	$2Fe \leftrightarrow Fe_2O_3$
Oxygen balance using water:	$2Fe + 3H_2O \leftrightarrow Fe_2O_3$
Hydrogen balance using H^+ ions:	$2Fe + 3H_2O \leftrightarrow Fe_2O_3 + 6H^+$
Charge balance:	$2Fe + 3H_2O \leftrightarrow Fe_2O_3 + 6H^+ + 6e^-$
Write as a reduction reaction:	$Fe_2O_3 + 6H^+ + 6e^- \leftrightarrow 2Fe + 3H_2O$

The sum of an element on one side of the reaction expression must balance it on the other. If there are two iron atoms as reactants, there will be two iron atoms as products. However, in some cases a compound may have several atoms of one kind. Thus, the balance must include stoichiometric coefficients and the number of atoms in each compound. A mathematical equation is useful to balance complex reactions. The mathematical balance is:

$$\sum v_{j,i}A_{i,h} = 0 \tag{2.15}$$

The number in front of species "i" in reaction "j" is the stoichiometric coefficient, $v_{j,i}$. The number of atoms of element "h" within species "i" is $A_{i,h}$. Species on the left-hand side of the equation have a positive stoichiometric coefficient. Species on the right-hand side have negative coefficients. Thus, products have positive stoichiometric coefficients. Reactants have negative stoichiometric coefficients. This organized mathematical approach can be particularly helpful in determining overall reaction balances.

It is easy to write balanced, chemical reactions that are not realistic. Thermodynamics allows for determination of reaction viability. In other words, calculations can show whether or not a reaction is favorable under specified conditions. Thermodynamic calculations can also be used to determine individual species concentrations when reactions are at equilibrium.

> *Thermodynamics can be used to determine if balanced chemical reactions are likely to occur.*

2.2 Chemical Potential

For an ideal solution, the chemical potential of species "i" can be expressed as:

$$\mu_i = \mu_i^o + RT \ln x_i \tag{2.16}$$

μ_i represents the chemical potential μ_i^o represents the standard reference chemical potential determined under standard conditions, R is the gas constant (see Appendices B and C), T is the temperature in Kelvin, and x_i is the mole fraction. For nonideal solutions that are more commonly encountered in industry,

$$\mu_i = \mu_i^o + RT \ln a_i \tag{2.17}$$

where: $a_i = k_i x_i$ (k is a correction factor for nonideality).

In solutions containing dissolved species, the mole fraction, x, is determined as:

$$x = \frac{n_i}{n_{solvent} + \sum n_{solutes}}, \tag{2.18}$$

where n = number of moles. In water the number of moles is generally large (55.51 mol per 1000 g H_2O) compared to the number of moles of all solutes. Consequently, the mole fraction can be approximated for aqueous chemistry:

$$x_i \approx \frac{n_i}{n_{\text{solvent}}} \tag{2.19}$$

This approximation introduces negligible error (0.2%) for typical solutions (<0.1 m). The error is less than 2% for solutions containing 1 mol of solute. Compensation for this error will be discussed later. Most dissolved species are not reported in terms of mole fractions. Thus, it is more convenient to use a common standard concentration unit that most accurately represents mole fraction. The most

Molality is the concentration form used for thermodynamic calculations because it is similar to mole fraction.

common units of concentration are molality, m, (moles of solute/1000 g of solvent), molarity, M, (moles of solute/1000 cm^3 of solution) and parts per million, ppm, (micrograms of solute/gram of solution or solid, or grams/metric ton of solution or solid). Molality can be used to approximate mole fraction using a constant. Molarity is based on solution volume. Solution volume is a function of temperature and solute solvation. Parts per million is based upon the total weight of solution. Solution weight is dependent on dissolved species as well as solvent. Parts per million (ppm) units represent accurately the mass fraction. However, this term is often used interchangeably with μg/ml or mg/l terms. These terms are only equal approximately to ppm in very dilute solutions. Molality is the reference concentration for most thermodynamic data because it is most like mole fraction. Conseqeuently, molality is the basis for thermodynamic derivations. Using molality, Eq. 2.12 can be rewritten as:

$$\mu_i = \mu_i^o + RT \ln\left(\frac{\gamma_{i(aq)} m_i}{m^o}\right) \tag{2.20}$$

where m^o is a molal reference added in the denominator to put concentration in terms of a scaled version of mole fraction, thus maintaining a dimensionless logarithmic term. $\gamma_{i(aq)}$ is the activity coefficient of species "i" in aqueous media. Note that in general the subscript (aq) will be dropped for most of the discussions involving ions because ions are generally found in aqueous media. The aqueous activity coefficient is only applicable to aqueous media on a molality basis.

At this point it should be clear that m_i/m_o is not the true mole fraction. Instead, it is a scaled version. 1000 g of water contain 55.5 mol of water. Consequently, a factor of 55.5 must be used for mole fraction/molal comparisons in aqueous media. Molality based activity is the aqueous standard in thermodynamic literature. The difference between mole fraction and molality-based activity is not a problem when thermodyamics are properly applied.

The chemical potential can be related to free energy using:

$$dG = VdP - SdT + \mu_i dn_i + \mu_j dn_j + \ldots, \tag{2.21}$$

For a constant pressure and temperature reaction this equation becomes:

$$dG = \mu_i dn_i + \mu_j dn_j + \ldots \tag{2.22}$$

Integration of Eq. 2.22 leads to Eq. 2.23:

$$G_{final} - G_{init.} = \mu_i(n_{ifinal} - n_{iinit.}) + \mu_j(n_{jfinal} - n_{jinit.}) + \ldots \tag{2.23}$$

For a reaction of x moles of pure B reacting to form y moles of pure C:
$xB \rightarrow yC$.
Equation 2.23 becomes

$$\Delta G_{BC} = \mu_B(0 - x) + \mu_C(y - 0). \tag{2.24}$$

After substituting in the chemical potentials for species B and C (as obtained using Eq. 2.16), Eq. 2.24 becomes:

$$\Delta G_{BC} = y(\mu_C^o + RT \ln a_C) - x(\mu_B^o + RT \ln a_B) \tag{2.25}$$

Rearranging leads to:

$$\Delta G_{BC} = y\mu_C^o - x\mu_B^o + RT \ln\left(\frac{a_C^y}{a_B^x}\right) \tag{2.26}$$

Chemical potential is equal to the partial derivative of the free energy with respect to the number of moles of that species. Consequently, the standard chemical potential can be replaced by the free energy. However, this replacement can only be made if the free energy is a molar quantity. It is also necessary that change in free energy with respect to the number of moles of the species involved is linear over all ranges of concentrations. These assumptions are generally accurate. Thus, it can be stated that:

$$\Delta G_{BC} = yG_C^o - xG_B^o + RT \ln\left(\frac{a_C^y}{a_B^x}\right) \tag{2.27}$$

or alternatively,

$$\Delta G_{BC} = \Delta G_{BC}^o + RT \ln\left(\frac{a_C^y}{a_B^x}\right) \tag{2.28}$$

where, $\Delta G^o{}_{BC}$, is equal to the standard free energies of the products (C) minus those of the reactants (B). Proper consideration of each species' stoichiometry (y for

C and x for B) as shown in Eq. 2.28 must be given. A more general form of Eq. 2.28 using a subscript "r" to denote the reaction is:

$$\Delta G_r = \Delta G_r^o + RT \ln Q, \qquad (2.29)$$

ΔG_r^o, the change in standard free energy for a reaction, can be expressed as:

$$\Delta G_r^o = \sum_k^{\text{products}(k-z)} v_k \Delta G_k^o - \sum_a^{\text{reactants}(a-j)} v_a \Delta G_a^o \qquad (2.30)$$

v is the stoichiometric coefficient for the species. ΔG_a^o is the reactant species standard free energy of formation. ΔG_k^o is the product species standard free energy of formation. Q is the mass action coefficient or the ratio of activities of products to reactants. Proper consideration of stoichiometry must be made. Q can be expressed as:

$$Q = \frac{\prod_p^{\text{products}(k-z)} \gamma_p^{v_j} m_p^{v_j}}{\prod_r^{\text{reactants}(a-j)} \gamma_r^{v_a} m_r^{v_a}} \qquad (2.31)$$

The subscript "p" denotes the product species. The subscript "r" represents the reactant species.

Equation 2.29 is a very important equation. It allows for the determination of free energy under varying concentrations. In order to utilize this equation, the standard free energy must be determined. The standard free energy can be calculated using Eq. 2.30. Example 2.3 demonstrates the use of Eq. 2.30. Data from Example 2.3 are used to demonstrate the application of Eq. 2.29 in Example 2.4.

> The standard free energy of a reaction is the amount of energy available to do useful work under standard conditions.

Example 2.3 Calculate the standard free energy change, $\Delta G^o{}_r$, of the reaction,

$$2Fe^{3+} + 3SO_4^{2-} \leftrightarrow Fe_2(SO_4)_3.$$

Standard free energy is calculated by taking the sum of the standard free energies of the products minus the sum of the standard free energies of the reactants. Proper consideration of stoichiometry and units must be made as shown below (see Appendices for data):

$$\Delta G^o_{Fe3+} \quad = -4600 \text{ J/mole}$$

$$\Delta G^o_{SO42-} = -744,630 \text{ J/mole}$$

$$\Delta G^o_{Fe2/SO4)3} = -2,249,555 \text{ J/mole}$$

$$\Delta G^o_r = [-2,249,555] - [2(-4600) + 3(-744,630)] = -6465 \text{ (J/mole)}$$

Note that this is the standard free energy change. It is the free energy change that would take place if the reaction occurred under standard conditions with standard unit activities of all compounds. More details will be explained in the next section. The actual free energy can be calculated using the standard free energy and the mass action coefficient, Q.

Example 2.4 Calculate the free energy for the reaction:

$$2Fe^{3+} + 3SO_4^{2-} \leftrightarrow Fe_2(SO_4)_3,$$

under standard atmospheric temperature and pressure (SATP) conditions (298.15 K, 1 bar of pressure) with unit activities using Eq. 2.29 with the data from Example 2.3.

$$\Delta G_r = \Delta G^o_r + RT \ln \left(\frac{a_{Fe_2(SO_4)_3}}{a^2_{Fe^{3+}} \, a^3_{SO_4^{2-}}} \right)$$

$$\Delta G_r = -6,465 \text{ J/mole} + RT \ln \left(\frac{1}{1^2 1^3} \right)$$

Because all activities under SATP conditions (i.e. 1 m ion concentrations in ideal solution and 1 bar for pure ideal gases) are one, the logarithm term becomes zero, and the change in free energy for the reaction under standard conditions is

$$\Delta G_r = -6,465 \text{ J/mole}.$$

This result is equal to the standard free energy calculated in Example 2.3 because the reaction occurs under standard conditions.

2.3 Free Energy and Standard Conditions

The reaction free energy, ΔG_r, is the change in usable or "free" energy of a reaction. The principle of free energy is illustrated in Fig. 2.1. A ball on a hill has significant potential energy as shown. If released, gravity will force the ball down the hill. The available potential energy of the ball depends on its position. Near the bottom the ball has less available energy than it has near the top. If released, the ball's potential energy will be reduced. Thus, the final energy will be less than the initial energy. The resulting change in energy for the process will be negative. If, however, the ball is initially at the bottom of the hill, it must acquire energy in order to go up the hill to a higher energy state. Consequently, if the energy change is negative, the event can occur spontaneously. Moreover, no external energy is needed if the energy change is negative.

In a similar way if a chemical reaction releases energy, the change in energy is negative. Reactions that release free energy are naturally favorable. Reactions that need external energy do not occur naturally. Reactions that require energy are naturally unfavorable reactions. Reaction free energy depends on the reaction conditions and constituents.

Reaction conditions have a significant influence on the free energy change. Gas molecules have more "free" energy at 1,000 kPa than at 100 kPa. Similarly, water has more available energy at 99 °C than at 1 °C. Thus, both temperature and pressure impact "free" energy. As a result, a set of standard conditions has

> *Temperature, pressure, phase, and concentration influence reaction free energy.*

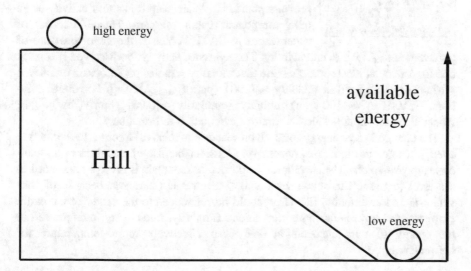

Fig. 2.1 Illustration of a ball on a hill with more available energy on top than the same ball at the bottom of the hill

been established. Standard conditions are needed to make uniform free energy comparisons.

The standard atmospheric temperature and pressure (SATP) conditions used internationally are 298.15 K (25.15 °C), and 1 bar (0.987 atm or 100 kPa). SATP is the basis for all thermodynamic reference states and standards. However, note that standard temperature and pressure (STP) definitions vary. STP in many references is 273.15 K and 1 bar. SATP is the most convenient and widely used standard condition for testing and thermodynamic data.

Pressure changes due to weather and elevation are significantly greater than the difference between 1 bar and 1 atm. Thus, the distinction between 1 bar and 1 atm is not usually considered in most engineering calculations for plant operations in the open atmosphere.

The phase of reaction constituents also affects free energy. For example, water vapor at 100 °C has much more energy than liquid water at 100 °C. The standard reference state is based on the most stable phase at SATP. For example, water's standard reference state at SATP is liquid water. However, at SATP water vapor coexists with liquid water. For the element hydrogen the standard state or phase is H_2 gas.

Another factor that affects the free energy of a reaction is activity. Activity is essentially the effective concentration. Reactions often require a threshold level of reactant activity.

The reference or standard activity of a pure solid compound is one. The standard activity is set at one for convenience. The reference activity for a pure liquid is also chosen as one. The reference activity of a pure ideal gas is one bar at SATP. Pure ideal gasses can include hydrogen, nitrogen, and oxygen. Gasses of compounds with liquid standard states have pure component vapor pressure standards. Water vapor has unit activity at its pure component vapor pressure. Thus, 0.026 bar of water vapor at SATP, which is the vapor pressure of pure water at SATP, has unit activity. The reference activity for dissolved species is one for a 1 molal ideal solution. Note that molality = moles per kilogram of solvent $\approx[(\text{molarity})(\rho_{\text{solution}})]$ or molality $\approx[0.001 \ (\text{ppm})(\rho_{\text{solution}})/\text{Mw}]$. For dilute solutions (<0.001 m or 100 ppm) molality $\approx$molarity $\approx[0.001 \ (\text{ppm})/(\text{Mw})]$. (see Appendices for more detailed laboratory calculation information.)

> *SATP is the basis for all thermodynamic reference states and standards.*

The change in free energy for a given reaction is relative. In order to define free energy, a reference system is necessary. Consider the use of elevation as an analogous measurement. The elevation of a particular mountain is always referenced to sea level (e.g. 6500 ft. above sea level). Sea level is referenced because of convenience and availability. Elevation could be referenced to the center of the earth. However, sea level is far more convenient. Similarly, free energy is referenced to the energy of pure elements at 298 K or effectively room temperature for convenience.

The standard free energy of an element or compound is the free energy of formation at a specified temperature using activities equal to one. Pure elements at SATP have zero free energy of formation. This standard is set for convenience. Pure elements can be readily obtained. Associated reaction free energies can be measured experimentally. Thus, the elements at 298 K and 1 bar of pressure are convenient free energy references.

> *The standard free energy of an element or compound is its free energy of formation using unit activities at a specified temperature .*

Finally, due to the judicious use of unit activities, the free energy of formation of any compound under standard conditions will be equal to the standard free energy of formation

> *Pure elements have zero free energies of formation under SATB conditions.*

according to Eq. 2.30. Based on Eq. 2.30, the logarithm term becomes zero under unit activity, standard conditions. However, if the activities are not equal to one, which is usually the case in real world applications, the outcome is different.

2.4 Free Energy and Nonstandard Activities

It is extremely unusual to have a reaction occur with all activities equal to one. Therefore, determining free energy under nonstandard conditions is critical. The most common and important nonstandard condition encountered in hydrometallurgy involves the activity or concentration of species. Activities generally deviate significantly from the reference standard activity of one. Accommodation for deviations from standard reference activities is made using the logarithm term of Eq. 2.30. Example 2.4 shows the logarithm term becomes zero for standard activities. Because activity or concentration can be orders of magnitude different from one, the logarithm term is usually not zero.

Example 2.5 Calculate the change in free energy for the reaction,

$$2Fe^{3+} + 3SO_4^{2-} \leftrightarrow Fe_2(SO_4)_3,$$

if the initial ferric ion (Fe^{3+}) and sulfate ion (SO_4^{2-}) activities are each 0.6 when the reaction takes place under SATP (standard atmospheric temperature of 298.15 K and pressure of 1 bar) conditions.

Equation 2.30 is used to solve this problem. Because the ferric sulfate ($Fe_2(SO_4)_3$) in the reaction is neutral and relatively insoluble as a non dissociated compound (the compound itself has significant solubility due to the dissociation of the ferric and sulfate ions, but the reaction equation assumes that it does not completely dissolve, leading to some residual solid, it is treated as a solid, which by definition has an activity of 1 (in some unusual cases neutral species have significant solubility and must be treated as dissolved species). The free energy is, therefore:

$$\Delta G_r = -6,465 \text{ J/mole} + RT \ln\left(\frac{1}{(0.6)^2(0.6)^3}\right)$$

$$\Delta G_r = -137 \text{ J/mole}$$

This value for the change in free energy for the reaction is almost negligible. In a real world context this means that if the reaction begins with the specified activities, there will be almost no free energy released.

If the reaction used for Examples 2.4 and 2.5 is utilized with activities greater than unity, the result is different. If the activity of the ions is 2.00, the resulting free energy will be –11,565 J/mole. If the reaction were initiated with ion activities of 2.00 and allowed to react until the activities reached 0.6, the free energy available for reaction would decrease as illustrated in Fig. 2.2. Consequently, it should be apparent that the initial and final conditions are critical to the calculation of free energy.

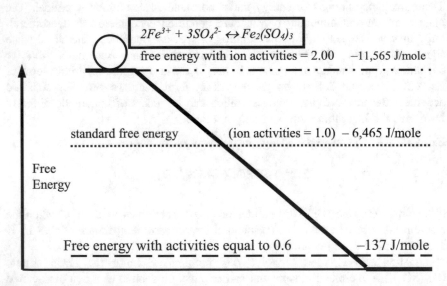

Fig. 2.2 Schematic diagram illustrating free energies at different activity levels for the reaction of ferric ions and sulfate ions to form ferric sulfate. (The activities listed are for the ferric and sulfate ions.)

2.5 Equilibrium

The special condition that occurs when the logarithm term equals the standard free energy is called equilibrium. At equilibrium $\Delta G_r = 0$. Physically, this special case is an important reference point because it means the system has no free energy to gain or lose during the reaction. Because there is no energy to be gained or lost at equilibrium, there is no driving force to produce a net change in the system. Equilibrium is particularly important when determining the final configuration or speciation of a given system. For many engineering calculations, the final state of the system is very close to true equilibrium, thus calculating the equilibrium speciation can provide valuable, practical information. Using the definition of thermodynamic equilibrium, which is $\Delta G_r = 0$, the standard free energy can be used to determine the final or equilibrium speciation in the system. At equilibrium, the substitution of $\Delta G_r = 0$ into Eq. 2.29 leads to:

$$\Delta G_r^o = -RT \ln K \tag{2.32}$$

where K is the equilibrium constant, which may be substantially different from Q, the mass action coefficient. Q represents nonstandard conditions that lie somewhere between the initial ratio and the equilibrium ratio that is realized at the end of the reaction, regardless of where the reaction begins. The equilibrium constant, K, is equal to the activity ratios of the products divided by those of the reactants—*only* at equilibrium.

If a system at equilibrium is perturbed, the system responds to counteract the perturbation effects and reestablish equilibrium. This is known as Le Chatelier's principle. Thus, if a reactant is suddenly added to a system at equilibrium, product forms in order to reestablish equilibrium. Conversely, if product is added, additional reactant will be produced to equilibrate the system.

Example 2.6 Calculate the equilibrium activity of ferric ions (Fe^{3+}) for the reaction of

$$2Fe^{3+} + 3SO_4^{2-} \leftrightarrow Fe_2(SO_4)_3,$$

when the initial sulfate ion (SO_4^{2-}) activity is 0.8.

Using Eq. (2.32), which is only valid for equilibrium conditions:

$$\Delta G_r^o = -RT \ln \left(\frac{a_{Fe_2(SO_4)_3}}{a_{Fe^{3+}}^2 \, a_{SO_4^{2-}}^3} \right) = -RT \ln \left(\frac{1}{a_{Fe^{3+}}^2 \, (0.8)^3} \right)$$

After using the standard free energy value of –6,465 J/mole (see Example 2.4) and rearranging, the expression becomes:

$$a_{Fe^{3+}} = \sqrt{\frac{1}{\exp[\frac{-(-6465)}{RT}](0.8)^3}}$$

$$a_{Fe^{3+}} = 0.379$$

This result, however, assumes that the ferric sulfate precipitate or solid existed.

In many cases, the activities of the dissolved species are insufficient to lead to the formation of a precipitate or solid. If a solid or precipitate does not form, the initial expression, which implicitly assumed its existence, becomes invalid.

> *Le Chatelier's Principle:*
>
> *If a system at equilibrium is perturbed, the system responds to counteract the perturbation and reestablish equilibrium.*

2.6 Solubility Product

Species often combine to form precipitates. Precipitation occurs when the reactant activities exceed the solubility product. The solubility product is generally written as K_{sp}. Reactions with the precipitate on the left-hand side of the equation are dissociation reactions. The equilibrium constant of a dissociation reaction equals the solubility product. Reactions with the precipitate on the right-hand side of the equation are formation reactions. Equilibrium constants of formation reactions are equal to $1/K_{sp}$. Consequently, dissociation reactions are more convenient for precipitation evaluations. Furthermore, the solubility product is only valid at equilibrium.

If the product of the species activities of the ions needed to form a precipitate exceeds the solubility product, K_{sp}. precipitation occurs. The ions in excess of the solubility product are utilized to form the precipitate. Thus, as precipitate forms, ion activities reduce to meet the solubility product.

If the solubility product is not exceeded, available precipitate will dissolve. Thus, the equilibrium solubilities of the associated reactants can be calculated from K_{sp}. In problems involving precipitate, the product of species activities should be used to determine if precipitation occurs. If the product of the activities does not exceed the solubility product, precipitation does not occur. In such cases the associated formation equation becomes invalid.

> *If the product of the species activities of the ions needed to form a precipitate exceeds the solubility product, K_{sp}, precipitation occurs.*

Example 2.7 Determine if ferric sulfate ($Fe_2(SO_4)_3$) precipitation,

$$2Fe^{3+} + 3SO_4^{2-} \leftrightarrow Fe_2(SO_4)_3,$$

will occur at 298.15 K if the ferric ion (Fe^{3+}) activity is 0.7 and the sulfate ion (SO_4^{2-}) activity is 0.3. Rearranging Eq. 2.32 to determine the equilibrium constant leads to:

$$K = \exp\left(\frac{-\Delta G_r^o}{RT}\right)$$

$$K = \exp\left[\frac{-(-6,465)}{(8.314)(298)}\right] = 13.59$$

Because the precipitate or solid is on the right-hand side of the equation, the solubility product is equal to $1/K$ (note that if the equation were written backwards, $K = K_{sp}$):

$$K_{sp} = 1/K = 0.07358$$

By comparison, the product of the activities of the ions needed to form the precipitate is:

$$\prod a_{products} = a_{Fe^{3+}}^2 \, a_{SO_4^{2-}}^3$$

$$\prod a_{products} = (0.7)^2 (0.3)^3 = 0.0132$$

Because the product of the activities is less than K_{sp}, the precipitate will not form. Therefore, the original equation for Example 2.4 is not valid for this problem with the activities specified in this example. However, it is valid at the levels given previously in Example 2.4. Because the precipitate does not form in this example, the final species activities (or concentrations) will be equal to the original values.

2.7 Relationships Amongst K, pK, pK$_a$, and pH

Consider the reaction:

$$HX^- \leftrightarrow H^+ + X^{2-} \tag{2.33}$$

In this reaction the hypothetical compound HX^- dissociates into free H^+ ions and free X^{2-} ions. The equilibrium constant for this reaction is expressed as:

$$K = \frac{(a_{H^+})(a_{X^{2-}})}{(a_{HX^-})} \tag{2.34}$$

Because this reaction involves acid or H^+ ions, the equilibrium constant K is often referred to as K_a. Most often the value of K or K_a will be extremely small.

Extremely small values are more easily compared on a logarithmic basis. Correspondingly, logarithm-based terms are used such as pK or pK_a. In these terms the "p" represents the negative logarithm. For example, pK_a is expressed as:

$$pK_a = -\log(K_a) \tag{2.35}$$

Consequently, if the K_a value for a reaction is 1×10^{-7}, the pK_a value is 7. Similarly, pH is a logarithmic representation of the hydrogen ion activity:

$$pH = -\log(a_{H^+}) \tag{2.36}$$

Example 2.8 Calculate the pH of an acidic solution in which the activity of hydrogen ions is 0.0001.
$$pH = -\log(0.0001) = 4$$

pH and pK_a are related but rarely equivalent.
In the important case of water dissociation:

$$H_2O \leftrightarrow H^+ + OH^- \tag{2.37}$$

The resulting equilibrium expression is:

$$K_a = \frac{(a_{H^+})(a_{OH^-})}{(a_{H_2O})} = 1 \times 10^{-14} \tag{2.38}$$

where $pK_a = 14$. If the activity of water is assumed to be equal to one, then:

$$pK_a = 14 = -\log[(a_{H^+})(a_{OH^-})] \tag{2.39}$$

If $a_{H^+} = a_{OH^-}$ due to dissociation with no other species present in solution, then:

$$pK_a = 14 = -\log[(a_{H^+})^2] \tag{2.40}$$

and,

$$\frac{1}{2}pK_a = \frac{1}{2}(14) = 7 = -\log[(a_{H^+})] = pH \tag{2.41}$$

Thus, if the water activity is one and the hydroxide and hydrogen ion activities are equal, then the pH equals half of the pK$_a$.

Note that for some systems such as ferric ions in water, several hydrolysis steps can occur ($Fe^{3+} + H_2O \leftrightarrow Fe(OH)^{2+} + H^+$, $Fe^{3+} + 2H_2O \leftrightarrow Fe(OH)_2^+ + 2H^+$ and $Fe^{3+} + 3H_2O \leftrightarrow Fe(OH)_3 + 3H^+$). The first reaction is referred to as the first hydrolysis reaction with its associated equilibrium constant K_{a1}. The second hydrolysis reaction has the equilibrium constant K_{a2}. The third is referred to as the third hydrolysis reaction with constant K_{a3}. Often, the first reaction constant will be referred to as K_a in connection with pK_a.

2.8 Free Energy and Nonstandard Temperatures

The effect of temperature on free energy is important to many solution equilibrium calculations. The standard free energy values at temperatures other than 298 K are often not available. Thus, it is useful to utilize data at 298 K and extrapolate to higher temperatures using heat capacity information. The standard free energy change at a given temperature, T, for reactions in which no phase changes occur, is given as:

$$\Delta G_T^o = \Delta H_T^o + T\Delta S_T^o \tag{2.42}$$

This equation applies to many systems. However, for ions in solution, another form will be discussed later.

Additional details regarding free energy as a function of temperature are found in reference 2.1. For most applications the change in free energy with change in temperature is small. The change is of the order of 50 J/mole per degree centigrade.

> *Modest changes in temperature result in small changes in the reaction free energy.*

The effect of temperature can, therefore, often be neglected near STAP. However, small temperature changes can be significant for reactions with small free energies (less than approximately 5,000 J/mole). Often, reactions such as precipitation can be significantly influenced by temperature. Entropic, enthalpic, and heat capacity data can be found in References 2.2–2.5. Also, the availability of necessary data at higher temperatures for dissolved ions in aqueous systems is poor, although some data is available (see Appendix F).

Effect of Temperature on Equilibrium Constant

The Van't Hoff Isocore Equation is [1]:

$$\frac{d(\ln K)}{dt} = \frac{\Delta H_r^o}{RT^2} \tag{2.43}$$

This equation can be rewritten as:

$$\frac{d(\ln K)}{d(\frac{1}{T})} = -\frac{\Delta H_r^o}{R} \tag{2.44}$$

Rearrangement of this equation, under the assumption that enthalpy is not a function of temperature for small temperature changes, leads to a common equation for determining K values at different temperatures:

$$\ln K' = \ln K_{298} - \frac{\Delta H_r^o}{R}\left(\frac{1}{T'} - \frac{1}{298}\right) \tag{2.45}$$

This equation gives a reasonable estimate, provided the change between the reference and desired temperature is generally less than 100 °C.

Exothermic reactions, or reactions that generate heat, are inhibited by increasing temperature. This observation is related to LeChatlier's principle. Correspondingly, precipitation reactions with negative reaction enthalpies are inhibited by increasing temperatures. Endothermic precipitation reactions are enhanced by increased temperature.

> *Exothermic reactions, or reactions that generate heat, are inhibited by increasing temperature.*

Effect of Temperature on Free Energy for Ions

Systems involving ions above 100 °C are often evaluated using a different approach. Ionic species free energies at elevated temperatures are often found using the equation [2]:

$$\Delta G_T^o = \Delta G_{298}^o - (T - 298)\Delta S_{298,\text{adj.}}^o + \int_{298}^{T} \Delta c_p\, dT - T \int_{298}^{T} \Delta c_p \frac{dT}{T} \tag{2.46}$$

in which

$$\Delta S_{298,adj.}^o = \Delta S_{298}^o - 20.92z \tag{2.47}$$

z is the charge of the ion. Based on correspondence principles, research by Criss and Coble, it can be shown that for ions the average molar heat capacity over a temperature range is [2]:

$$c_{p,\text{ions}}\big|_{298}^{T} = \alpha_T + \beta_T \Lambda S_{298,\text{adj.}}^o \tag{2.48}$$

The assumption that the molar heat capacity can be averaged as a constant, allows for it to be moved outside of the integral. Thus, when determining the change in molar heat capacity, this constant can be used for ions. Thus, in practice for the formation of ions from pure metals, the integrals are:

$$\int_{298}^{T} \Delta c_p dT = \left(c_{p,\text{ion}}\big|_{298}^{T} - \frac{\int_{298}^{T} \Delta c_{p,\text{solid}} dT}{\int_{298}^{T} dT}\right)(T - 298) \qquad (2.49)$$

$$T\int_{298}^{T} \Delta c_p \frac{dT}{T} = T\left(c_{p,\text{ion}}\big|_{298}^{T} - \frac{\int_{298}^{T} \Delta c_{p,\text{solids}} dT}{\int_{298}^{T} \Delta c_p}\right) \ln \frac{T}{298} \qquad (2.50)$$

Values for α and β are given in Table 2.1.

Example 2.9 As an example of using this approach, determine the free energy of formation for Ni^{2+} from Ni at 150 °C if the standard entropy of Ni and Ni^{2+} at 298 K are 31.35 and −106.7 J/molK, respectively. The molar heat capacity of Ni is 16.99 + 0.0295 T and the standard free energy of formation of Ni^{2+} is −45,600 J/mole.

$$\Delta S^o_{298,\text{adj.Ni}^{2+}} = \Delta S^o_{298,\text{Ni}^{2+}} - 20.92 z_{\text{Ni}^{2+}} = -106.7 - 20.92(2) = -148.5\,\text{J/molK}$$

$$\Delta S^o_{298,\text{adj.Ni}} = \Delta S^o_{298,\text{Ni}} - 20.92 z_{\text{Ni}} = 31.4 - 0 = 31.4\,\text{J/molK}$$

$$\Delta S^o_{298,\text{adj.}} = \Delta S^o_{298,\text{adj.}} - \Delta S^o_{298,\text{adj.}} = -149 - 31.4 = 180\,\text{J/molK}$$

z is the charge of the ion. Based on correspondence principles research by Criss and Coble, it can be shown that for ions [2]:

$$c_{p,\text{Ni}} = \frac{\int_{298}^{T} c_{p,\text{Ni}} dT}{\int_{298}^{T} dT} = \frac{\int_{298}^{T} (16.99 + 0.0295T)dT\big|_{298}^{T}(16.99T + 0.0295\frac{T^2}{2})}{T - 298}$$

$$= 27.6\,\text{J/molK}$$

$$c_{p,\text{Ni}^{2+}}\big|_{298}^{T} = \alpha_T + \beta_T \Delta S^o_{298,\text{adj.}} = 192 + -0.59(-148.54) = 280\,J/moleK$$

$$\Delta c_p(280 - 27.6) = 252.4\,J/mole$$

Table 2.1 Values of α and β as a function of temperature and ions based on data in Ref. [2]

Temp (°C)	Cations		Simple anions		Oxyanions		hydroxyanions	
	α	B	A	β	A	β	α	β
50	146	−0.41	−192	−0.28	−532	1.96	−510	3.44
100	192	−0.55	−243	0	−577	2.24	−565	3.97
150	192	−0.59	−255	−0.03	−556	2.27	−598	3.95
200	209	−0.63	−272	−0.04	−607	2.53	−636	4.24

$$\Delta G_T^o = \Delta G_{298}^o - (T - 298)\Delta S_{298,adj.}^o + \int_{298}^{T} \Delta c_p dT - T\int_{298}^{T} \Delta c_p \frac{dT}{T}$$

$$\int_{298}^{T} \Delta c_p dT = (c_{p,ion}|_{298}^{T} - \frac{\int_{298}^{T}\Delta c_{p,solid}dT}{\int_{298}^{T}dT})(T-298)$$

$$\int_{298}^{T} \Delta c_p\, dT = (280 - 27.6)(T - 298) = 31,550\,\text{J/mole}$$

$$T\int_{298}^{T} \Delta c_p \frac{dT}{T} = T(c_{p,ion}|_{298}^{T} - \frac{\int_{298}^{T}\Delta c_{p,solids}dT}{\int_{298}^{T}\Delta c_p})\ln\frac{T}{298}$$

$$T\int_{298}^{T} \Delta c_p \frac{dT}{T} = 423(280 - 27.6)\ln\frac{423}{298} = 37,398\,\text{J/mole}$$

$$\Delta G_T^o = -45,600 - (423 - 298)180 + 31550 - 37398 = -73,948\,\text{J/mole}$$

2.8.1 Effect of Temperature on Gas Solubility

Equilibrium gas solubilities are also affected strongly by temperature. As an example, the solubilities of oxygen and hydrogen in water decrease until the boiling point of water is reached. Above the boiling point, further increases in temperature result in rapid increases in solubilities as indicated by the data in Table 2.2.

2.9 Heat Generation Due to Reactions

Another important aspect relating to temperature is the generation of heat due to reaction. Most hydrometallurgical systems operate at constant pressure and do not involve useful work performed on or by the system. Consequently, the heat

Table 2.2 Oxygen and hydrogen gas solubilities (ml gas/gram of water) in Water at 300 PSI, 20.6 bar or 20.3 atm. (calculated using data from Wadsworth [3])

Temp (K)	300	422	435	472	475	533	589
Oxygen (m)	0.024		0.018		0.023	0.048	0.088
Hydrogen (m)	0.017	0.017		0.023		0.047	0.084

generation or consumption due to reactions can often be determined using reaction enthalpy changes. The associated temperature changes can be determined from the heat generation or consumption data together with heat capacity information. The heat of reaction can be expressed as [4]:

$$\Delta H = \Delta H_{298}^o + \int_{298}^{T} \Delta c_p dT \tag{2.51}$$

The temperature rise due to reaction can be calculated using this equation. Often, simplifying assumptions are made. In leaching applications the temperature change is modest and the enthalpy of reaction can be estimated to be the same at 298 K as it is at the reaction temperature. Consequently, an estimate of the temperature rise can be made using the standard enthalpy and heat capacities.

Example 2.10 Estimate the change in temperature associated with bacterial leaching of a sulfide heap containing 0.7 wt% pyrite (FeS_2), 69.3 wt% quartz, and 30 wt% water under ambient (SATP) conditions assuming the leaching reaction is $FeS_2 + H_2O + 3.5O_2 = Fe^{2+} + 2H^+ + 2SO_4^{2-}$. Assume the standard enthalpies of formation at 298 K for the reaction species are $-171,544$, $-285,830$, 0, $-89,100$, 0, and $-909,270$ J/mole, respectively for the species in the order they appear in the reaction equation. The heat capacity of quartz and water are 44 and 75.19 J/moleK, respectively. Assume the enthalpies and heat capacities remain constant over the estimated temperature change. The molecular weights of quartz, pyrite, and water are 60.09, 119.97, and 18.015 g/mole, respectively. Neglect the effect of pyrite, oxygen, and nitrogen heat capacities.

$$\Delta H_{r,298} = \sum_{products} \Delta H_{f,298} - \sum_{reactants} \Delta H_{f,298}$$

$$\Delta H_{r,298} - \sum_{products} 2(-909,270) + 0 + (-89,100)$$

$$- \sum_{reactants} (-171,544) + (-285,830)$$

$$\Delta H_{r,298} = -1,450,266 \text{ J/mole}$$

$$\Delta T_{estimate} = \frac{m_{reactant}\Delta H_{r,298}}{m_{pyrite}Cp_{pyrite} + m_{quartz}Cp_{quartz} + m_{water}Cp_{water}}$$

$$\Delta T_{estimate} = \frac{\frac{0.7}{119.97}\text{mole}(-1,450,266 \text{ J/mole})}{\frac{69.3\,\text{mole}}{60.09}(44 \text{ J/mK}) + \frac{30\,\text{mole}}{18.015}(75.19 \text{ J/mK})} = 48.1 \text{ K}$$

Thus, the temperature in such a heap leaching operation has the potential of rising 48.1 K. Although this value will not be achieved due to heat losses associated with gas and solution flows. In addition, evaporative losses and incomplete reaction were not considered. However, this type of estimation shows that the heat of reaction can be important. Heat generation is of particular importance in closed vessels during concentrate leaching. Closed vessels may require heat removal equipment for proper temperature control.

2.10   Free Energy and Nonstandard Pressures

Free energy changes as a function of pressure due to alterations in molar volumes (V_m) that occur during the reaction. Increasing pressure increases stored energy. The additional energy stored upon compression is analogous to the energy stored by compressing a spring. The change in free energy with respect to pressure can be expressed as:

$$\Delta G_p^o - \Delta G_{1\text{atm}}^o = \int_{1\text{atm}}^{p} \Delta V_m \, dP \tag{2.52}$$

Typically, free energy changes by 0.1 J/mole per bar. Therefore, unless the change in pressure is extreme, the effect of pressure on free energy is generally negligible. However, for equilibria involving very small changes in free energy, it can be significant.

2.11   Equilibrium Concentration Determinations

Using the free energy data for a typical hydrometallurgical reaction at equilibrium,

$$bB + hH^+ \leftrightarrow cC + dH_2O, \tag{2.53}$$

Species activities can be obtained using the expression:

$$\Delta G_r^o = -RT \ln\left[\frac{a_{H2O}^d a_C^c}{a_B^b a_{H+}^h}\right] \tag{2.54}$$

Converting this expression to base 10 logarithm leads to:

$$\Delta G_r^o = -2.303RT \log\left[\frac{a_{H2O}^d a_C^c}{a_B^b a_{H+}^h}\right] \tag{2.55}$$

Because $pH = -\log(a_{H+})$, Eq. 2.55 can be rearranged to:

$$\Delta G_r^o = -2.303RT(\log[a_{H_2O}^d a_C^c] - b\log a_B + hpH) \tag{2.56}$$

After further rearranging, and assuming the activity of water is one, it can be observed that a generalized equation can be expressed as:

$$\frac{\Delta G_r^o}{2.303RT} = b\log a_B - c\log a_C - hpH, \tag{2.57}$$

Example 2.11 Determine the equilibrium or final ferrous ion (Fe^{2+}) activity for the reaction,

$$FeO + 2H^+ \leftrightarrow Fe^{2+} + H_2O,$$

in water if the final pH is 8. Assume that the solution behaves ideally. In other words, assume the activity coefficients are equal to one.

The reaction is of the same form as Eq. 2.53, therefore, Eq. 2.56 applies, and:

$$\frac{\Delta G_r^o}{2.303RT} = \log(a_{FeO}) - \log(a_{Fe^{2+}}) - 2pH$$

Using the data in Appendix D, leads to

$$\Delta G_r^o = [-78,870 + -237,141] - \left[-251,156 + (0)^2\right] = -64,855 \text{ J/mole}$$

After substitution of free energy (and $a_{FeO} = 1$) and rearrangement the equation becomes:

$$\log(a_{Fe^{2+}}) = \frac{-(-64,855)}{(2.303)(8.314)(298)} + \log(1) - 2(8.0)$$

$$aFe^{2+} = 2.33 \times 10^5.$$

2.12 Activities and Activity Coefficients

Activity is a measure of thermodynamic availability of a species. Activity is sometimes viewed as the escaping tendency of a species from its environment. In fact, activity is nearly the same as fugacity. Fugacity comes from the Latin word, fuga, which means flight or escape. If a species interacts unfavorably with its

> *Activity is a measure of a species "activity" or availability in an environment or its tendency to leave its environment.*

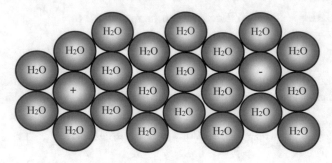

Fig. 2.3 Schematic diagram showing ions and water molecules in dilute solution

environment, its activity increases. Conversely, if a species interacts favorably with its environment its activity decreases. Thus, a species activity is linked to its concentration. The probability of individual molecules leaving or reacting increases as the population density or concentration increases. In ideal solutions activity is directly proportional to concentration.

If a species interacts unfavorably with its environment it seeks to leave. In such a case, the activity of that species is greater than would be found in an ideal solution; therefore, the activity is no longer equal to concentration and a correction factor, known as the activity coefficient must be applied. The relationship between activity and activity coefficient is given as:

$$a_j = \gamma_j \frac{m_j}{m^o} \tag{2.58}$$

If the species does not like its environment, the activity coefficient is greater than one. If the species is fond of its environment the activity coefficient is less than one. Ions interact favorably with other ions in water. Consequently, the activity coefficient of ions in water is less than one. The presence of other ions increases favorable interactions.

Dilute solutions generally behave as ideal solutions because the ions are generally far from the influence of neighboring ions. A dilute aqueous solution is represented in Fig. 2.3.

Favorable interactions can occur with different ions in aqueous media. Thus, increasing concentration of other ions decreases activity. The decrease is due to additional favorable interactions in solution that make ions less likely to leave. For example, if a salt is added to solution, its positively charged "cations" are likely to be associated with negatively charged "anions" in solution to form

Ionic environments are favorable for ions, making them less active and less likely to leave, thereby lowering activity.

electrostatic associations in solution, making them less likely to react or leave the system. This effect is illustrated in Fig. 2.4.

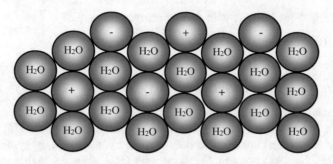

Fig. 2.4 Schematic diagram showing ions and water molecules in concentrated solution

For aqueous solutions ion activity is related to ionic strength. The ionic strength is a measure of the charged nature of the solution. Since ions are charged, their activity is influenced by the electrical fields of surrounding ions. In addition, ion activity is affected by the separation distance between ions. The separation distance is related to concentration. Thus, both concentration and charge of ions contribute to activity.

The ionic strength, I, is defined as [5],

$$I = \frac{1}{2} \sum_{i}^{n} \frac{m_i z_i^2}{m^o},$$ (2.59)

m_i = molality (moles per 1000 g of water) of species "i", z_i = charge of species "i", and m^o = reference molality (1 molal) for aqueous systems.

If the ionic strength is less than 0.00001, the solution is effectively ideal. Ideal solutions have activity coefficients, γ_i, of unity. If the ionic strength is greater than 0.00001, the activity coefficient will differ from one.

One common method of calculating activity coefficients is the Debye–Huckel Equation [6]:

$$-\log \gamma_i = \frac{A z_i^2 \sqrt{I}}{1 + d_i B \sqrt{I}},$$ (2.60)

in which,

$A = 0.2409 + 9.01 \times 10^{-4}$ T, (valid from 273 to 333 K).

(equal to 0.5094 at 298 K),

$B (10^8/cm) = 0.280 + 1.62 \times 10^{-4}$ T (equal to 0.3283 at 298 K),

and typical values for the effective diameter d_i are:

2.5 to 4×10^{-8} cm for ions with single charge

4.5 to 8 $\times$ 10^{-8} cm for ions with double charge

9 to 11 $\times$ 10^{-8} cm for ions with more charge

Notable exceptions are 9 $\times$ 10^{-8}cm for H^+ and

6 $\times$ 10^{-8}cm for Li^+. (for additional information

see Solutions, Minerals, and Equilibria [6]).

For pure water, the ionic strength is low enough (less than 0.00001) to assume the activity coefficients are equal to one. In most hydrometallurgy applications the activity coefficients are not one. Solutions with ionic strengths higher than 0.1 are often treated using the mean salt method, Meissner correlations, or other high concentration correlations (see Reference 2.10 for additional details).

Example 2.12 Calculate the activity coefficient of $SO_4{}^{2-}$ ions in a solution containing 0.01 m.

K_2SO_4. Assume complete dissolution at SATP and an effective ionic diameter of 4.3 $\times$ 10^{-8} cm:

$$I = \frac{1}{2}\sum \frac{2(0.01\,\text{m})(+1)^2}{1\,\text{m}} + \frac{(0.01\,\text{m})(-2)^2}{1\,\text{m}} = 0.03$$

$$A = 0.2409 + 9.04 \times 10^{-4}(298) = 0.5094$$

$$B(10^8/\text{cm}) = 0.280 + 1.62 \times 10^{-4}(298) = 0.3283 \times 10^8/\text{cm}$$

$$-\log(\gamma_{SO_4^{2-}}) = \frac{0.5094(-2)^2\sqrt{0.03}}{1 + (4.3 \times 10^{-8})(0.3283 \times 10^8)\sqrt{0.03}}$$

$$\gamma_{SO_4^{2-}} = 0.520$$

Activity coefficients can also be determined using the Davies Equation:

$$-\log \gamma_i = Az_i^2\left(\frac{\sqrt{I}}{1+\sqrt{I}} - 0.2I\right) \tag{2.61}$$

Surprisingly, the Davies equation can provide reasonable activity coefficients for some ions up to 1 M. However, caution should be exercised when applying this equation for $I > 0.1$.

Activity coefficients of nonionic entities dissolved in aqueous media can be approximated using expressions such as [7]:

$$\log(\gamma_{i(\text{nonionic})}) = bI \tag{2.62}$$

where b is often close to 0.15 [7] and I is the ionic strength.

Activities of gases are usually calculated using partial pressures. In general, the activity of individual species in the gas phase is analogous to that in solution. The activity for gas species is given as:

$$a_{i(gas)} = \gamma_{i(gas)} \frac{P_i}{P_i^o} \tag{2.63}$$

where P_i is the partial pressure of the gas in the gas phase, and P_i^o is the partial pressure of pure "i" under standard conditions. Note that the partial pressure of pure "i" is not the same for oxygen as it is for compounds such as chloroform or ethanol. The reason for this phenomenon is that oxygen is typically a gas under typical conditions. In contrast, chloroform and ethanol are typically liquids. However, ethanol has a vapor pressure of 0.1 atm at standard atmospheric temperature and pressure (SATP) conditions (i.e. 298.15 K, and 1 bar or 0.987 atm). Thus, it is also prevalent as a gas at SATP. Other compounds that are liquids at SATP, such as mercury have very low vapor pressures (2.43×10^{-6} atm). Consequently, the value of P_i^o can vary widely. However, P_i^o remains 1 bar for compounds whose standard states are gases.

> *The activity coefficients of dissolved gasses and neutral species in aqueous media are not strongly influenced by ionic strength.*

Following the definition of activity for species in solution,

$$a_{\text{"}i\text{" gas (aq)}} = \gamma_{\text{"}i\text{" gas (aq)}} \frac{m_{\text{"}i\text{" gas}}}{m^o} \tag{2.64}$$

In a typical reaction of a gas such as oxygen dissolving into aqueous media:

$$O_{2(g)} \leftrightarrow O_{2(aq)} \tag{2.65}$$

the equilibrium constant, K, can be written as:

$$K = \frac{a_{O_2(aq)}}{a_{O_2(g)}} \tag{2.66}$$

which after substitution of Eqs. 2.63 and 2.64, becomes:

$$K = \frac{\gamma_{O2(aq)} \frac{m_{O2(aq)}}{m^o}}{\gamma_{O2(g)} \frac{P_{O2(g)}}{P_{O2}^o}} \tag{2.67}$$

The activity coefficients in the gas phase can generally be assumed to be one. Also, the activity coefficient of dissolved gas in water is approximately one. The near unit activity of gas in water is due to charge neutrality of the gas. Thus, Eq. 2.67 can be rewritten in terms of the dissolved concentration of oxygen as:

$$m_{O2(aq)} = K \frac{P_{O2(g)} m^o}{P^o_{O2}} \tag{2.68}$$

which can be reduced to a more familiar form for gases. The standard pressure gasses at SATP (thus $P^o = 1$ atm), keeping in mind the partial pressure of the gas, P_i needs to be reported in bars (or atm) and the final solution concentration is in molality ($m^o = 1$ molal):

$$m_{i(aq)} = K P_{i(g)} \tag{2.69}$$

This is very similar, but not the same as Henry's law which states that:

$$P_i = x_i K_H \tag{2.70}$$

where K_H is Henry's constant. The number of moles of water, 55.5 mol, in 1000 g of water is large. Comparitively, the number of moles of solute is usually much smaller. Thus, the total number of moles per 1000 g of water is approximately 55.5. Consequently, it is clear that $x_i \approx m_i/55.5$. Furthermore, it can be readily shown that:

$$m_i = \frac{55.5 P_i}{K_H} = K P_i \tag{2.71}$$

This assumes that pressure is reported as a dimensionless partial pressure. Furthermore, $K = 55.5/K_H$. However, it should be noted that some incorrectly report the Henry's constant as the equilibrium constant. Moreover, caution should be exercised when using Henry's constant. If the units are in terms of pressure, then the constant is the true Henry's constant. If the units are dimensionless, the constant is really the equilibrium constant.

Finally, it should be pointed out that the activity of a liquid such as water or ethanol in any mixture at low to moderate pressures (<10 atm) is given by the vapor pressure of that component divided by the pure component vapor pressure. Thus, the activity of water in a concentrated electrolyte can be determined from its vapor pressure. The vapor pressure divided by the pure water vapor pressure equals its activity.

2.13 Practical Equilibrium Problem Solving

In real world applications of solution speciation, there is complexity. There are more unknown parameters than in the theoretical problems shown. Final speciation is often solved based on initial quantities. However, free energy information is based on the final speciation. Thus, a connection between the initial and final conditions must be established. The connection between initial and final concentrations is made using stoichiometry.

By combining information in the form of a system of equations, the resulting equilibria can be determined. For the generic hydrometallurgical reaction:

$$aA + bB \leftrightarrow cC + dD \tag{2.72}$$

the equilibrium expression can be written as:

$$K = \frac{a_{C_E}^c a_{D_E}^d}{a_{A_E}^a a_{B_E}^b} = \frac{\gamma_{C_E}^c \gamma_{D_E}^d}{\gamma_{A_E}^a \gamma_{B_E}^b} \frac{m_{C_E}^c m_{D_E}^d}{m_{A_E}^a m_{B_E}^b} \tag{2.73}$$

in which the subscript "E" indicates equilibrium values. The stoichiometric mass balances are then invoked to form equations. These equations express molalities in terms of initial and equilibrium values. Initial values are denoted by the subscript "I". Equilibrium concentrations of species are tracked in terms of one species. Tracking is facilitated using the fraction, "x", of species "A" that underwent reaction to achieve equilibrium. The mass balance expressions are:

$$m_{A_E} = m_{A_I} - x m_{A_I} \tag{2.74}$$

$$m_{B_E} = m_{B_I} - x \frac{b}{a} m_{A_I} \tag{2.75}$$

$$m_{C_E} = m_{C_I} + x \frac{c}{a} m_{A_I} \tag{2.76}$$

$$m_{D_E} = m_{D_I} + x \frac{d}{a} m_{A_I} \tag{2.77}$$

Substitution of these expressions into the expression for the equilibrium constant leads to:

$$K = \frac{\gamma_{C_E}^c \gamma_{D_E}^d}{\gamma_{A_E}^a \gamma_{B_E}^b} \frac{\left(m_{C_I} + \frac{c}{a} x m_{A_I}\right)^c \left(m_{D_I} + \frac{d}{a} x m_{A_I}\right)^d}{\left(m_{A_I} - x m_{A_I}\right)^a \left(m_{B_I} - \frac{b}{a} x m_{A_I}\right)^b} \tag{2.78}$$

The resulting equation is in terms of initial concentration values. The previous expression was in terms of equilibrium values. The new expression can be solved to determine the value of x. The value of x can be used to determine equilibrium molalities. The expression for x is often not trivial to solve. Such equations are often solved using the quadratic equation (if appropriate). Alternatively, such equations can be solved numerically using a computer.

Example 2.13 Calculate the final concentration of calcium ions when 2 g of calcium chloride are mixed (at SATP) with 0.5 g of sodium fluoride in 1000 g of pure water given that the activity coefficients are 0.4 and 0.75 for calcium and fluoride ions, respectively. (Assume that calcium chloride and sodium fluoride completely dissolve and dissociate.)

The reaction of interest here is:

$$Ca^{2+} + 2F^- \leftrightarrow CaF_2$$

The other reaction with sodium and chloride ions will not lead to the formation of a precipitate due to the high solubility of sodium chloride or table salt. (The calculation to show this is not presented because it is quite obvious.)

The next step is to determine the molar quantities of the involved species.

$$\text{Moles of } Ca = 2g/MwCaCl_2 = 2/110.99 = 0.0180 \, \text{moles}$$

$$\text{Moles of } F = 0.5g/MwNaF = 0.5/41.99 = 0.0119 \, \text{moles}$$

Because molality is defined as the number of moles per 1000 g of solvent,

$$m_{initCa} = 0.0180 \, m$$

$$m_{initeE} = 0.0119 \, m$$

According to the specified stoichiometry, the molar consumption of fluoride ions is twice that of the calcium ions, so the final molalities are given as:

$$m_{fuca} = 0.0180 \, m - m_{consumpCa}$$

$$m_{fuF} = 0.0119 \, m - 2m_{consumpCa}$$

Using free energy considerations for the final equilibrium condition (see Eq. 2.29) along with the substitution of $a_i = \gamma m_i/m^o$), leads to.

$$\Delta G_r^o = -2.303RT \log\left(\frac{\gamma_{CaF_2} m_{CaF_2}}{\gamma_{Ca^{2+}} m_{Ca^{2+}} \gamma_{F^-}^2 m_{F^-}^2}\right)$$

After substitution for the final molalities, the equilibrium free energy expression becomes:

$$\Delta G_r^o = -2.303RT \log\left(\dfrac{1}{\gamma_{Ca^{2+}}(0.0180 - m_{consumpCa})\gamma_{F^-}^2(0.0119 - 2m_{consumpCa})^2}\right)$$

Calculation of the free energy value for the reaction leads to:

$$\Delta G_r^o = [-1,162,000] - [-553,540 + 2(-276,500)] = -55,460 \text{ J/mole}$$

$$-55,460 = -2.303RT \log\left(\dfrac{1}{\gamma_{Ca^{2+}}(0.0180 - m_{consmpCa})\gamma_{F^-}^2(0.0119 - 2m_{consumpCa})^2}\right)$$

$$9.72 = \log\left(\dfrac{1}{0.4(0.0180 - m_{consumpCa})(0.75)^2(0.0119 - 2m_{consumpCa})^2}\right)$$

$$10^{9.72}(0.4)(0.75)^2\left[(0.0180 - m_{consumoca})(0.0119 - 2m_{conswmpCa})^2 - 1 = 0\right.$$

Solving for m_{consCa} by iteration leads to:

$$m_{consCa} = 0.005818$$

Using this value for the molal consumption, the final equilibrium values are calculated.

$$m_{f_{fn}Ca} = 0.0180 - 0.005818 = 0.01218 \text{ m}$$

$$m_{fnF} = 0.0119 - 2(0.005818) = 0.000264 \text{ m}$$

Often, the activity coefficient must be calculated without knowing the ionic strength. In such cases, an initial estimate of the ionic strength must be made. The initial estimate is based on given information. After making an initial guess for the ionic strength, activity coefficients can be determined. The activity coefficients can be used to solve for molalities. Using the calculated molalities, the ionic strength can then be recalculated. The calculated ionic strength is then compared to the guessed value. If the initial guess and the calculated ionic strengths are within reasonable tolerance, the final solution has been reached. If the guessed and calculated ionic strengths are significantly different, the new calculated value of the ionic strength is used to recalculate the activity coefficients. The new activity coefficients are used to recalculate molalities. New molalities are used to recalculate ionic strength. If the new ionic strength is close to the previous one the solution is appropriate. If there is a significant difference, the process is repeated until the value of the previous ionic strength is within the specified tolerance of the new calculated value.

Example 2.14 Consider the same information used in Example 2.12 without knowledge of activity coefficients: Calculate the final concentration of calcium ions when 2 g of calcium dichloride are mixed (at SATP) with 0.5 g of sodium fluoride in 1,000 g of pure water. (Assume that calcium chloride and sodium fluoride completely dissolve and dissociate.)

The initial part of the problem is the same as Example 2.12, so using the results:

$$Ca^{2+} + 2F^- \leftrightarrow CaF_2$$

$$\text{Moles of } Ca = 2g/MwCaCl_2 = 2/110.99 = 0.0180 \text{ moles}$$

$$\text{Moles of } F = 0.5\,g/MwNaF = 0.5/41.99 = 0.0119 \text{ moles}$$

$$m_{fnCa} = 0.0180m - m_{consumpCa}$$

$$m_{fvuF} = 0.0119m - 2m_{consumpCa}$$

$$\Delta G_r^o = -2.303RT \log\left(\frac{1}{\gamma_{Ca^{2+}}(0.0180 - m_{consumpCa})\gamma_{F^-}^2(0.0119 - 2m_{consumpCa})^2}\right)$$

$$-55,460 = -2.303RT \log\left(\frac{1}{\gamma_{Ca^{2+}}(0.0180 - m_{consumpCa})\gamma_{F^-}^2(0.0119 - 2m_{consCa})^2}\right)$$

The activity coefficient calculations require the ionic strength. The best estimate for an initial guess for the ionic strength is made using the available information:

$$I = \frac{1}{2}\sum \frac{2(0.018m)(-1)^2}{1m} + \frac{(0.018m)(+2)^2}{1m} + \frac{(0.0119m)(-1)^2}{1m} + \frac{(0.0119m)(+1)^2}{1m}$$

$$I = 0.057$$

$$-\log \gamma_{Ca2+} = 0.5094(-2)^2\left(\frac{\sqrt{0.057}}{1+\sqrt{0.057}} - 0.2(0.057)\right) = 0.3695$$

$$\gamma_{Ca2+} = 0.427$$

$$-\log \gamma_{F-} = 0.5094(1)^2(\frac{\sqrt{0.057}}{1+\sqrt{0.057}} - 0.2(0.057)) = 0.0924$$

$$\gamma_{F-} = 0.808$$

Substitution and rearrangement leads to:

$$5.25x10^9 = (\frac{1}{0.427(0.0180 - m_{consumpCa})(0.808)^2(0.0119 - 2m_{consumpCa})^2})$$

$$1.46 \times 10^9 = \left(\frac{1}{(0.0180 - m_{consumpCa})(0.0119 - 2m_{consumpCa})^2}\right)$$

Solving for m_{consCa} by iteration leads to:

$$m_{cassCa} = 0.005832$$

Using this value for the molal consumption, the final equilibrium values are calculated.

$$m_{finCa} = 0.0180 - 0.005832 = 0.0122m$$

$$m_{finF} = 0.0119 - 2(0.005832) = 0.000236m$$

The new value of ionic strength is:

$$I = \frac{1}{2}\sum 2(0.018)(-1)^2 + (0.0122)(+2)^2 + (0.000236)(-1)^2 + (0.0119)(+1)^2$$

$$I = 0.0395$$

$$-\log \gamma_{Ca2+} = 0.5094(-2)^2\left(\frac{\sqrt{0.0395}}{1 + \sqrt{0.0395}} - 0.2(0.0395)\right) = 0.3217$$

$$\gamma_{Ca2+} = 0.4767$$

$$-\log \gamma_{F-} = 0.5094(1)^2\left(\frac{\sqrt{0.0395}}{1 + \sqrt{0.0395}} - 0.2(0.0395)\right) = 0.804$$

$$\gamma_{F-} = 0.831$$

Using these new values for activity coefficients the molalities are recalculated:

$$m_{consCa} = 0.005841$$

Using this value for the molal consumption, the final equilibrium values are:

$$m_{finCa} = 0.0180 - 0.005841 = 0.01216\,m$$

$$m_{fnF} = 0.0119 - 2(0.005841) = 0.000218\,m$$

These values are very close to the values from the previous step. Consequently, the ionic strength will be nearly the same as the previous iteration. Thus, no further iteration is needed. These values are appropriate estimates of the correct molality values.

2.14 Electrochemical Reaction Principles

For reactions involving electron transfer, such as:

$$Fe^{3+} + e^- \leftrightarrow Fe^{2+} \tag{2.79}$$

the free energy of the reaction is linked to electron transfer. Electron transfer, in turn, is linked to an electrochemical potential or voltage. The potential is equivalent to energy per unit of charge transferred (1 V = 1 J/Coulomb). The relationship between free energy and electrochemical potential, E, is given as:

$$\Delta G_r = -nN_A eE = -nFE, or, \Delta G_r^o = -nFE^o, \tag{2.80}$$

E equals electric potential or voltage, E^o is the standard potential or voltage, n is the number of electrons transferred per mole of reaction, N_A is Avogadro's number (6.022 × 10²³ atoms or molecules/mole), e is the charge per electron (1.602 × 10⁻¹⁹ Coulombs/electron), and F is the Faraday constant (96,485 Coulombs/mole of electrons). Free energy can be converted to electrochemical potential for reactions involving electron transfer. Dividing the free energy equation by $-nF$ results in:

> *Electrochemical potential is directly related to free energy.*

$$\frac{\Delta G_r}{-nF} = \frac{\Delta G_r^o}{-nF} + \frac{RT}{-nF}\ln Q \tag{2.81}$$

Substitution using the relationship between potential and free energy results in:

$$E = E_o - \frac{RT}{nF}\ln Q \tag{2.82}$$

This equation is known as the Nernst Equation.

The Nernst Equation can be rewritten for a reaction with ferric and ferrous ions as:

$$E = E^o - \frac{2.303RT}{F} \log \left[\frac{a_{Fe+2}}{a_{Fe+3}} \right], \tag{2.83}$$

Thus, as product concentration increases, the solution potential decreases. On the other hand, if the concentration of the reactant increases, the potential increases. Note that the ferric ion $(Fe^{3+})/(Fe^{2+})$ ferrous ion reaction is written cathodically. In other words, it is written as a reduction reaction. The oxidized species is the reactant and the reduced species is the product. Using this convention, the resulting potential will correspond to measured voltages. If the reactions are written

> *The Nernst Equation can be used to determine the electrochemical potential of half-cell oxidation/reduction reactions.*

anodically, the calculated voltages will be the opposite of the measured voltages. (Some electrochemists write the equations the opposite way, then compensate by changing signs with respect to the relationship between free energy and potential.) Consequently, the common convention is to write reactions cathodically. Cathodic reactions have electrons on the left-hand side of the equation.

The standard free energy of the $(Fe^{3+})/(Fe^{2+})$ reaction is -74.32 kJ/mole. Substituting this value into Eq. 2.68 leads to a standard potential of 0.770 V. This reaction is known as a half-cell reaction. In order for this reaction to occur there must be a complimentary half-cell reaction. The complimentary reaction provides the electrons to complete the reaction. In some cases the complimentary reaction will receive the electrons. Suppose the complimentary half-cell reaction is:

$$Cu^{2+} + 2e^- \leftrightarrow Cu \tag{2.84}$$

for which the standard free energy is $-65,520$ J/mole. The two half-cell reactions can be combined to give:

$$Cu + 2Fe^{3+} \leftrightarrow Cu^{2+} + 2Fe^{2+} \tag{2.85}$$

The direction of the reaction can be determined from free energy. If the activities of the species are unknown, the direction can be determined based on standard free energy. The standard free energy of the copper/iron reaction in Eqs. 2.79 and 2.84 is:

$$\Delta G_r^\circ = 2\left(\Delta G_{r1}^\circ\right) - \Delta G_{r2}^\circ = 2(-74,320) - (-65,520) = -83,120 \, \text{J/mole} \tag{2.86}$$

The negative standard free energy the reaction indicates a tendency to proceed as written. The actual free energy will depend on species activities. Note that the free energy for the first reaction (Fe^{3+}/Fe^{2+}) was multiplied by 2 because 2 electrons were needed for the copper reaction. A simpler comparison of the reactions could be made using electrochemical potentials.

The electrochemical potential provides an easy comparison of reaction tendencies. Electrochemical potentials are simply free energies normalized by charge. The potentials of the iron and copper half-cell reactions are $E^o_{Fe} = 0.770$ V and $E^o_{Cu} = 0.339$ V. These potentials reveal that the iron half-cell reaction is more likely to proceed cathodically than the copper reaction. Higher potentials indicate greater tendencies to proceed cathodically. By coupling the half-cell reactions, the reaction with the highest potential will go forward. Correspondingly, the lower potential reaction will be forced in the opposite direction. The same comparison can be made among any half-cell reactions. A list of common half-cell reactions is given in Appendix G.

Half-cell reactions with higher potentials have more available (negative reaction free energy) free energy per electron. Consequently, higher potential reactions will drive lower potential reactions in the opposite direction. Thus, for example, gold is nearly always found in an elemental, reduced state in nature. This phenomenon is due to the high half-cell potential for gold reduction from a gold ion. In contrast, elemental iron tends to dissolve (or rust) due to its low half-cell potential.

For more complex hydrometallurgical reactions involving electron transfer, such as:

$$bB + hH^+ + ne^- \leftrightarrow cC + dH_2O, \tag{2.87}$$

the corresponding electrochemical potential expression is:

$$E = E^o - \frac{2.303RT}{nF}\log[\frac{a_C^c a_{H2O}^d}{a_B^b a_{H+}^h}] \tag{2.88}$$

which after assuming unit activity for water and substituting pH for $-\log a_{H+}$ becomes:

$$E = E^o + \frac{2.303RT}{nF}(b\log \gamma_B[B] - hpH - c\log \gamma_C[C]) \tag{2.89}$$

which is a useful mathematical relationship between solution potential, species concentrations, and pH for a typical hydrometallurgical reaction.

For water, several important reactions, most of which involve electron transfer, can be written. One possible reaction with water is:

$$O_2 + 4H^+ + 4e^- \leftrightarrow 2H_2O \tag{2.90}$$

This reaction is an important step in converting food to energy for respiring (breathing) organisms—Wow! The reverse reaction shows that if electrons, oxygen, or hydrogen ions can be removed from the system, water will be consumed. In other words, water will break down into oxygen, hydrogen ions, and electrons under proper conditions. Electrons must have a place to go for this to occur. Thus, a counter reaction is needed. The water reaction is critical in hydrometallurgy. It is difficult to perform industrial reactions when water breaks down. However, some industrial processing leads to water splitting into oxygen and hydrogen ions. Some processing under different conditions breaks water into hydrogen gas and hydroxide ions. The water splitting reaction marks an important boundary for practical hydrometallurgy reactions. At equilibrium, this boundary condition for water decomposing into oxygen gas and hydrogen ions can be expressed using Eq. 2.89 as:

$$E(V) = 1.228 - 0.0591\text{pH} + 0.0147\log P_{02} \quad (\text{at } 25\,^{\circ}\text{C}), \qquad (2.91)$$

An equivalent reaction, which tends to occur at high pH, can be written as:

$$O_2 + 2H_2O + 4e^- \leftrightarrow 4OH^- \qquad (2.92)$$

Other important water reactions involving hydrogen are:

$$2H^+ + 2e^- \leftrightarrow H_2, \text{ and} \qquad (2.93)$$

$$2H_2O + 2e^- \leftrightarrow H_2 + 2OH^- \qquad (2.94)$$

for which,

$$E = 0 - 0.0591\text{pH} - 0.0295\log P_{H_2} \quad (\text{at } 25\,^{\circ}\text{C}). \qquad (2.95)$$

The relationship between ion activity and potential is often used in electrochemical sensors and selective ion electrodes. This relationship allows for the determination of concentration based on a potential or voltage measurement. The application of this relationship will be demonstrated in the next section.

2.15 Equilibrium and Electrochemical Equations

As discussed in connection with free energy, equilibrium occurs when the free energy goes to zero. Consider a hypothetical set of two half-cell reactions:

$$Y^{P+} + je^- \leftrightarrow Y^{m+} \qquad (2.96)$$

$$Z^{Q+} + ke^- \leftrightarrow Z^{S+} \qquad (2.97)$$

for which the coupled overall reaction is:

$$Y^{P+} + \frac{j}{k}Z^{S+} \leftrightarrow Y^{m+} + \frac{j}{k}Z^{Q+} \qquad (2.98)$$

The free energy of the overall reaction can be calculated from the individual reactions. At equilibrium the free energy will be zero as shown:

$$\Delta G_r = \Delta G_{r(Y)} - \frac{j}{k}\Delta G_{r(Z)} = 0 \tag{2.99}$$

This equation can be rearranged to:

$$\Delta G_{r(Y)} = \frac{j}{k}\Delta G_{r(Z)} \tag{2.100}$$

which after substitution of $\Delta G = -nFE$ (n is "j" or "k" here) leads to:

$$-jFE_Y = -kF\left(\frac{j}{k}E_Z\right) \tag{2.101}$$

Further rearranging leads to the equilibrium expression:

$$E_Y = E_z \tag{2.102}$$

Thus, at equilibrium the potentials for the half-cell reactions are equal. The same concept is true for multiple half-cell reactions. This is also intuitive because reactions generally occur on a conductive substrate. Thus, all half-cell reactions occurring on a common conductive substrate are exposed to the same potential. Although free energy of an overall reaction is zero at equilibrium, the equilibrium potentials will rarely be zero. These are important concepts that deserve additional attention.

> At equilibrium, all half-cell potentials in a system are equal, but rarely zero.

The Nernst equation indicates that electrochemical potentials are related to species activities. Coupled Nernst equations can be used to determine equilibrium concentrations. Coupled Nernst equations will acquire the same potential at equilibrium. This relationship, shown as Eq. 2.89, allows for the equilibrium calculation. Consequently, electrochemical reaction equilibria can be calculated by equating the respective Nernst Equations. As an example, consider the following reactions.

$$Sn^{2+} + 2e^- \leftrightarrow Sn \quad E^\circ = -0.136\,V \tag{2.103}$$

$$Ni^{2+} + 2e^- \leftrightarrow Ni \quad E^\circ = -0.250\,V \tag{2.104}$$

The expanded Nernst Equation for each of these reactions is given as:

$$E_{Sn} = -0.136 - \frac{2.303RT}{2F}\log\left[\frac{a_{Sn}}{a_{Sn+2}}\right], \text{ and} \tag{2.105}$$

$$E_{Ni} = -0.250 - \frac{2.303RT}{2F}\log\left[\frac{a_{Ni}}{a_{Ni+2}}\right] \tag{2.106}$$

When these reactions are coupled, the reaction with the higher potential (Sn) will go in the forward direction. Consequently, the lower potential reaction (Ni) will go in the opposite direction. Tin metal will be produced as nickel metal is dissolved. Thus, the solution species concentrations will change as the reactions proceed. The reactions will proceed until equilibrium is achieved. Equilibrium occurs when the potentials of the reactions are equal. Equilibrim will be obtained when Eq. 2.105 equals Eq. 2.106. Note that this approach to determine equilibrium speciation is an alternative to the free energy approach. Clearly, the two approaches are directly related.

The equilibrium for the nickel-tin system will depend on the species activities of Ni^{2+} and Sn^{2+}. The equilibrium assumes that Ni and Sn are present at equilibrium. If one of the metals is entirely consumed its activity will no longer be one. Consequently, if one of the components is eliminated its equation will not be valid. Assuming the metals are both present at equilibrium ensures both equations are valid. Setting the two Nernst Equations equal to each other leads to:

$$-0.136 - \frac{2.303RT}{2F} \log\left[\frac{a_{Sn}}{a_{Sn+2}}\right] = -0.250 - \frac{2.303RT}{2F} \log\left[\frac{a_{Ni}}{a_{Ni+2}}\right], \quad (2.107)$$

which simplifies to:

$$\log(a_{Sn+2}) = \left[\frac{-0.114}{\left(\frac{2.303RT}{2F}\right)}\right] + \log(a_{Ni+2}), \quad (2.108)$$

Thus, the final activity and concentration of Sn^{2+} can be calculated. The calculation requires the final concentration of Ni^{2+}. The final concentration determination requires an additional stoichiometric balance. If the equilibrium potential were known the Nernst Equation (Eq. 2.83) could be used to determine equilibrium activities. In principle, the value of the electrochemical potential of a solution can be measured using a redox electrode. A redox electrode is an electrode that measures electron withdrawing tendencies of dissolved species. The electrode typically consists of a platinum disk and an internal reference system. Species at the platinum surface exert an electron withdrawing potential. This potential is compared to the internal reference. The difference in potential between the platinum and reference is the resulting redox potential. The redox potential is also referred to as the oxidation–reduction potential (ORP). When the potential, E, is measured relative to a standard hydrogen cell it is sometimes referred to as Eh. Eh and ORP are often used interchangeably. In many cases references other than hydrogen are associated with them. However, it is important to understand that redox measurements are made relative to a reference reaction and potential. The value of the reference is critical to the interpretation of the measurement. In addition, it is important to realize that kinetic effects can affect redox

> *Electrochemical potentials are measured relative to a reference potential.*

Table 2.3 Comparison of reference half-cell reaction potentials [8–10]

Reaction	Conc. specification	Potential (E)
$CuSO_4 + 2e^- \leftrightarrow Cu + SO_4^{2-}$	Standard conditions	0.339
$CuSO_4 + 2e^- \leftrightarrow Cu + SO_4^{2-}$	Saturated $CuSO_4$ soln	0.318
$Hg_2Cl_2 + 2e^- \leftrightarrow 2Hg + 2Cl^-$	Standard conditions	0.268
$Hg_2Cl_2 + 2e^- \leftrightarrow 2Hg + 2Cl^-$	Saturated KCl soln	0.241
$AgCl + e^- \leftrightarrow Ag + Cl^-$	Standard conditions	0.222
$AgCl + e^- \leftrightarrow Ag + Cl^-$	Saturated KCl solution	0.199
$2H^+ + 2e^- \leftrightarrow H_2$	Standard conditions	0.000

measurements. The measured potential is actually the mixed potential of all adsorbed species. Consequently, the measured ORP or Eh is affected by a combination of activities and interaction kinetics. Details of the kinetics and mixed potential will be discussed later. A list of common reference reactions and potentials are presented in Table 2.3.

One reaction that has been tracked by such measurements is the ferrous/ferric half-cell reaction. Clearly, some caution must be used in interpreting ORP/Eh results for reasons already discussed. Another issue that affects measurements is

Reference potentials are usually based on half-cell reactions with stable potentials near zero.

solution complexation. The solution complexation issues can, however, be easily reconciled. The first step in compensation requires a calculation of solution speciation. In the case of the ferrous/ferric half-cell reaction, only 6% of the added ferric ions in a 0.01 m solution of ferric sulfate are found as free Fe^{3+} ions. In contrast, the remainder is complexed in other forms, primarily $FeSO_4^+$. In the case of ferrous ions, 70% are found as free Fe^{2+} ions in a similar solution (0.01 m ferrous sulfate). Thus, a mixture of 0.01 m of ferrous ($FeSO_4$) and 0.005 m ferric sulfate ($Fe_2(SO_4)_3$) will not result in equivalent Fe^{3+} and Fe^{2+} species concentrations. The ideal theoretical Eh would be the standard potential for the half-cell reaction (0.770 V). However, the solution complexation greatly lowers the available ferric ion concentration. In addition, the ferric ion activity coefficient is significantly

The standard hydrogen ion / hydrogen gas half-cell reaction provides a zero reference potential.

less than one. As a result of complexation and activity coefficient reduction the measured Eh is 0.665 V (0.01 m $FeSO_4$, 0.05 m $Fe_2(SO_4)_3$). The ORP for this system is 0.466 V using a common silver-silver chloride reference based redox electrode. Thus, it is critical to know the free ion concentrations and activity coefficients. Alternatively, redox measurements can be used to determine these values under appropriate conditions.

The following equation can be used to isolate activity coefficients from ion concentrations for the ferric/ferrous reaction:

$$E = E_o - \frac{RT}{nF} \ln \frac{\gamma_{Fe2+}}{\gamma_{Fe3+}} - \frac{RT}{nF} \ln \frac{m_{Fe2+}}{m_{Fe3+}} = E_{eff} - \frac{RT}{nF} \ln \frac{m_{Fe2+}}{m_{Fe3+}} \quad (2.109)$$

This equation shows the potential is equal to an effective potential minus the logarithmic contribution of concentration. The combination of activity coefficients with standard potential is constant if the activity coefficients do not change. Activity coefficients are constant if the ionic strength is stable. Free or uncomplexed ion concentrations can be calculated using thermodynamics.

The Nernst equation is also used extensively to determine species concentrations. Examples include hydrogen or metal ion concentrations using ion selective electrodes. Selective species electrodes typically operate based on electrochemical reactions. A general reaction form is:

$$M^{n+} + ne^- + X \leftrightarrow MX \quad (2.110)$$

Assuming unit activities of X and MX, the associated Nernst equation can be written. Assuming a dilute system in units of ppm (molality ≈ 0.001(ppm) $(\rho_{solution})/$(molecular weight $[Mw]$)) the expression is:

$$E = E_o - \frac{RT}{nF} \ln \frac{Mw}{0.001 ppm_{M^{n+}} \rho_{soln} \gamma_{M^{n+}}} = E_{eff} + \frac{2.303RT}{nF} \log(ppm_{M^{n+}}) \quad (2.111)$$

This expression merges the constants with the standard potential to form an effective potential, E_{eff}. Consequently, the measured potential, E, should change in direct proportion to the logarithm of concentration. This assumes the electrode responds selectively to the desired species. The equation form can be modified for other concentration units. The logarithmic proportionality with concentration remains the same if ionic strength is constant. An example plot of potential versus cupric ion concentration (ppm) is presented in Fig. 2.5. Figure 2.5 results from data

Fig. 2.5 Comparison of potential versus logarithm of cupric ion concentration (ppm) using a cupric ion selective electrode in 0.1 M sodium nitrate aqueous media at 22 °C

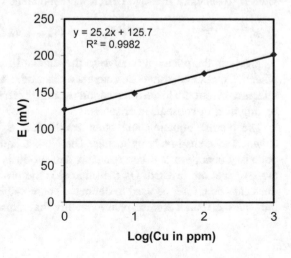

measured in constant ionic strength medium (0.1 M sodium nitrate). In practice, however, there is a finite range of accurate electrode response. Most electrodes experience strong interference by similar or antagonistic species. Thus, electrode manuals need to be consulted to ensure use within proper conditions.

All electrochemical potential measurements are made relative to a reference. Ideally, potentials would be measured with respect to a standard hydrogen half-cell reaction. The hydrogen half-cell reaction has a zero potential. Thus, such a reference would be ideal mathematically. However, this reference would require 1 bar of hydrogen gas, and a pH of zero. This is not practical. Therefore, measurements are made with other reference systems enclosed in electrodes.

Electrodes designed around reference systems, called reference electrodes, have a salt bridge connection to solution. The salt bridge connects the internal reference reaction to the external solution. The bridge completes the electrical circuit for the measurement system. Common reference reactions involve calomel or silver/silver chloride. If the electrochemical potential of the solution is zero (referenced to the standard hydrogen electrode (SHE)), an ORP electrode with a calomel internal reference (saturated KCl solution) will measure a voltage of −0.241 V. This measured potential would be relative to the internal reference. The internal calomel reference is 0.241 V above the standard hydrogen reaction [11]. Thus, the difference between the solution potential of 0 V (versus SHE) and the reference against which it is compared, is −0.241 V. This potential is reported for a standard calomel electrode (saturated KCl solution) in V (versus standard calomel electrode or SCE), although it is often converted to V (versus SHE) by adding 0.241 V to the SCE reading. The electrode with the internal Ag/AgCl reference cell (saturated KCl solution) would give a reading of −0.199 V [11]. Thus, the addition of 0.199 V will convert the measurement into the more meaningful standard hydrogen electrode cell voltage of 0 V. The SHE value can be correlated directly with thermodynamic data. A schematic representation of the application of a reference electrode is presented in Fig. 2.6. The general equation for converting reference electrode measurement valuess to standard hydrogen electrode equivalent values is:

$$E_{(SHE)} = E_{meas} + E_{ref} \qquad (2.112)$$

$E_{(SHE)}$ is the potential relative to the standard hydrogen electrode, E_{meas} is the potential measured using the reference electrode, and E_{ref} is the potential of the reference electrode. More information regarding electrochemical measurements will be provided in subsequent chapters.

The Nernst Equation and other equilibrium expressions are useful in many chemical systems involving metals. The Nernst Equation provides the equilibrium boundary condition between reactants and products. On one side of the boundary the reactants are favored. On the other side the products are favored. These equations can, therefore, be used to delineate phase stability boundaries. Consequently, they are the foundation for phase stability diagrams.

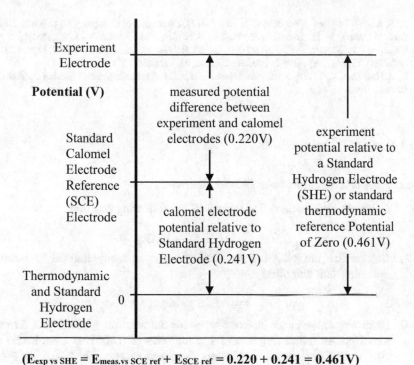

$$(E_{exp\ vs\ SHE} = E_{meas.vs\ SCE\ ref} + E_{SCE\ ref} = 0.220 + 0.241 = 0.461V)$$

Fig. 2.6 Illustration of relative relationship between experimental, calomel reference, and standard hydrogen potentials

References

1. D.J.G. Ives, *Chemical Thermodynamics* (University Chemistry, Macdonald Technical and Scientific, 1971)
2. K. Han, *Fundamentals of Aqueous Metallurgy*, SME, Littleton (2002)
3. M.E. Wadsworth, *Hydrometallurgy Course Notes*
4. H.S. Fogler, *Elements of Chemical Reaction Engineering* (Prentice-Hall, Englewood Cliffs, 1986)
5. J.F. Zemaitis Jr., D.M. Clark, M. Rafal, N.C. Scrivner, *Handbook of Aqueous Electrolyte Thermodynamics* (Design Institute for Physical Property Data, New York, 1986)
6. R.A. Robie, B.S. Hemingway, J.R. Fisher, Thermodynamic properties of minerals and related substances at 298.15 K and 1 Bar (10^5 Pascals) pressure and at higher temperatures. Geol. Surv. Bulletin 1452, U.S. Department of the Interior, Washington, D.C.
7. J.N. Butler, *Ionic Equilibrium* (Wiley, New York, 1998), p. 49
8. J.M. West, *Basic Corrosion and Oxidation*, 2nd edn. (Ellis Horwood Limited, New York, 1980)
9. L. Pauling, *General Chemistry*, 3rd edn. (Dover Publications Inc., New York, 1970)
10. D.A. Jones, *Principles and Prevention of Corrosion* (Macmillan Publishing Company, New York, 1992)
11. American Society for Testing and Materials (ASTM), D 1498-76 (Reapproved, 1981)
12. D.R. Gaskell, *Introduction to Metallurgical Thermodynamics*, 2nd edn. (Hemisphere Publishing Corporation, New York, 1981)

13. P.W. Atkins, *Physical Chemistry*, 3rd edn. (W. H. Freeman and Company, New York, 1986)
14. D.D. Wagman, W.H. Evans, V.B. Parker, R.H. Schumm, I. Halow, S.M. Bailey, K.L., Churney, R.L. Nuttall, The NBS tables of chemical thermodynamic properties. J. Phys. Chem. Ref. Data **11**, supplement # 2, National Bureau of Standards, Washington, D.C.
15. R.M. Garrels, C.L. Christ, *Solutions, Minerals, and Equilibria*, (Jones and Bartlett Publishers, Boston, 1990)

Problems

(Assume SATP conditions unless stated otherwise.)

(2-1) Calculate the standard free energy of the following reaction:

$$2H_2 + O \rightarrow H_2O$$

(2-2) Determine if the following reaction is favored thermodynamically as written assuming unit activities:

$$3Fe^{+2} \rightarrow Fe + 2Fe^{+3}$$

(2-3) Determine the change in free energy for the reaction of ferric and ferrous ions from an initial ferrous to ferric ion ratio of 0.0001 to a final ratio of 100, assuming activity coefficients are equal to one. ($Fe^{3+} + e^- \leftrightarrow Fe^{2+}$, electron free energies are neglected) (answer: +34,229 J/mol)

Calculate the equilibrium potential as well as the equilibrium concentration of nickel if excess metallic cobalt is added to a 0.019 m nickel dichloride solution. Do not use free energy data to solve this problem. (Assume that cobalt will tend to form Co^{2+} ions if it dissolves. Assume nickel has only one oxidized form (Ni^{2+}). Assume chloride does not play a role in the reaction. Assume all nonspecified species have negligible activity.)

(2-5) Calculate the measured slope and intercept (when potential is plotted versus the natural logarithm of concentration of cupric ions (**in ppm**)) for a cupric ion selective electrode measuring dilute concentrations of cupric (Cu^{2+}) ions in a sulfuric acid (0.0333 m H_2SO_4) solution using a silver-silver chloride reference electrode that is saturated with respect to KCl (0.199 V SHE). Assume that the electrode performs as expected based upon the reaction $Cu^{2+} + 2e^- = Cu$ according to the Nernst Equation.

(2-6) Calculate the solubility product for calcium dihydroxide if the products are calcium ions (Ca^{2+}) and hydroxide ions (OH^-).

(2-7) Determine whether a KCl precipitate will be present at equilibrium if the initial activities of K^+ and Cl^- are 4.

(2-8) Determine the equilibrium concentration of Ag^+ and the equilibrium pH when 0.05 mol of $AgNO_3$ are added to 1000 g of water with an initial pH of 12.3, assuming the only product is Ag(OH). Assume $I = 0.1$.

(2-9) Calculate the pKa for the hydrolysis reaction: $Fe^{3+} + 2H_2O = Fe(OH)_2^+ + 2H^+$, and what will the pH be if the activities of Fe^{3+} and $Fe(OH)_2^+$ are equal?

(2-10) A copper solution containing 7 g/l copper at pH 3 is to be reduced in an autoclave with hydrogen (5 atm constant pressure). If the reaction proceeds to completion, what will be the final copper concentration and pH? (Assume an ionic strength of 0.1) (answer: $m_{Cu} = 5.45 \times 10^{-14}$ m)

(2-11) What is the electrochemical potential of a solution containing 0.0001 M Fe^{+2} with 0.0001 M Fe^{+3} assuming only those ions are involved and unit activity coefficients? (answer: 0.770 V)

(2-12) A copper solution containing initially 7 g/l copper at pH 3 is to be reduced in an autoclave (pressure vessel) operated in batch mode with hydrogen (maintained at 5 atm pressure). If the reaction in the autoclave proceeds to completion, what will the final copper concentration and pH be? (Assume an ionic strength of 0.1) (answer: 5.43×10^{-14} m).

Chapter 3
Speciation and Phase Diagrams

> *Speciation and phase diagrams quantify the effects of changing conditions on species concentrations.*

Key Learning Objectives and Outcomes:
Determine equilibrium species concentrations in complex systems
Create a speciation diagram
Create a simple Eh pH diagram

3.1 Speciation (or Ion Distribution) Diagrams

Speciation diagrams are useful in determining the effects of conditions on the distribution of species. For example in the sulfate/water system it is useful to know the available sulfate ion concentration as a function of pH. At low pH, much of the sulfate is complexed with hydrogen. The complexed sulfate is not available for reaction as sulfate. At high pH, nearly all sulfate is dissociated from hydrogen ions. Dissociated sulfate is readily available to react with other species. A speciation diagram can provide a visual and quantitative representation of species availability. Speciation and phase diagrams can be made as a function of a variety of parameters.

Producing a speciation diagram requires thermodynamic data and calculations. The details of these calculations were discussed previously. The first step in producing a diagram is to determine the species that need to be evaluated. In the sulfuric acid/water system, the sulfate, hydrogen sulfate, and sulfuric acid need to be evaluated. The common atom in each of these species is sulfur. It is, therefore, useful to construct the diagram with each species specified in terms of a fraction of the total sulfur. For the sulfate/water system, the reactions of greatest interest are:

> *Speciation and phase diagrams are based on thermodynamic data.*

© The Minerals, Metals & Materials Society 2022
M. L. Free, *Hydrometallurgy*, The Minerals, Metals & Materials Series,
https://doi.org/10.1007/978-3-030-88087-3_3

$$H^+ + SO_4^{2-} \leftrightarrow HSO_4^- \tag{3.1}$$

$$H^+ + HSO_4^- \leftrightarrow H_2SO_4 \tag{3.2}$$

Note that the reaction $2H^+ + SO_4^{2-} \leftrightarrow H_2SO_4$ could also have been included. It was not included because it is not independent. In other words, it could have been derived using the previous equations.

The free energies and equilibrium constants for Eqs. 3.1 and 3.2 are:

$$\Delta G_r^o = -755.91 - [0 + -744.63] = -11.28 \text{ kJ/mol}; \quad K_{H1} = 94.9 \tag{3.3}$$

$$\Delta G_r^o = -689.995 - [-755.910 + 0] = 65.915 \text{ kJ/mol}; \quad K_{H2} = 2.79 \times 10^{-12} \tag{3.4}$$

Rewriting these expressions by substituting in the species activities results in:

$$\frac{a_{HSO_4-}}{a_{H+}a_{SO_4^{2-}}} = 94.9 \tag{3.5}$$

$$\frac{a_{H_2SO_4}}{a_{H+}a_{HSO_4-}} = 2.79 \times 10^{-12} \tag{3.6}$$

The H_2SO_4 concentration will be small based on Eq. 3.6 because the other activities will be near or less than one. Therefore, H_2SO_4 will be neglected. Thus, Eq. 3.3 is sufficient to adequately describe the effect of pH on the sulfur-containing species in this simple analysis. Taking the logarithm of Eq. 3.3 leads to:

$$\log\left(\frac{a_{HSO_4-}}{a_{SO_4^{2-}}}\right) = \log(94.9) + \log(a_{H+}) = 1.977 - pH \tag{3.7}$$

Equation 3.7 is used along with the assumption that the HSO_4^- and SO_4^{2-} activity coefficients are equal. Additionally, the total sulfur molality/activity is assumed to be one. These assumptions and the following equations result in Table 3.1.

$$\left(\frac{\gamma_{HSO_4-}m_{HSO_4}}{\gamma_{SO_42-}m_{SO_42-}}\right) = 10^{(1.977-pH)} \tag{3.9}$$

$$m_{HSO4-} = (S_{tot} - m_{HSO4-})10^{(1.977-pH)} \tag{3.9}$$

$$m_{HSO_4-} = \frac{m_{Stotal}10^{(1.977-pH)}}{1 + 10^{(1.977-pH)}} \tag{3.10}$$

The accompanying plot of the data in Table 3.1 is presented in Fig. 3.1.

The data in Fig. 3.1 show that sulfate concentration increases dramatically above pH 1. The sulfate ion activity reaches a maximum at approximately pH 4. At pH 2, HSO_4^- and SO_4^{2-} ions have nearly equal activities. Although Fig. 3.1 is useful, more practical speciation diagrams include more species. More practical diagrams are more complicated to produce.

The phosphate/water system is more complex than the previous example. For the phosphate/water system, there are three dominant equilibria:

$$H^+ + PO_4^{3-} \leftrightarrow HPO_4^{2-} \ (\Delta G_r^o = -75{,}000 \ \text{J/mol}; \ K_{H1} = 1.40 \times 10^{13}) \quad (3.11)$$

$$H^+ + HPO_4^{2-} \leftrightarrow H_2PO_4^- \ (\Delta G_r^o = -41{,}000 \ \text{J/mol}; \ K_{H2} = 1.54 \times 10^7) \quad (3.12)$$

$$H^+ + H_2PO_4^- \leftrightarrow H_3PO_4 (\Delta G_r^o = -12{,}000 \ \text{J/mol}; \ K_{H3} = 126.9) \quad (3.13)$$

This results in three equations and five unknowns. However, the variable H^+ will be set by varying the pH over a range of values. Thus, H^+ is considered as a known variable. The other equation needed is a mass balance equation. The relevant equation is $H_3PO_4 + H_2PO_4^- + HPO_4^{2-} + PO_4^{3-} = P_{total}$. P_{total} will be considered to be one for convenience. However, it should be noted that in practice, P_{total} is often plotted with reference to 100% rather than one.

The system of equations for the phosphate/water system includes three nonlinear equations together with one linear equation. Therefore, obtaining a solution is not as simple as it was for the sulfate/water system. However, the solution can be easily obtained using modern spreadsheet software packages.

Many species, such as sulfate and phosphate, form complexes with hydrogen ions. Thus, the final equilibrium speciation is often pH dependent. Similarly, other complexes and related speciation are influenced by other ions.

The spreadsheet is set up with one column dedicated to each of the species activities as well as the pH. An additional column is set up for the total phosphorus. Each cell is then set up in terms of the other cells based on appropriate constants. Equilibrium or mass balance constants are established from chemical and thermodynamic principles. One of the phosphate species will be the basis for the other cell calculations. Its value will be determined by the spreadsheet solving tool. Its value will be optimized to best satisfy all of the equations. The spreadsheet software solution to the phosphate/water system is presented in Fig. 3.2.

Using data from Figs. 3.1 and 3.2, the fraction of any of the relevant species can be easily obtained. This is a very important tool in hydrometallurgy. Most hydrometallurgical reactions depend on individual species concentrations. Usually

Table 3.1 Comparison of HSO$_4^-$ and SO$_4^{2-}$ as a Function of pH

pH	$S_{HSO_4^-}$	$S_{SO_4^{2-}}$
0.0	9.90E−01	1.04E−02
1.0	9.05E−01	9.53E−02
2.0	4.87E−01	5.13E−01
3.0	8.67E−02	9.13E−01
4.0	9.40E−03	9.91E−01
5.0	9.48E−04	9.99E−01
6.0	9.49E−05	1.00E+00
7.0	9.49E−06	1.00E+00
8.0	9.49E−07	1.00E+00
9.0	9.49E−08	1.00E+00
1.0	9.49E−09	1.00E+00
1.1	9.49E−10	1.00E+00
1.2	9.49E−11	1.00E+00
1.3	9.49E−12	1.00E+00
1.4	9.49E−13	1.00E+00

only the total quantity of a given atom in the system can be measured. Speciation information also provides pH values of important ion dissociation events. The same principles used to generate the pH-based diagrams can also be used to generate Eh-based, partial pressure-based, or other similar ion distribution diagrams.

Fig. 3.1 Speciation or ion distribution diagram for the sulfate/water system

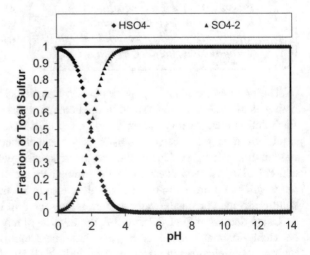

Fig. 3.2 Speciation or ion distribution diagram for the phosphate/water system based on activities rather than molar quantities. Note that the total fraction is based upon the assumption that concentration is equivalent to activity

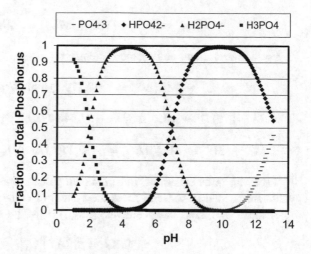

3.2 Metal–Ligand Speciation Diagrams

The approach to determining metal speciation is similar to the previous examples. However, some simplifications are often possible. Metal ions are often treated differently because they do not complex readily with H^+ ions. Consequently, resulting diagrams are often plotted versus the counter ion or ligand with which they associate. Some terminology is needed to explain these methods.

β_n is the formation constant from the fundamental ions. It is the same as the formation equilibrium constant. In contrast, equilibrium constants can represent dissociation reactions as well as formation reactions with varying coordination numbers and transitions. If used properly, β_n is equal to $1/K_{sp}$. β_n is applicable for a coordination number of "n" from n units of the ligand.

$$\beta = \frac{a_{ML_n}}{a_L^n a_M} \approx \frac{[ML_n]}{[L]^n[M]} \tag{3.14}$$

L is the ligand. A ligand is a species than is capable of combining with another species to form a complex. Ligand is often used in association with metals as the other species. As an example consider $Cu(Cl)_2^-$ in which Cu^+ is the metal ion of interest and Cl^- is the ligand. A general chemical formulation that uses a ligand and a metal is ML_n where n is the coordination number, and L represents the ligand species.

α_n is the species fraction. It is defined as the fraction of species ML_n relative to M_{total} or

$$\alpha_n = \frac{[ML_n]}{[M_{total}]} \tag{3.15}$$

$$M_{tot} = [M^+] + [ML] + [ML_2^-] + \ldots [ML_n^{(n-1)-}] \tag{3.16}$$

For a model metal and ligand system:

$$M^+ + L^- \leftrightarrow ML(\Delta G_r^o = -38{,}570 \text{ J/mol}; \beta_1 = 5.77 \times 10^6) \tag{3.17}$$

$$M^+ + 2L^- \leftrightarrow ML_2^- (\Delta G_r^o = -27{,}540 \text{ J/mol}; \beta_2 = 6.72 \times 10^4) \tag{3.18}$$

$$M^+ + 3L^- \leftrightarrow ML_3^{2-} (\Delta G_r^o = -32{,}170 \text{ J/mol}; \beta_3 = 4.36 \times 10^5) \tag{3.19}$$

There are three equations, five unknown values. Thus, one equation and one set value are needed in order to solve the system of equations.

A metal (M) mass balance gives: $[M_{total}] = [M^+] + [ML] + [ML_2^-] + [ML_3^{2-}]$

$$[ML] = \beta_1 [M^+][L^-] \tag{3.20}$$

$$[ML_2^-] = \beta_2 [M^+][L^-]^2 \tag{3.21}$$

$$[ML_3^{2-}] = \beta_3 [M^+][L^-]^3 \tag{3.22}$$

The fraction of total M found in each of the individual species is given by:

$$\alpha_n = \frac{\beta_n [M^+][L^-]^n}{[M^+] + \beta_1 [M^+][L^-] + \beta_2 [M^+][L^-]^2 \ldots \beta_{total}[M^+][L^-]^{total}} \tag{3.23}$$

For the specific example given, the individual fractions can be given in simplified forms. The simplified forms result from the cancellation of the M^+ concentration as follows:

$$\alpha_o = \frac{1}{1 + \beta_1 [L^-] + \beta_2 [L^-]^2 + \beta_3 [L^-]^3} \tag{3.24}$$

$$\alpha_1 = \frac{\beta_1 [L^-]}{1 + \beta_1 [L^-] + \beta_2 [L^-]^2 + \beta_3 [L^-]^3} \tag{3.25}$$

$$\alpha_2 = \frac{\beta_3 [L^-]^2}{1 + \beta_1 [L^-] + \beta_2 [L^-]^2 + \beta_3 [L^-]^3} \tag{3.26}$$

$$\alpha_3 = \frac{\beta_3 [L^-]^3}{1 + \beta_1 [L^-] + \beta_2 [L^-]^2 + \beta_3 [L^-]^3} \tag{3.27}$$

Note that for this scenario that involves only one metal oxidation state and one ligand, the resulting metal species fractions can be calculated by simple solution methods that do not require the use of solver routines in spreadsheets or calculators.

However, for many realistic problems, the speciation is more complicated. More complicated scenarios generally cannot be simplified sufficiently to avoid numerical solutions. As an example consider the following set of equations:

$$M^+ + L^- \leftrightarrow ML(\Delta G_r^o = -38{,}570 \text{ J/mol; } K_1 = 5.77 \times 10^6) \tag{3.28}$$

$$M^+ + 2L^- \leftrightarrow ML_2^-(\Delta G_r^o = -27{,}540 \text{ J/mol; } K_2 = 6.72 \times 10^4) \tag{3.29}$$

$$M^{2+} + L^- \leftrightarrow ML^+(\Delta G_r^o = -2450 \text{ J/mol; } K_3 = 2.688) \tag{3.30}$$

$$M^{2+} + e^- \leftrightarrow M^+(E^o = 0.1611 \text{ V }; K_4 = 0.0926 \text{ at } E_{\text{soln}} = 0.3 \text{ V}) \tag{3.31}$$

There are four equations and six unknowns, so one more equation and one set value are needed.

The metal M mass balance gives $[M_{\text{total}}] = [M^+] + [ML] + [ML_2^-] + [ML^+] + [M^{2+}]$. The equilibrium equations are:

$$[ML] = K_1[M^+][L^-] \tag{3.32}$$

$$[ML_2^-] = K_2[M^+][L^-]^2 \tag{3.33}$$

$$[ML^+] = K_3[M^{2+}][L^-] \tag{3.34}$$

$$[M^+] = K_4[M^{2+}] \tag{3.35}$$

The fraction of total M found in each of the individual species is given by:

$$\alpha_o = \frac{[M^+]}{[M^+] + K_1[M^+][L^-] + K_2[M^+][L^-]^2 + K_3[M^{2+}][L^-] + K_4[M^{2+}]} \tag{3.36}$$

$$\alpha_1 = \frac{K_1[M^+][L^-]}{[M^+] + K_1[M^+][L^-] + K_2[M^+][L^-]^2 + K_3[M^{2+}][L^-] + K_4[M^{2+}]} \tag{3.37}$$

$$\alpha_2 = \frac{K_2[M^+][L^-]^2}{[M^+] + K_1[M^+][L^-] + K_2[M^+][L^-]^2 + K_3[M^{2+}][L^-] + K_4[M^{2+}]} \tag{3.38}$$

$$\alpha_3 = \frac{K_3[M^{2+}][L^-]}{[M^+] + K_1[M^+][L^-] + K_2[M^+][L^-]^2 + K_3[M^{2+}][L^-] + K_4[M^{2+}]} \tag{3.39}$$

$$\alpha_4 = \frac{[M^{2+}]}{[M^+] + K_1[M^+][L^-] + K_2[M^+][L^-]^2 + K_3[M^{2+}][L^-] + K_4[M^{2+}]} \quad (3.40)$$

Note that K has been used in place of β because both M^+ and M^{2+} ions are present. Consequently, the β term cannot be applied.

Using increasing set values for the ligand concentration, the individual fractions can then be plotted with respect to set ligand concentration values when the equations are solved simultaneously. One effective method of simultaneously solving such a system of equations involves utilizing spreadsheet solver and macro routines.

The basic format for spreadsheet solutions is as follows:

Spreadsheet Columns

A	B	C	D	E	F	G	H
$[L^-]$	$[M^+]$	$[M^{2+}]$	$[ML]$	$[ML_2^-]$	$[ML^+]$	$[M_{total}]$	Target
Set	solve	f(B,G)	f(A,B)	f(A,B)	f(A,C)	set	f(all M)

Target cell value = [M_{total}] - column B - column C - column D - column E - column F

The system of equations is solved using a spreadsheet solver routine. This process requires a target cell value that is targeted by the calculation. It is often useful to set the target value as the difference between a calculated and known value. As the simulation reaches the solution the target cell value will approach zero. In this example the computer will change the value of M^+ until the target cell approaches zero. As the value of M^+ is changed, the other column values, which are functions of M^+, will also change. Thus, column B will be varied until the value in column H is lower than a specified limit. Often the user needs to check to make sure the minimization will go to less than 1×10^{-15}. Consequently, the user must often change default calculation options.

Values for the ligand concentration and the total metal concentration must be set in the appropriate cells. Often it is useful to set the ligand concentration at a desired level in the first row of the first column. A multiplier, such as $A2 = A1 * k$ can be used in subsequent cells to naturally increment the ligand concentration. The functions used for the other columns are determined by the equilibrium constants and other appropriate columns as indicated. As an example the result for Eq. 3.29, which solves for the concentration of ML, would be placed in the first available cell in column D. The cell would look something like: $D4$ = (value of equilibrium constant $K1)*A4*B4$. Other values would be calculated in a similar manner. Macros are often helpful in generating automated calculation routines.

The resulting plot for the equations shown is presented in Fig. 3.3.

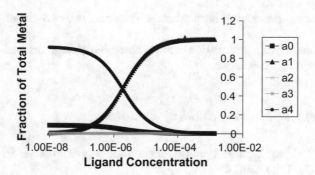

Fig. 3.3 Comparison of metal ion fraction and ligand concentration as determined using a spreadsheet for the example presented previously. a0, a1, a2, a3, and a4 represent the fraction of metal associated with the respective number of ligands

3.3 Phase Stability Diagrams

Phase stability diagrams, such as Fig. 3.4, provide useful species information. They relate conditions to equilibrium species. However, as with speciation diagrams, stability diagrams are only valid under equilibrium conditions. Additionally, phase diagrams provide only regions for the dominant species. They provide no information about nondominant species concentrations, which may be significant. Also, if the system is not at equilibrium (sometimes true equilibrium takes years to achieve due to slow kinetics), the diagram may give misleading information. They can be made to include metastable species. Metastable species are species that are thermodynamically favored but not as favorable as their decomposition into another species. Most diagrams do not include metastable species. Despite their limitations,

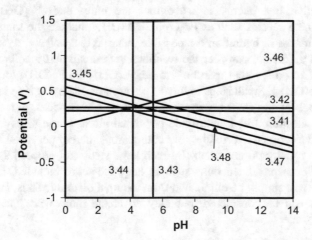

Fig. 3.4 Phase stability plot of copper system given in Examples 3.1 (parts a–g)

phase diagrams are useful in predicting dominant species under equilibrium conditions.

The accompanying information shows how phase diagrams are constructed. Copper is used as an example.

3.3.1 Steps to Making Stability Diagrams

Example 3.1 Make a phase stability diagram for copper in water including only Cu^+, Cu^{2+}, Cu, CuO, CuO_2.

(1) Write out the species to be considered:

Example 3.1a Cu^+, Cu^{2+}, Cu, CuO, CuO_2 (note that $Cu(OH)_2$ is not stable, $Cu(OH)^+$ and CuO_2^{2-} were neglected here to reduce the number of species and simplify the construction of the diagram. Additional species will be shown in a similar diagram in Chap. 5.)

(2) Write Out Chemical Equilibria

   (a) Make sure all of the possible equilibria are listed. There may be more or fewer equations than there are species, since phase boundaries, not species concentrations, are being calculated.
   (b) Make sure each species is represented in at least one equation. Most species will be involved in more than one equation.
   (c) Make sure that no metastable species are considered. (As an example, when $Fe(OH)_3$ and Fe_2O_3 are considered, only Fe_2O_3 is thermodynamically stable in water at room temperature (i.e. $2Fe(OH)_3 \leftrightarrow Fe_2O_3 + 3H_2O$ has a negative free energy and activities are unity until $Fe(OH)_3$ vanishes) Examine species such as Fe_2O_3 and $Fe(OH)_3$ that are the same except for differences in hydration for possible metastable species.
   (d) As a guideline, consider the reactions with small and medium steps first. (Fe^{3+} to Fe^{2+}, rather than Fe^{3+} to Fe; $Fe(OH)_2^+$ to $Fe(OH)_3$, rather than Fe to $Fe(OH)_3$). Small steps involve 1 electron or one hydroxide ion per metal atom. Medium steps involve 2 electrons or 2 hydroxides, or a combination of 1 electron and 1 hydroxide per metal atom. Large steps involve 3 or more electrons, hydroxides, or combinations thereof per metal atom. Large steps usually do not occur. Instead, large steps can proceed as a series of smaller steps. If reactions such as large steps are included, the resulting lines will simply be eliminated from the final diagram. Thus, large reaction steps can be considered, but they do create more work than is usually necessary.

(e) Balance chemical equilibria by first balancing oxygen using H_2O, then balancing hydrogen using H^+, then balancing the charge by adding electrons. (for $Cu^+ \leftrightarrow CuO$, the steps would be: $Cu^+ + \mathbf{H_2O} \leftrightarrow CuO$, then. $Cu^+ + H_2O \leftrightarrow CuO + \mathbf{2H^+}$, then finally $Cu^+ + H_2O \leftrightarrow CuO + 2H^+ + \mathbf{e^-}$).

(f) Always write oxidation-reduction reactions cathodically with the electron on the left hand side of the expression $(Fe^{3+} + e^- \leftrightarrow Fe^{2+})$. If there is a reaction with water, hydrogen, and an electron, the electron takes priority in terms of how the reaction is written - always cathodically with the electron on the left hand side. (Note that the European community often uses the reverse notation.)

(g) The expressions should contain all logical relationships between species.

Example 3.1b

$$Cu^+ + e^- \leftrightarrow Cu \tag{3.41}$$

$$Cu^{2+} + e^- \leftrightarrow Cu^+ \tag{3.42}$$

$$CuO + 2H^+ \leftrightarrow Cu^{2+} + H_2O \tag{3.43}$$

$$Cu_2O + 2H^+ \leftrightarrow 2Cu^+ + H_2O \tag{3.44}$$

$$2CuO + 2H^+ + 2e^- \leftrightarrow Cu_2O + H_2O \tag{3.45}$$

$$2Cu^{2+} + H_2O + 2e^- \leftrightarrow Cu_2O + 2H^+ \tag{3.46}$$

$$Cu_2O + 2H^+ + 2e^- \leftrightarrow 2Cu + H_2O \tag{3.47}$$

$$CuO + 2H^+ + 2e^- \leftrightarrow Cu + H_2O \tag{3.48}$$

(3) Calculate the Free Energy for the Reactions:

(Note that kJ (kilojoules) were used instead of J (Joules) to save space).

Example 3.1c

$$\Delta G_r^o = 0 - [49.98] = -49.98 \text{ kJ/mol} \tag{3.49}$$

$$\Delta G_r^o = 50.16 - [65.52] = -15.36 \text{ kJ/mol} \tag{3.50}$$

$$\Delta G_r^o = 65.52 - 237.14 - [-129.56] = -42.06 \text{ kJ/mol} \tag{3.51}$$

$$\Delta G_r^o = 2(49.98) - 237.14 - [-146.03] = 8.85 \text{ kJ/mol} \tag{3.52}$$

$$\Delta G_r^o = -146.03 - 237.14 - 2[-129.56] = -124.05 \text{ kJ/mol} \qquad (3.53)$$

$$\Delta G_r^o = -146.03 - [2(65.52) - 237.14] = -39.93 \text{ kJ/mol} \qquad (3.54)$$

$$\Delta G_r^o = -237.14 - [-146.03] = -91.11 \text{ kJ/mol} \qquad (3.55)$$

$$\Delta G_r^o = -237.14 - [-129.56] = -107.58 \text{ kJ/mol} \qquad (3.56)$$

(4) Decide what needs to be plotted and which plot will work best.

   (a) For a reaction of the type: $bB + hH^+ \leftrightarrow cC + dH_2O$, it will probably be best to plot the logarithm of the activity of B versus pH. This applies even if the reaction is written in terms of OH^- instead of H^+.
   (b) If the reaction is of the type: $bB + hH^+ + ne^- \leftrightarrow cC + dH_2O$, it will be best to plot E (or Eh) versus pH.
   (c) If a reaction involves gases as either species B or C, it may be best to plot the logarithm of the gas pressure versus pH.
   (d) If a reaction involves gases and does not involve either pH or E, a plot of the logarithm of the gas pressure versus the logarithm of the activity (or pressure) of other species may be needed.
   (e) If a reaction involves metal ions and H^+ with no change in oxidation state it may be best to plot ion concentration or activity versus pH.

Example 3.1d In this case electrons are transferred and pH is an important variable, so it will generally be most useful to make an Eh, pH diagram.

(5) Calculate Equations for Equilibrium Lines:

Example 3.1e For Eq. 3.41, the general expression is:

$$E = E_o - \frac{2.303RT}{nF} \log\left[\frac{a_{Cu}}{a_{Cu^+}}\right]$$

The value of E^o can be calculated from the free energy (see Chap. 2) or by obtaining the value from Appendix G. For this reaction, the E^o value is 0.520 V. Using this value along with some rearranging and substituting leads to:

$$E = 0.520 + \frac{2.303RT}{F}\left[\log a_{Cu^+}\right]$$

Because the plot is potential (E) versus pH, activities cannot be variables if the plot is to remain 2-dimensional. Thus, the activities must be assigned. Different activities will lead to different diagrams. In this example, the activity of Cu^+ will be assigned a value of 1×10^{-5}. Using this value, the equilibrium line for the Eq. 3.41 becomes:

$$E = 0.224 \text{ V}$$

For Eq. 3.42, the general equation is:

$$E = 0.153 + \frac{2.303RT}{F} \left[\log \frac{a_{Cu+2}}{a_{Cu+}} \right]$$

which, after substitution of the activities ($a_{Cu+2}a_{Cu+2} = 1 \times 10^{-3}$), becomes:

$$E = 0.271 \text{ V}$$

Equation 3.43 is not a half-cell reaction. Therefore, the Nernst Equation is not used. Thus, the free energy equation is used in the form:

$$\Delta G_r^o = -2.303RT \log \left[\frac{a_{H_2O}a_{Cu^{2+}}}{a_{CuO}a_{H^+}^2} \right],$$

or

$$\frac{\Delta G_r^o}{2.303RT} = \frac{-42,060}{2.303(8.314)(298)} = -\log a_{Cu^{2+}} - 2pH,$$

which for the specified cupric ion activity ($a_{Cu^{2+}} = 1 \times 10^{-3}$) and reaction free energy becomes:

$$pH = 5.19$$

Equation 3.44 leads to the following line expression.

$$\frac{\Delta G_r^o}{2.303RT} = \frac{8850}{2.303(8.314)(298)} = -2 \log a_{Cu^+} - 2pH,$$

which upon substitution for $a_{Cu^+} = 1 \times 10^{-5}$ becomes:

$$pH = 4.22$$

The equations needed for Eq. 3.45 are:

$$E = E_o - \frac{2.303RT}{nF} \log \left[\frac{a_{Cu_2O} a_{H_2O}}{a_{CuO}^2 a_{H^+}^2} \right]$$

$$E_o = \frac{-\Delta G_r^o}{nF} = \frac{-(-124,050 \text{ J/mole})}{2(96,485 \text{ J/Coul.mol})(\frac{J}{\text{Coul.V}})} = 0.643 \text{ V}$$

$$E = E_o - \frac{2.303RT}{nF} \log \left[\frac{(1)(1)}{(1)^2 a_{H^+}^2} \right] = E_o + \frac{(2)2.303(8.314)(298)}{2(96,485)} \log(a_{H^+})$$

$$E = 0.643 - 0.0591 \text{pH}$$

Equation 3.46 can be expressed in the Nernst Equation format as:

$$E = E_o - \frac{2.303RT}{nF} \log \left[\frac{a_{Cu_2O} a_{H^+}^2}{a_{Cu^{2+}}^2 a_{H_2O}} \right]$$

$$E_o = \frac{-\Delta G_r^o}{nF} = \frac{-(-39,930 \text{ J/mole})}{2(96,485 \text{ J/Coul.mol})(\frac{J}{\text{Coul.V}})} = 0.207 \text{ V}$$

The combination of the previous two expressions and rearrangement give:

$$E = 0.207 + 0.0591 \text{pH} + 0.0591 \log a_{Cu+2} = 0.0296 + 0.0591 \text{pH}$$

For Eq. 3.47, the equation is:

$$E = E_o - \frac{2.303RT}{nF} \log \left[\frac{a_C^c a_{H_2O}^d}{a_B^b a_{H^+}^h} \right]$$

$$E_o = \frac{-\Delta G_r^o}{nF} = \frac{-(-91,110 \text{ J/mole})}{2(96,485 \text{ J/Coul.mol})(\frac{J}{\text{Coul.V}})} = 0.472 \text{ V}$$

$$E = 0.472 - \frac{2.303(8.314)(298)}{2(96,485)} \log \left[\frac{a_{Cu}^2 a_{H2O}^1}{a_{Cu2O} a_{H^+}^2} \right]$$

which reduces to:

$$E = 0.472 - 0.0591 \text{pH}$$

For Eq. 3.48, the equations are:

$$E = E_o - \frac{2.303RT}{nF} \log \left[\frac{a_{Cu}a_{H_2O}}{a_{CuO}a_{H^+}^2} \right]$$

$$E_o = \frac{-\Delta G_r^o}{nF} = -(-107,580 \text{ J/mole})2(96,485 \text{ J/Coul.mol]})\left(\frac{\text{J}}{\text{Coul.V}}\right) = 0.558\text{V}$$

$$E = 0.558 - \frac{2.303(8.314)(298)}{2(96,485)} \log \left[\frac{a_{Cu}^2 a_{H_2O}^1}{a_{Cu_2O}a_{H^+}^2} \right]$$

which reduces to:

$$E = 0.558 - 0.0591\text{pH}$$

(6) Plot the equilibrium lines with equation labels.

Example 3.1f See Fig. 3.4.

Note that in Example 3.2, some of the intersections of three lines do not meet exactly at one point as they should. These discrepancies are usually due to inaccuracies in either the plot or the free energy data.

(7) Determine the Species on Each Side of Each Line

(a) The easiest way to decide which species goes on each side of the equilibrium line is look at the line, the axes, and the reaction. Next, ask questions such as: "If the pH increases, will the product or reactant be favored according to the reaction?" or "If the ion concentration increases, will the product or reactant become dominant?" or "If the solution potential, *E*, increases (i.e. the more oxidized species form becomes more dominant), will the product or reactant be favored?" (If the reaction is written cathodically, as it should be, the reactants will be favored at higher potentials).

Example 3.1g The diagram was not included here for reasons of clarity (too many species).

(8) Helpful hints to determine the stability regions.

(a) Only one phase combination will satisfy at least a portion of most or all lines.
(b) If lines intersect, at least one portion of each intersecting line must be erased either before or after the intersection. (Some lines may be erased entirely.)

(c) Start assigning regions and erasing line segments using the following logic:

    (1) Ask questions such as: Will the Cu^{2+}/CuO equilibrium line (3.43) be valid below the Cu^{2+}/Cu^+ line (3.41)?—The answer would be no in this case because Cu^{2+} would not exist below the Cu^{2+}/Cu^+ equilibrium. Therefore, the portion of the Cu^{2+}/CuO line (3.43) that is below the Cu^{2+}/Cu^+ line (3.41) should be erased. As another example consider the Cu/CuO line (3.48), which is between the line for Cu/Cu_2O (3.47) and Cu_2O/CuO (3.44) and cannot exist between those lines because all the Cu would have converted to Cu_2O and all of the CuO would have converted to Cu_2O before the line is reached by the species. Thus, Cu would not exist above the line for Cu^+ formation, making the Cu/CuO line invalid throughout the diagram. As another example consider the Cu/Cu^+ line (3.41) which would not be valid after crossing to the right of the Cu^+/Cu_2O (3.44) line because the Cu^+ would be converted to Cu_2O after crossing that line. Thus, the Cu/Cu^+ line (3.41) would be erased to the right of the Cu^+/Cu_2O (3.44) line.

    (2) Noncomplexed metal ions tend to be stable on the acidic side with the lowest oxidation states at lower potentials and lower ion activities.

    (3) Hydroxi/de precipitates tend to be found at high pH values with a vertical line separating them from each succeeding ionized metal hydroxide complex.

    (4) Species with a high oxidation state are generally found in the upper portion of Eh-pH diagrams.

    (5) When removing line segments, it is helpful to note that the line segments with reactions involving less hydroxide (e.g. $M^{2+} + OH^- \leftrightarrow M(OH)^+$) tend to appear at lower pH values than those involving more hydroxide (e.g. $M^{2+} + 2OH^- \leftrightarrow M(OH)_2$) at high pH values.

    (6) When three lines nearly intersect, and should intersect based upon the speciation, the data used to produce them are not completely accurate (or the plotting is not accurate), so some adjustment should be made to allow the lines to intersect.

    (7) When the diagram is finished check to make sure that the lines represent the equilibria between the species shown on the diagram.

Example 3.2h See Fig. 3.5.

(9) Draw in the water stability lines ($4H^+ + O_2 + e^- \leftrightarrow 2H_2O; 2H^+ + 2e^- \leftrightarrow H_2$) as a reference, since water is only stable between these lines. Also, make sure to label any fixed activity values.

Example 3.2i See Fig. 3.6.

A more complete and typical diagram for metal that includes more oxides and one oxyanion is shown in Fig. 3.7.

Figure 3.7 shows many of the transitions that can occur as pH and potential (Eh) change. Increases in potential lead to more oxidized species. Transitions to

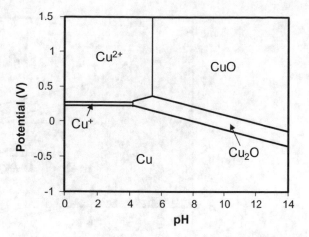

Fig. 3.5 Equilibrium line plot for the copper system given as Example 3.1

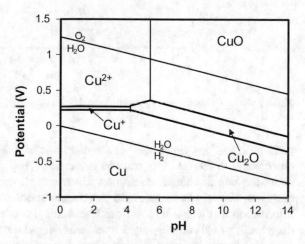

Fig. 3.6 Final phase stability diagram for the copper system given in Examples 3.1 (a–h) with cupric ion activity at 10^{-3} and the cuprous ion activity at 10^{-5}. Note that many possible species were neglected to improve clarity

high pH lead to more oxygen rich species. This figure shows that metal oxides can be converted to soluble species at high and low pH levels. It also shows that metals can be converted to metal oxides by increasing potential, or under some conditions by simply changing pH. The diagram also shows that increased potentials at intermediate pH levels will convert most metals to metal oxides or hydroxides, which are generally not soluble, and, therefore, not leachable. Eh–pH diagrams can also be used to determine the pH levels needed to remove metals from solution or to separate one oxide from another based on dissolution tendencies. Thus, these diagrams are very helpful in determining conditions needed to transform minerals into metals or to purify solutions by hydrometallurgical processing. The diagrams also show the conditions needed for metal corrosion.

Phase diagrams give useful information to the wise user. Improper interpretation of phase diagrams leads to incorrect conclusions and potentially serious errors. The

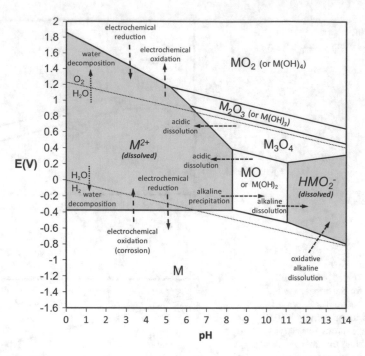

Fig. 3.7 Eh–pH diagram for a hypothetical metal, M, showing typical phases relative to potential and pH

main problems associated with phase diagrams include the fact that the diagrams are specific to the set of conditions presented. They do not provide information about the rate of species transitions. Even though one phase is shown in phase diagrams, several phases are often present at various concentrations. (Consider the information that would be lost from Fig. 3.2 if it was simply converted to a phase diagram. It would contain only 3 lines and four phases as shown in Fig. 3.8.) Most often the conditions used to derive the diagram do not match the conditions of application, thus caution is always advised. Additional useful information about phase stability diagrams can be found in Ref. [5] in Chap. 2.

3.3.2 Precipitation Diagrams

The solubility of many metals in aqueous media is pH related. Metals tend to form hydroxides at specific pH levels. The general formula is:

$$M^{y+} + yOH^- \leftrightarrow M(OH)_y$$

Fig. 3.8 Comparison of information from both a speciation and phase diagram for the phosphoric acid water system. The dashed lines represent phase boundary lines, and the arrows indicate which species would be given as the dominant phase in that region

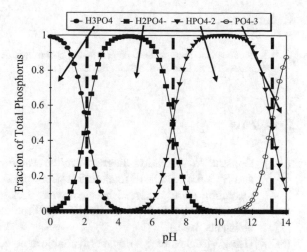

The associated solubility products can be used to construct a simplified phase stability diagram for a variety of metals and their hydroxides as shown in Fig. 3.9.

> *Solubility product data can be used to construct a precipitation diagram that identifies pH conditons leading to precipitation.*

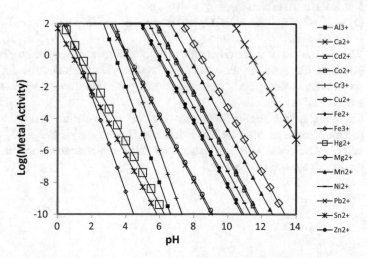

Fig. 3.9 Precipitation diagram based on metal hydroxide solubility products based on data in appendices and Ref. [1]

Reference

1. A.J. Monhemius, Precipitation diagrams for metal hydroxides, sulphides, arsenates, and phosphates. Trans. Inst. Min. Metall. **86**, C202 (1977)

Problems

(3-1) Construct a speciation diagram for the water-sulfide system (H_2S_{aq}, HS^-, and S^{2-}) for 0.0001 m total sulfide from pH 0 to pH 14 with 0.1 pH unit resolution using a spreadsheet software package of your choice. Assume unit activity coefficients to simplify the problem. (show the equations and constants used.)

(3-2) Using Visual Minteq software (available at no cost at http://www.lwr.kth.se/english/OurSoftware/Vminteq/) or a spreadsheet software program determine the concentrations of SO_4^{2-} and HSO_4^- at pH 1, 2, and 3. Assume 1 m total sulfate at 25 °C.

(3-3) Using Visual Minteq software produce a speciation diagram (Fraction of total phosphorus versus pH from pH 1 to pH 13 with 0.1 pH unit resolution) for the phosphorous/water system (PO_4^{3-}, HPO_4^{2-}, $H_2PO_4^-$, and H_3PO_4 species need to be evaluated).

(3-4) Calculate the equilibrium lines (Pourbaix diagram lines in terms of Eh and pH) for the following reactions with an oxygen partial pressure of 1 atmosphere (note that $a_{H+}a_{OH-} = 10^{-14}$):
$O_2 + 2H_2O + 4e^- \leftrightarrow 4OH^-$ $O_2 + 4H^+ + 4e^- \leftrightarrow 2H_2O$ (answer: E = 1.23 − 0.0591 pH).

(3-5) Without the aid of thermodynamic computer software produce an Eh-pH or Pourbaix diagram for the zinc/water system that includes the following species: Zn^{2+}, ZnO, ZnO_2^2, and Zn assuming the dissolved species activities are 1×10^{-6}.

(3-6) Construct a potential-pH diagram for the water-sulfur system using a software package such as HSC. Include $H_2S(aq)$, S^{2-}, HS^-, H^+, SO_4^{2-}, HSO_4^-, OH^-, and H_2SO_4. (Use a total sulfur concentration of 1 m, and a total gas pressure of 1 atm.)

Chapter 4
Rate Processes

> *Rates of hydrometallurgical reactions determine if processing is technically and economically viable.*

Key Learning Objectives and Outcomes:
Understand what factors determine overall reaction rates
Understand basic reaction kinetics
Understand basic biological oxidation kinetics
Understand basic electrochemical kinetics
Determine kinetics parameters from experimental data
Understand basic mass transport equations
Understand and apply relevant specialized mass transport equations
Apply combined kinetics and mass transport equations to solve relevant problems

4.1 Chemical Reaction Kinetics

Understanding the thermodynamic stability of species is critical. However, the rate of reactions is crucial to practical applications. Reactions that require millions of years are of little industrial interest. Reactions that require only milliseconds are potentially dangerous. Consequently, the rate of reactions is very important. Reaction kinetics describes both the rate and mechanism associated with dynamic chemical systems.

4.1.1 Reaction Rate Concepts

Reaction rates are most often determined by reacting species concentrations. Consider a reaction such as:

$$aA + bB \leftrightarrow cC + dD \tag{4.1}$$

© The Minerals, Metals & Materials Society 2022
M. L. Free, *Hydrometallurgy*, The Minerals, Metals & Materials Series,
https://doi.org/10.1007/978-3-030-88087-3_4

The rate of reaction will be related to reactant concentrations and stoichiometry. The associated reaction rate (far from equilibrium) is commonly written as:

$$R = k_f C_A^\chi C_B^\xi \tag{4.2}$$

k_f is the forward reaction rate constant. C_A is the concentration of reactant A. C_B is the concentration of reactant B. The superscripts "χ" and "ξ" reflect the reaction order. If the superscript χ is one, the reaction is first order with respect to the concentration of A. If the superscript ξ is two, the reaction is second order with respect to the concentration of B. The overall reaction order is the sum of χ and ξ. Often, the superscripts "χ" and "ξ" are equal to the coefficients "a" and "b" in the reaction equation. However, due to intermediate steps they may not be equal. Therefore, "χ" and "ξ" may not be the same as "a" and "b".

4.1.2   Determination of the Reaction Order and the Forward Reaction Rate Constant

The reaction order with respect to individual species concentrations can be determined by performing reaction experiments. These experiments measure the change in rate associated with a change in species concentration. Each species concentration is varied independently of the other(s). The effect of concentration change on reaction rate can be compared to theory. Modification of the initial rate equation can facilitate an evaluation. A logarithmic version of the rate equation (Eq. 4.2) is:

$$\ln(R) = \ln(k_f) + \chi \ln(C_A) + \xi \ln(C_B) \tag{4.3}$$

A logarithm-based plot of rate versus concentration reveals reaction order. The reaction order is the slope for each species. Figure 4.1 illustrates such a plot for the reaction of $iI + aAr \leftrightarrow 0.5iI_2 + aAr$. The corresponding slope for argon is one. This slope indicates the reaction is first order with

> *Reaction order is determined as the sum of the powers associated with the species in the rate expression.*

respect to argon. Similarly, the slope for iodine is two. The iodine slope indicates second order iodine kinetics. The sum of the slopes for iodine and argon is three. Thus, the overall reaction order is three. In other words, the overall reaction is a third order reaction. This example of argon and iodine was used because of the clearly defined kinetics. Many hydrometallurgical reactions are less ideal in their kinetics behavior.

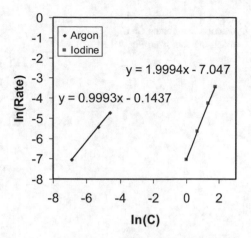

Fig. 4.1 Comparison of the logarithm of the reaction rate versus the logarithm of concentration for argon and iodine for the reaction of argon and iodine to form iodine gas. Data obtained from Ref. [1]

The reaction rate constant can be determined in multiple ways. The rate constant can be determined from Fig. 4.1. The intercept in the plots can be used together with the concentration of the species. Equation 4.3 is used to convert slope and intercept information to obtain k_f. It can also be determined using alternative methods. A plot of rate versus $C_I^2 \, C_{Ar}$ should have a slope of k_f. The corresponding intercept should be zero according to Eq. 4.2. These evaluations assume the reaction follows the kinetics shown. Figure 4.1 can be used to verify the reaction orders. Note that the relationships in Figs. 4.1 and 4.2 are linear. The linearity indicates the kinetics model adequately describes the reaction.

> Rate constants can be found using plots involving the logarithm of concentrations, or if the reaction orders are known, appropriate plots of rates and concentrations can be used to determine rate constants.

Example 4.1 Determine the measured or apparent reaction order for a reaction of acid and a copper mineral if the concentration of copper at the end of one minute of reaction is 6.3, 11.6, and 18.5 ppm for vessels containing 0.015, 0.03, and 0.05 m H^+, respectively. Neglect the effect of mass transport limitations.

The rates of reaction can be measured in ppm per minute to make it easy to evaluate. Next the natural logarithm of the concentrations can be plotted as a function of the natural logarithm of the rate (ppm/min) as shown in Fig. 4.2.

The resulting slope, which is indicative of the reaction order with respect to the reacting species, H^+, is 0.89 or approximately 1. However, the effect of mass transport has not been considered. Consequently, this is the measured or apparent reaction order, but it may not indicate the true reaction order due to the neglect of mass transport, which will be discussed in a subsequent section (Fig. 4.3).

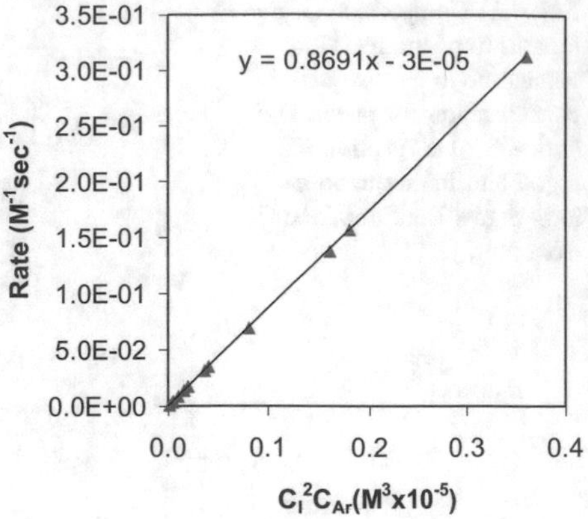

Fig. 4.2 Comparison of reaction rate versus $C_I^2 C_{Ar}$ using the data from Fig. 4.1

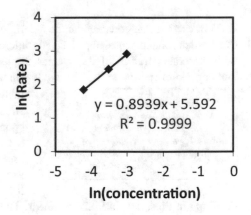

Fig. 4.3 Comparison of the natural logarithm of rate versus the natural logarithm of the concentration for Example 4.1 data

4.1.3 Equilibrium and Reversibility

The rate expressions shown previously are appropriate for reactions far from equilibrium. However, many reactions occur near equilibrium. In such cases the backward reaction rate must be considered. For a simple reaction such as:

$$jX \leftrightarrow mY \tag{4.4}$$

The net rate of a reaction is the difference between forward and backward rates. The associated rate of reaction can be expressed mathematically as:

$$R_{net} = R_{forward} - R_{backward} \tag{4.5}$$

The equation can be expressed in more detail as:

$$R_{\text{total}} = k_f C_X^j - k_b C_Y^m \tag{4.6}$$

V is the volume of the reaction. n is the number of moles. k_f is the forward reaction rate constant. C_X^j is the concentration of species "X" to the j th power. k_b is the backward reaction rate constant. C_Y^m is the concentration of species "Y" to the mth power.

At equilibrium the forward and backward rates are equal by definition. Thus, for first order reactions it can be stated that:

$$k_f C_X = k_b C_Y \tag{4.7}$$

and by rearranging,

$$\frac{k_f}{k_b} = \frac{C_Y}{C_X} \tag{4.8}$$

Equation 4.8 is similar to the equilibrium constant. It is a ratio of the product concentration over the reactant concentration. However, the equilibrium constant is the ratio of activities, not concentrations. Therefore, if concentrations are approximately equal to activities ($a_i \approx C_i$):

$$K = \frac{C_Y}{C_X} = \frac{k_f}{k_b} \tag{4.9}$$

Otherwise, the conversion to concentration needs to be made. Eq. 4.6 can then be equated to free energy ($\Delta G_r^o = -RT\ln K$). The ratio of forward and backward rate constants can then be determined. Equating Eq. 4.9 to free energy leads to:

$$k_b = k_f \exp\left(\frac{\Delta G^o}{RT}\right) \tag{4.10}$$

This equation can be substituted back into Eq. 4.6 to arrive at:

$$R_{\text{total}} = k_f C_X^j - k_f C_Y^m \exp\left(\frac{\Delta G_r^o}{RT}\right) \tag{4.11}$$

The free energy can be approximated using concentrations for a first-order reaction:

$$\Delta G_r = \Delta G_r^o + RT \ln \frac{C_Y}{C_X} \tag{4.12}$$

Rearrangement leads to:

$$\Delta G_r = \Delta G_r^o + RT \ln C_Y - RT \ln C_X \tag{4.13}$$

Substitution of Eq. 4.13 into Eq. 4.11 assuming a first order reaction leads to:

$$R_{\text{total}} = k_f C_X - k_f C_Y \exp\left(\frac{\Delta G_r - RT \ln C_Y + RT \ln C_x}{RT}\right) \tag{4.14}$$

Further rearrangement leads to:

$$R_{\text{total}} = k_f C_X - k_f C_Y \exp\left(\frac{\Delta G_r}{RT}\right) \exp\left(\frac{-RT \ln C_Y}{RT}\right) \exp\left(\frac{RT \ln C_x}{RT}\right) \tag{4.15}$$

Additional rearranging leads to a significant conclusion regarding the reaction rate:

$$R_{\text{total}} = k_f C_X \left(1 - \exp\left(\frac{\Delta G}{RT}\right)\right) \tag{4.16}$$

Equation 4.16 indicates reaction rate is related to free energy. As free energy goes to zero, the reaction rate goes to zero. As the free energy becomes very negative, the reaction can be considered as a forward reaction. In other words, the backward reaction can be neglected if the free energy is sufficiently negative. If the free energy of reaction is −20,000 J/mole, the backward reaction rate, which is given by the exponential term in the $(1 - \exp(\Delta G/RT))$ term, is 0.031% of the forward reaction rate. Under such conditions the backward reaction can be neglected. However, this value is relative. It is relative to the forward reaction rate. Thus, it gives no direct information about the absolute rate. The forward reaction rate constant and reactant concentration must be known to determine absolute rates.

> *Backward reactions are negligible when reaction free energies are greater in magnitude (more negative) than − 20,000 J/mole.*

Example 4.2 Determine the fraction of backward reaction (relative to the forward rate) for a reaction with a free energy of −4,500 J/mole.

The backward portion of the reaction is found as:

$$\text{backward(fraction)} = \exp\left(\frac{\Delta G}{E}\right) = \exp\left(\frac{-4500\,\dfrac{\text{J}}{\text{mole}}}{8.314\,\dfrac{\text{J}}{\text{mole K}}(323\ \text{K})}\right) = 0.187$$

4.1.4 Effect of Temperature on Chemical Reaction Kinetics

Reactions occur only when molecules collide with sufficient energy. Thus, reaction rates are related to collision frequency. Rates are also related to collision energy. The reaction rate constant is proportional to the reaction rate. The rate constant can be expressed in terms of a collision frequency factor, v, and the probability that the reaction collision will have sufficient energy. The energy term is equivalent to exp $(-E_a/RT)$. Mathematically, this relationship is:

$$k_f = ve^{(-E_a/RT)} \tag{4.17}$$

This equation is known as the Arrhenius equation. E_a is the activation energy, and v is the frequency factor. Thus, the frequency factor and the activation energy of the reaction can be determined using an appropriate plot. The intercept (intercept = $\ln(v)$), and slope (slope = $-E_a/R$) of a plot of $\ln(k_f)$ versus $1/T$ provide the needed constants. An example plot is shown in Fig. 4.4.

> *Activation energies are often determined using Arrhenius plots of the natural logarithm of the rate constant versus the inverse of the the absolute temperature.*

The Arrhenius equation can provide valuable information in designing hydrometallurgical processes. For example, the activation energy can be used to predict the rate response to a change in temperature. In other words, the Arrhenius equation can be used to determine the rate as a function of temperature if the constants are known.

The activation energy provides information about temperature sensitivity. The larger the activation energy is, the greater the impact of temperature on rate. A 10 ° C increase in temperature increases reaction rate by 93% when the activation energy is 50,000 J/mole.

Activation energies can also be used to evaluate kinetics control. In many reactions, the overall rate is determined by mass transport. Mass transport processes

Fig. 4.4 Logarithm of reaction constant versus inverse absolute temperature for the reaction of hydrogen with iodine. Data from Reference [2]

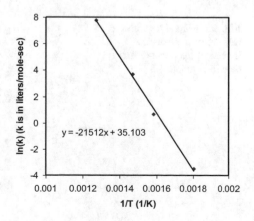

y = -21512x + 35.103

such as diffusion have low activation energies. Aqueous diffusion controlled reactions often have low activation energies less than 15,000 J/mole. Reactions controlled by the chemistry have much larger activation energies. Consequently, diffusion-controlled reactions are not influenced as strongly by temperature as most reaction-controlled processes. Thus, activation energy is often used to help identify mass transport and reaction control in processes.

Example 4.3 A student measures gold adsorption at different temperatures, then determines the adsorption rate constant. The following data is obtained:

k	T (°C)
2620	24
3150	42
3900	62
5330	81.5

4.1.5   Determine the Activation Energy for the Adsorption Process

Determination of activation energy can be made from rate constant and temperature data by plotting the $\ln(k)$ versus $1/T$ and from the slope, which is equal to $-E_a/R$, the activation energy can be determined as the –slope times R (Fig. 4.5).

The slope from the plot is -1274.8. Thus, the activation energy is found as:

$$E_a = -\text{slope}(R) = -(-1274.8)8.314 \text{ J/mole} = 10{,}599 \text{ J/mole}$$

Fig. 4.5 Natural logarithm of rate constant versus the inverse of temperature for data in Example 4.3

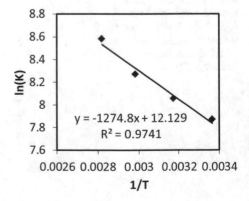

4.2 Biochemical Reaction Kinetics

Biochemical reactions typically involve enzymes together with nutrients. Usually, the nutrient species is first adsorbed at a cell site. At that site there is access to enzymes, proteins, and solution that facilitate transport or conversion. The nutrient either enters the bacterium or remains adsorbed for processing. The bacteria convert

> *Bacteria can enhance hydrometallurgical reaction kinetics substantially by catalyzing reactions such as ferrous ion oxidation to produce ferric ion, which is an oxidant needed for some leaching.*

nutrients for either metabolism or cell components. In aqueous metal processing, bacteria often oxidize dissolved species to obtain energy. *Acidithiobacillus ferrooxidans* (note that the name was changed from *Thiobacillus* to *Acidithiobacillus*) oxidize ferrous ions to ferric ions. They accomplish this by coupling the ferrous-ferric half-cell reaction with water

production. This reaction ($4H^+ + O_2 + 4e^- \leftrightarrow 2H_2O$) can only be utilized if oxygen is provided to the cell. The reaction of oxygen and hydrogen ions in respiring (breathing) organisms facilitates metabolism in these organisms. Metabolic energy is obtained through a series of reactions that are coupled to the water formation reaction.

Ferrous oxidizing bacteria have other needs as well. Bacteria (and other organisms) need other nutrients (i.e. K, Mg, PO_4^{3-}, NH_3, CO_2 and O_2) in order to survive. Assuming that all of the basic needs are met, the oxidation rate is associated with ferrous ions. The ferrous biooxidation rate can be determined using traditional Michaelis–Menten or Monod kinetics [3]:

$$R_{Fe+2ox.} = \frac{C_{cells}\mu_{max}C_{Fe+2}}{Y_c(C_{Fe+2} + K_m)} \tag{4.18}$$

In this equation, C_{cells} is the cell concentration. μ_{max} is the maximum specific growth rate. Y_c is the cell yield coefficient or cell mass pro-

> *Biooxidation rates often follow Michaelis-Menton kinetics.*

duced per substrate mass consumed. K_m is the Michaelis constant. Michaelis–Menten kinetics is based on the assumption of reaction-limiting site availability. Example values for these variables in stirred reactors with sulfide mineral concentrates are: $C_{cells} = 3$ g/l, $\mu_{max} = 0.03$ h^{-1}, $Y_c = 0.07$, $K_m = 0.025$ g/l, $C_{Fe+2} = 0.03$ g/l, based on data in the research literature [4, 5]. The rate of ferrous oxidation is often made using a redox or ORP electrode. The ORP information can be related to concentrations as disucussed in Chap. 2. Values for heap leaching are likely similar, except that the cell concentration is much lower. Heap leaching is also more likely to be controlled by oxygen and carbon dioxide availability. The same form of Eq. 4.18 can be used to describe the effect of oxygen on bacterial activity. Because *Acidithiobacillus ferrooxidans* are autotrophic, they utilize carbon dioxide to build cell mass. If carbon dioxide is not readily available, the mass of

Fig. 4.6 Batch ferrous
biooxidation rate data from
Reference [4]

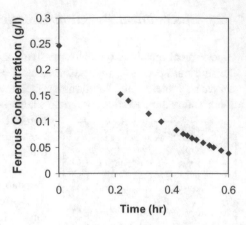

cells will be limited. In bioreactors, the feed gas can be supplemented with CO_2 (1% in air) in order to maximize the cell mass.

In most bioleaching operations there are several strains of bacteria that perform different oxidation functions. The most common bacteria associated with mineral leaching are *Leptospirillum ferrooxidans, Acidithiobacillus ferrooxidans, Acidithiobacillus thiooxidans, Acidithiobacillus acidophilus*, and *Sulfolobus*-like bacteria. Typical ferrous oxidation data in a batch test taken from a steady-state bioreactor with a mixed bacterial culture are presented in Fig. 4.6.

In order to determine the reaction parameters for a given biooxidation scenario, test data can be evaluated. An example of test data was shown in Fig. 4.6. Equation 4.18 can be best used for data evaluation in a rearranged form:

$$\frac{1}{R_{Fe+2ox}} = \frac{Y_c K_m}{\mu_{max} C_{cells} C_{Fe+2}} + \frac{Y_c}{\mu_{max} C_{cells}} \qquad (4.19)$$

Thus, a plot of $1/R$ versus $1/C_{Fe2+}$ can be used to determine the needed constants. Such a plot, referred to as a Lineweaver-Burke Plot, is shown in Fig. 4.7. The data from Fig. 4.6 were used to obtain Fig. 4.7. The data in Fig. 4.7 are linearly related as predicted using the Michaelis–Menten or Monod kinetics model. Therefore, it is likely that the bacterial activity is linked to the ferrous ion concentration and associated enzymatic activity within the bacteria.

The rate of bacterial leaching is also affected by the level of nutrients, contaminants, and flow rates. If necessary nutrients are not available, the bacteria will not be able to reproduce or perform at optimal levels. Toxins such as organic compounds and heavy metals can limit or eliminate growth. Finally, if the bacteria-containing medium is exchanged faster than the bacteria can reproduce, the bacterial population will eventually be "washed out". The wash out condition is limited to low residence time bioreactors.

Fig. 4.7 Lineweaver-Burke plot of batch ferrous biooxidation rate data from Reference [4]

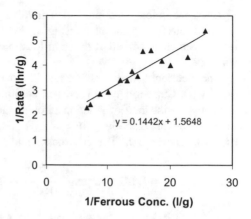

4.3 Electrochemical Reaction Kinetics

4.3.1 Electrochemical Reaction Rate Theory (Butler-Volmer Kinetics)

Electrochemical reaction kinetics is very similar to chemical reaction kinetics.

> Electrochemical kinetics are similar to traditional chemical reaction kinetics except that electron transfer is involved, and, therefore, the electrical potential is very influential.

However, in addition to common parameters, there is a charge transfer probability. The charge transfer probability is related to the potential at the reaction interface. For an electrochemical reaction given as:

$$aXp^{p+} + ne^- \leftrightarrow bX^{m+} \tag{4.20}$$

The resulting equation for the rate of reaction is:

$$R_{\text{electr}} = k_b C_{Xm+}^b \exp\left(\frac{\alpha_a FE}{RT}\right) - k_f C_{Xp+}^a \exp\left(\frac{-\alpha_c FE}{RT}\right) \tag{4.21}$$

a is the order of the reaction with respect to X^{p+}. b is the order of the reaction with respect to X^{m+}. α_a is the anodic charge transfer coefficient. α_c is the cathodic charge transfer coefficient. E is the electrochemical potential. F is the Faraday constant. The other constants are equivalent to previous definitions. The specified concentrations are surface concentrations. However, initially they will be assumed to be bulk concentrations.

Note that forward and backward rate constants have reversed positions. This reversal accommodates the electrochemical convention of positive electron flow for anodic reactions. This convention requires reverse positions relative to chemical kinetics equations.

Electrochemical reactions often involve conductive substrates. Consequently, the substrate is capable of transferring electrons between species. Thus, two different species, adsorbed at different locations, can exchange electrons through the substrate.

The electrochemical reaction rate is most often expressed in terms of a current density (i). Current density is current per unit of area ($i = I/A$). The current density is related to the rate of reaction through the number of electrons transferred per mole of overall reaction ($i/nF = R_{electr}$). The number of electrons transferred per mole of reaction is n. The electrochemical reaction rate is expressed as:

$$i = k_b C_{Xm+}^b \, nF \exp\left(\frac{\alpha_a FE}{RT}\right) - k_f C_{Xp+}^a \, nF \exp\left(\frac{-\alpha_c zFE}{RT}\right) \qquad (4.22)$$

This equation is a form of the Butler-Volmer equation. A more common, simpler form can be derived.

At equilibrium, which occurs at the equilibrium potential, E_{eq}, the net current density is zero:

$$0 = k_b C_{Xm+}^b \, nF \exp\left(\frac{\alpha_a FE_{eq}}{RT}\right) - k_f C_{Xp+}^a \, nF \exp\left(\frac{-\alpha_c zFE_{eq}}{RT}\right) \qquad (4.23)$$

However, the forward and backward rates are not zero at equilibrium. Instead, these rates are equal to the equilibrium exchange current density, i_o:

$$i_o = k_b C_{Xm+}^b \, nF \exp\left(\frac{\alpha_a FE_{eq}}{RT}\right) = k_f C_{Xp+}^a \, nF \exp\left(\frac{-\alpha_c FE_{eq}}{RT}\right) \qquad (4.24)$$

As indicated, i_o is a function of concentration, temperature, and rate constant. The associated rate constant depends strongly on the substrate. Consequently, i_o varies widely with substrate as indicated in Table 4.1. Platinum has a much higher i_o value than tin for the hydrogen reaction shown. When the equilibrium exchange current density is high, a slight perturbation from equilibrium results in a significant reaction rate. If i_o is small, a small perturbation from equilibrium results in a small rate. Thus, i_o represents an important characteristic parameter for determining the rate.

Table 4.1 Selected exchange current densities for (2H$^+$ + 2e$^-$ $\leftrightarrow$ H$_2$) in 1 N HCl [6]

Metal	Exchange current density (A/cm^2)
Pt	10^{-2}
Au	10^{-5}
Fe	10^{-5}
Sn	10^{-7}

Rearrangement of the current density equation leads to:

$$
\begin{aligned}
i = {} & k_b C_{Xm+} nF \exp\left(\frac{\alpha_a F E_{eq}}{RT}\right) \exp\left(\frac{\alpha_a FE}{RT}\right) \exp\left(\frac{-\alpha_a F E_{eq}}{RT}\right) \\
& - k_f C_{Xp+} nF \exp\left(\frac{-\alpha_c F E_{eq}}{RT}\right) \exp\left(\frac{-\alpha_c FE}{RT}\right) \exp\left(\frac{\alpha_c F E_{eq}}{RT}\right)
\end{aligned}
\tag{4.25}
$$

> **The Butler-Volmer equation describes the rate of electrochemical reactions.**

Substitution for the equilibrium exchange current density and rearrangement result in:

$$
i = i_o \left[\exp\left(\frac{\alpha_a F(E - E_{eq})}{RT}\right) - \exp\left(\frac{-\alpha_c F(E - E_{eq})}{RT}\right) \right]
\tag{4.26}
$$

Equation 4.26 is a common form of the Butler-Volmer equation. The collective term "$E - E_{eq}$" is often called the overpotential. The overpotential is represented by the symbol η. It represents the potential "over" the equilibrium potential. A negative overpotential indicates a potential below the equilibrium potential. If the system potential, E, is above E_{eq}, the first, anodic exponential term in the Butler-Volmer equation is more influential. If the potential is below E_{eq}, the second, cathodic term becomes more influential. In other words, if the potential is below E_{eq}, the current flow is negative or cathodic. Negative currents indicate a net reduction reaction.

The Butler-Volmer equation provides the current density for one half-cell reaction. Complete reactions involve at least two coupled half-cell reactions. Coupled half-cell reactions must have balanced currents. In other words, the cathodic current must balance the anodic current. If more than two reactions are coupled, the sum of currents must balance. Mathematically, the current balance can be expressed as:

$$
\sum I_{cathodic} = \sum i_{cathodic} A_{cathodic} = -\sum i_{anodic} A_{anodic} - \sum I_{anodic}
\tag{4.27}
$$

If the area of reaction is the same for all reactions, current densities can replace currents.

As an example, consider a reaction between hydrogen ions, H^+, and metal, M. The application of Eq. 4.26 would give the data as shown in Fig. 4.8. In Fig. 4.8, the production of hydrogen gas is the cathodic reaction (Reaction 1 in Fig. 4.8) and the dissolution of metal (Reaction 2 in Fig. 4.8) is the anodic reaction. The hydrogen and metal reaction currents must balance unless external current is applied. The reaction currents balance at the mixed potential, E_{mix}. The mixed

Fig. 4.8 Plot of current density versus potential for two half-cell reactions (1 and 2). The plot shows the equilibrium potentials ($E1_{eq}$ and $E2_{eq}$) of the individual half-cell reactions

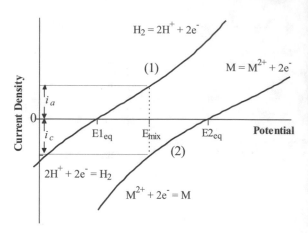

potential is also known as the corrosion or reaction potential. The difference between the mixed potential and the equilibrium potential ($E_{mix}-E_{eq}$; where $E = E_{mix}$ in this comparison) for each half-cell reaction is the "overpotential" for the respective reaction. In the case of the hydrogen reaction the overpotential is negative. The metal reaction has a positive overpotential in this scenario.

Figure 4.8 also shows the resulting mixed potential (E_{mix}) of the combined reactions. The mixed potential occurs when the cathodic and anodic current densities (i_c and i_a) are equal in magnitude, but opposite in sign.

A more typical format for plotting current density and potential is potential (y-axis) versus logarithm of the absolute value of the current density (x-axis) as shown in Fig. 4.9. The resulting plot illustrates the exchange current density values for the half-cell reactions. In addition, the plot shows the mixed potential and reaction current density.

Fig. 4.9 Comparison of potential and (log-scale) $|i|$ for hypothetical reactions (1 and 2). This plot is based on Butler-Volmer electrochemical reaction kinetics

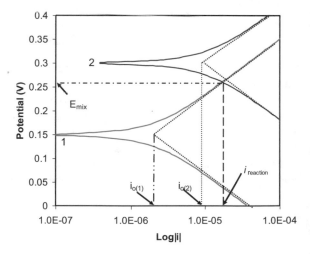

4.3.2 Electrochemical Rate Measurements

Electrochemical reactions usually occur on conductive substrates between ions and the surface. It is not feasible to measure individual electron transfer events with a probe between species. However, it is possible to modify the substrate potential. The conductive substrate is generally referred to as the working electrode. Modifying the electrode potential from equilibrium requires an applied current. The current applied for the potential modification can be measured. The relationship between measured current and potential is very useful. Extrapolations from such data provide reaction information that cannot be directly measured. For example, exchange current densities and reaction rates can be obtained by data extrapolation. However, this approach requires that the individual equilibrium potentials can be determined using the necessary available data and the Nernst Equation.

A typical electrochemical cell is shown in Fig. 4.10. The basic components include a working electrode, a reference electrode, a counter electrode, a vessel, and the desired solution. Often the vessel is sealable and sparged with a gas such as air, oxygen, or argon to provide the desired environment. The working electrode is often mounted in a rotatable fixture. The electrode has electrical contact to a controlled power supply through an insulated wire. The controlled power supply is usually referred to as a potentiostat.

> *Electrochemical measurements are typically made using a 3-electrode cell using working, counter, and reference electrodes.*

At the reaction or mixed potential there is no measured current. In order to change the working electrode potential, a current must be applied. The applied current comes from the counter electrode reaction. The counter electrode potential is changed by a potentiostat to supply the needed current.

> *The mixed potential is the potential at which anodic and cathodic currents balance. It is also known as the reaction or corrosion potential.*

The amount of needed current is determined by working electrode potential and reactions. The working electrode potential is set relative to the reference electrode. The working electrode can be held at a set value if the counter electrode current meets the working electrode current balance. The necessary counter electrode current that must be applied is:

$$I_{\text{applied}} = -\left(\sum I_{\text{anodic}} - \sum I_{\text{cathodic}}\right) \tag{4.28}$$

If all reactions occur on the same homogeneous surface:

$$i_{\text{applied}} = -\left(\sum i_{\text{anodic}} - \sum i_{\text{cathodic}}\right) \quad \text{(for equal cathodic and anodic areas)} \tag{4.29}$$

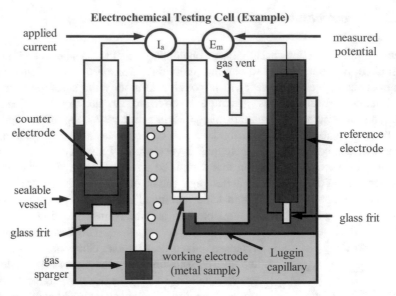

Fig. 4.10 Schematic diagram of a typical electrochemical testing reactor. I is the current. E is the electrochemical potential

> *The potential can be moved from the mixed potential using a counter electrode, which must supply current needed to balance anodic and cathodic currents at the applied potential.*

As the potential is moved from the natural reaction or mixed potential, applied current is needed. The quantity of applied current depends on the difference between the anodic and cathodic currents. The farther the electrode potential is from the mixed potential, the greater the need for applied current.

Figure 4.11 illustrates the applied current needed at a potential $E1$. The potential $E1$ is below the mixed potential given by the intersection of M and X lines. Because $E1$ is less than E_{mix}, reaction X consumes more electrons than can be provided by M. Consequently, a counter electrode must supply the electron deficit. At $E1$, the current density for M is 3.2 mA/cm^2. X has a current density of -6 mA/cm^2 at $E1$. Thus, at $E1$ the counter electrode must provide 2.8 mA/cm^2. Figure 4.12 shows the scenario at $E2$. $E2$ is lower than $E1$. Consequently, the current deficit is greater (8 mA/cm^2) than at $E1$. As the potential is lowered, the deficit increases. Eventually, M supplies an insignificant fraction of the total current. Thus, at large overpotentials the applied current approaches the dominant reaction line. At large overpotentials the potential versus Log$|i|$ data become linear. A comparison of applied current and potential is shown in Fig. 4.13 relative to theoretical values. The linear range of applied current data can be used for extrapolation. The natural reaction rate can be found where the two linear sections intersect. The reaction or mixed potential is found where the Log$|i|$ values approach zero. Because the values never reach zero, the mixed potential is found at minimum $|i|$.

Fig. 4.11 Illustration of the applied current density at a potential, $E1$, which is lower than the natural reaction potential. The anodic reaction rate at $E1$ is 3.2 mA/cm^2 and the cathodic reaction rate is -6 mA/cm^2. Thus, the applied current density is 2.8 mA/cm^2

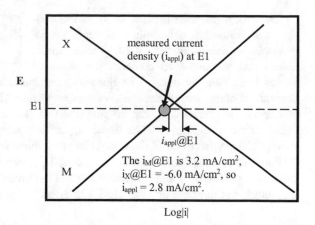

Fig. 4.12 Illustration of the applied current density at a potential, $E2$, which is lower than the mixed or reaction potential and the potential $E1$ shown in Fig. 4.11. The anodic reaction rate at $E2$ is 2 mA/cm^2 and the cathodic reaction rate is -10 mA/cm^2. Therefore, the applied current density is 8 mA/cm^2

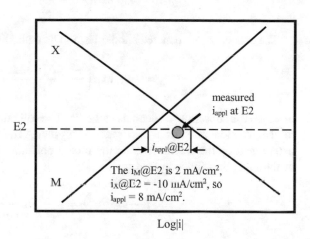

Fig. 4.13 Illustration of the applied current density values above and below the natural reaction potential, shown as the circles, relative to the half-cell reaction data for X and M. The dotted lines show the half-cell reaction data beyond the natural reaction

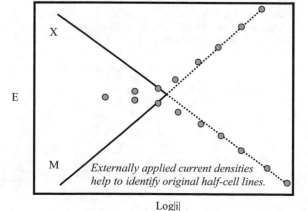

The mixed potential is often far from the anodic and cathodic equilibrium potentials. An
example is shown in Fig. 4.14. The example
shows that at the mixed potential the potential

> *Electrochemical measurement data can be used to determine rates at the mixed potential.*

versus log|i| data are linear. The linearity near the mixed potential is due to the
dominance of one term in the Butler-Volmer equation. Consequently, the respective
Butler-Volmer expressions can be simplified to one exponential term when analyzed far from equilibrium. The linear region is usually well-established approximately 100 mV from equilibrium potentials for typical reactions. In Fig. 4.10 M and
X illustrate the anodic and cathodic reactions. The intersection of M and X occurs at
the mixed potential. M and X reaction rates at the mixed potential are equivalent in
magnitude but opposite in sign. Thus, the anodic term is:

$$i_M = i_{oM} \exp\left(\frac{\alpha_{Ma}F(E - E_{oM})}{RT}\right) \tag{4.30}$$

The subscript "a" denotes anodic term. The cathodic term is:

$$i_X = -i_{oX} \exp\left(\frac{-\alpha_{Xc}F(E - E_{oX})}{RT}\right) \tag{4.31}$$

in which the subscript "c" denotes cathodic. The cathodic and anodic rates are equal
to the reaction rate for a two reaction system. The rate of a reaction can be
determined by the reaction rate at the mixed potential. Rearrangement and the use
of the logarithm function lead to:

$$E - E_{oM} = \eta = \frac{RT}{\alpha_{Ma}F} \ln \frac{i_M}{i_{oM}} \tag{4.32}$$

Fig. 4.14 Plot illustrating
Tafel slopes (dashed lines)
relative to Butler-Volmer
lines (solid lines) for reactions
involving X and M

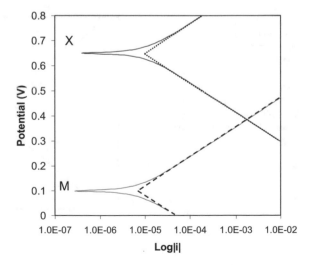

Further rearranging leads to the equation:

$$\eta = \frac{2.303RT}{\alpha_{Ma}F} \log \frac{i_M}{i_{oM}} \tag{4.33}$$

The revised form of the equation, known as the Tafel equation, is

$$\eta_a = \beta_a \log \frac{i_M}{i_{oM}} \tag{4.34}$$

> *The Tafel Equation is a linear form of the Butler-Volmer Equation that can be applied if reaction potentials are far (> 200mV) from their equilibrium.*

This expression is for the anodic reaction as denoted by the subscript "a". The corresponding Tafel slope, β_a, is expressed as:

$$\beta_a = \frac{2.303RT}{\alpha_a F} \tag{4.35}$$

The corresponding equations for the cathodic reaction, which has a negative slope that must be treated appropriately in the logarithmic expression, are:

$$E - E_{oX} = -\frac{RT}{\alpha_{Xa}F} \ln \frac{|i_X|}{i_{oX}} \tag{4.36}$$

$$\eta_c = -\frac{2.303RT}{\alpha_{Xa}F} \log \frac{|i_X|}{i_{oX}} \tag{4.37}$$

$$\eta_c = \beta_c \log \frac{|i_X|}{i_{oX}} \tag{4.38}$$

$$\beta_c = -\frac{2.303RT}{\alpha_c F} \tag{4.39}$$

> *The Tafel equations indicate the overpotential is linearly related to the logarithm of the current density or reaction rate.*

These equations show that the logarithm of the current density is directly proportional to overpotential. The latter forms, which include overpotential, η, are known as Tafel equations. The slopes of the resulting lines are β_a and β_c. These slopes are known as Tafel slopes. The associated lines, which are dashed in Fig. 4.14 are Tafel lines. The Tafel lines are often used in place of the Butler-Volmer lines to simplify plots. The Tafel lines are only appropriate when the analysis is performed far from equilibrium (often > 100 mV from E_o for each reaction) and the rate is controlled electrochemically, rather than by mass transport.

Reactions that are close to equilibrium (generally < 10 mV from E_{mix}) can be analyzed using an alternative approach. This approach, which applies near equilibrium, is referred to as polarization resistance. The term polarization resistance implies resistance to change in polarity. In other words it is resistance to change in potential. Resistance to a change in potential is related to the reaction rate at the reaction potential. This method is derived from the Butler-Volmer equation. It is based on the applied or measured change in current in response to a change in potential. Beginning with the Butler-Volmer equation for applied current:

$$i_{applied} = i_{measured} = -\left[i_{oa}\exp\left(\frac{\alpha_a F(E - E_{ao})}{RT}\right) - i_{oc}\exp\left(\frac{-\alpha_c F(E - E_{co})}{RT}\right)\right] \quad (4.40)$$

The overpotential is the potential "over" the equilibrium potential or the difference between an equilibrium half-cell reaction and the applied potential.

Following the same logic used to establish the equilibrium exchange current density, it can be shown that:

$$i_{measured} = -i_{reaction}\left[\exp\left(\frac{\alpha_a F(E - E_{mix})}{RT}\right) - \exp\left(\frac{-\alpha_c F(E - E_{mix})}{RT}\right)\right] \quad (4.41)$$

If E is within 10 mV of E_{mix} (<10 mV) a substitution can be made. When x is small the following expansion substitution can be made:

$$\exp(x) = 1 + x + \frac{x^2}{2!} + \frac{x^3}{3!} + \cdots \cong 1 + x, \quad \text{if} \quad (x < 0.1) \quad (4.42)$$

Thus, as long as the potential is within about 10 mV of the natural reaction potential E_{mix}:

$$\frac{i_{measured}}{i_{reaction}} = -\left[1 + \left(\frac{\alpha_a F(E - E_{mix})}{RT}\right) - 1 - \left(\frac{-\alpha_c F(E - E_{mix})}{RT}\right)\right] \quad (4.43)$$

which can be rearranged to:

$$\frac{i_{measured}}{2.303 i_{reaction}} = -\left[\left(\frac{\alpha_a F(E - E_{mix})}{2.303RT}\right) - \left(\frac{-\alpha_c F(E - E_{mix})}{2.303RT}\right)\right] \quad (4.44)$$

After substitution for the respective Tafel slopes the equation becomes:

$$\frac{i_{measured}}{2.303 i_{reaction}} = -\left[\frac{1}{\beta_a}(E - E_{mix})\left(\frac{|\beta_c|}{|\beta_c|}\right) + \frac{1}{|\beta_c|}(E - E_{mix})\left(\frac{\beta_a}{\beta_a}\right)\right] \quad (4.45)$$

The Tafel slopes β_a and β_c can be determined from the respective slope of a plot of E versus $\log|i|$ for each half-cell reaction. The potential needs to be more than about 100 mV from each of the half-cell equilibrium potentials for this measurement. Alternatively, β_a and β_c can be estimated by assuming a symmetry factor of 0.5. In other words, the charge transfer coefficients α_a and α_c are each equal to 0.5. When the symmetry factor is 0.5, the Tafel slopes of $\beta_a = 0.118$ V and $\beta_c = -0.118$ V.

Rearrangement of the equation for the measured and natural reaction current density leads to:

$$i_{measured} = -2.303 i_{reaction} \left[\frac{(E - E_{reaction})(|\beta_c| + \beta_a)}{|\beta_c|\beta_a} \right] \quad (4.46)$$

Consequently, a plot of $i_{measured}$ versus potential should result in a slope equal to:

$$\text{slope}(i_{meas.} vs. E_{(near.} E_{reaction})) = \frac{\Delta i_{measured}}{\Delta E_{(nearEmixed)}} = -2.303 i_{rxn} \left[\frac{(|\beta_c| + \beta_a)}{|\beta_c|\beta_a} \right] \quad (4.47)$$

> The term polarization indicates an electrode is "polarized" or is made more polar due to a change in its potential above or below its equilibrium.

Thus, $i_{reaction}$, which is relatively constant during a polarization (change in potential) test can be calculated from the slope:

$$i_{reaction} = \frac{|\Delta i_{measured}||\beta_c|\beta_a}{2.303|\Delta E_{nearEmix}|(|\beta_c| + \beta_a)} \quad (4.48)$$

This assumes the test is performed near (within 10 mV) of the natural reaction potential For applications near mass transport limitations, which will be discussed later, the associated equation is:

$$i_{reaction} = \frac{\Delta i_{measured} + i_{Lcath}}{1 + \frac{2.303\Delta E_{nearEmix}}{\beta_a}}; (\text{mass} - \text{transport} - \text{controlled}) \quad (4.49)$$

Partial or full mass transport limitations exist when changing fluid flow rate changes the reaction rate. Additional details for mass transport evaluations are discussed in a subsequent section.

Example 4.4 A sample plot of the measured current density versus potential for a piece of metal in a salt solution is shown in Fig. 4.15. Assume Tafel slope values of $\beta_a = 0.12$ V and β_c −0.12 V. Assume the reaction is not mass transport controlled.

The change in current is $[1 \times 10^{-5}$ A/cm$^2 - (-1 \times 10^{-5}$ A/cm$^2)] = 2 \times 10^{-5}$ A/cm^2, and the corresponding change in potential is $[-0.6969 - (-0.7047)] = 0.0078$ V (The resulting polarization resistance slope of 389 Vcm^2/A is given in the

Fig. 4.15 Comparison of potential versus the current density for a sample of metal in a salt solution. The reported potential is relative to a silver/ silver-chloride (saturated in KCl) reference electrode

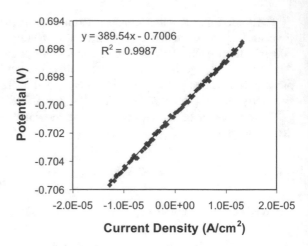

graph.). Thus, substituting this information into the equation for the natural reaction current density or rate of reaction at the mixed potential leads to:

$$i_{reaction} = \frac{|2 \times 10^{-5} \text{ /cm}^2||-0.12 \text{ V}|0.12V}{2.303|0.0078V|(|-0.12|+0.12 \text{ V})} = 6.68 \times 10^{-5} \text{ A/cm}^2$$

The rate of an electrochemical reaction is related to mass. Faraday's Law states current is directly proportional to reacted mass:

$$\text{Mass} = \frac{ItA_w}{nF} \tag{4.50}$$

A_w is the atomic weight of the reacting species undergoing reduction. Time is represented by t. The current is represented by I. The number of electrons transferred per mole of reaction is n. F is the Faraday constant. Faraday's law only applies under ideal conditions. Often, more than one species reacts at a surface. If the current utilization at a surface occurs from the reduction of two species, Faraday's law can only be applied if the current used by each species is known. Thus, Faraday's Law must be carefully applied in complex systems.

> *Faraday's Law states that the mass of an electrodeposit is directly proportional to the current applied to that reaction.*

Example 4.5 Calculate the mass of copper deposited using 1.000 amp of current for 1.000 h assuming that all of the applied current is used by the reduction of copper ($Cu^{2+} + 2e^- \leftrightarrow Cu$).

$$\text{Mass} = \frac{ItA_w}{nF} = \frac{1.000 \text{ amp}\left(\frac{\frac{C}{sec}}{1 \text{ amp}}\right)(1 \text{ h})\frac{3600 \text{ S}}{1.000 \text{ h}}\left(63.55\frac{gCu}{moleCu}\right)}{2\frac{e\text{-mole}}{moleCu}\left(96485\frac{C}{e\text{-mole}}\right)} = 1.243 \text{ g}$$

4.4   Mass Transport

Many hydrometallurgical processes involve reactions at solid–liquid interfaces. The overall rate for such reactions is complicated by transport of reactants and products. Transport restrictions can become more limiting to the overall rate than the chemical reaction. Although some important hydrometallurgical applications involve mass transport of neutral species, ion mass transport is more common.

4.4.1   Ion Mass Transport

Mass transport of ions in aqueous media often involves three processes. These processes are convection, diffusion, and migration. Advection is simply bulk mass transport of ions by fluid flow. Diffusion is mass transport involving movement of species in response to a concentration gradient. Migration is mass transport in response to an electrical potential gradient. Migration completes the flow of charge in the electrical circuit. These processes are

> *Ion mass transport can occur by advection, diffusion, and migration.*

also associated with reactions. The rates of these processes must be considered in a mass balance. Thus, a general ion mass balance for species "i" in a small element of fluid leads to:

$$\frac{\partial c_i}{\partial t} = \vec{U}\nabla c_i + z_i F\nabla(u_i c_i \nabla E) + \nabla(D\nabla c_i) + R_i \tag{4.51}$$

U is the velocity vector, u_i is ion mobility, and R_i is reaction rate. The gradient operator, ∇, represents the partial derivatives with respect to the coordinate system. Other symbols are the same as defined previously. The first term on the right accounts for convection. The second term on the right represents migration. The third term on the right is for diffusion. The last term represents the reaction rate. The left hand side provides the rate of change of concentration.

The processes of convection, diffusion, and migration are illustrated in Fig. 4.16. Figure 4.16 shows an electrochemical cell that produces hydrogen and chlorine gas from hydrochloric acid. It illustrates the quantitative dimensions for ion transport mechanisms. In the case of hydrochloric acid, hydrogen ions are 4 times more mobile than chloride ions. Consequently, the hydrogen ions transport more net charge than chloride ions. The migration of chloride and hydrogen ions is necessary to complete the flow of charge in the circuit. Migration is based on ion availability and mobility as indicated in the mass balance equation. The fraction of the migration current flux carried by a specific ion is proportional to its concentration and mobility relative to that of the system, and it is calculated by the transference number:

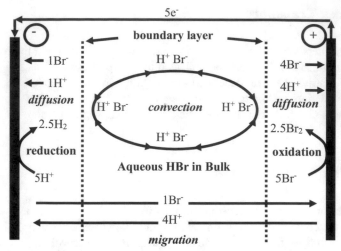

H⁺ ions are at about 4 times more mobile than Br⁻ ions, so migration of H⁺ is
about four times greater than migration of Br⁻ in pure HBr.

Fig. 4.16 Ion transport illustration for an electrochemical cell. Hydrobromic acid is used as the
example medium

$$t_j = \frac{z_j F(-u_j c_j \nabla E)}{-[F \sum_j z_j u_j c_j] \nabla E} = \frac{z_j u_j c_j}{\sum_j z_j u_j c_j} \tag{4.52}$$

Example 4.6 Calculate the transference number for sodium ions in solution containing 0.1 M Na$_2$SO$_4$ assuming complete dissociation of the ions and no solution
complexation. Assume the mobility of sodium and sulfate ions are 0.519×10^{-3}
and 0.827×10^{-3} cm^2/Vsec, respectively.

$$t_{Na+} = \frac{1 \left(\frac{0.519 \times 10^{-3} \frac{cm^2}{Vsec}}{RT} \right) (0.2 \text{ M})}{1 \left(\frac{0.519 \times 10^{-3} \frac{cm^2}{Vsec}}{RT} \right) (0.2 \text{ M}) + -2 \left(\frac{-0.827 \times 10^{-3} \frac{cm^2}{Vsec}}{RT} \right) (0.1 \text{ M})} = 0.399$$

The transference number is the fraction of migration current carried by a specific ion.

Diffusion is needed to provide the ions not
transported by migration. Diffusion is also needed to remove ions used for migration that are not
consumed by reaction. In the case of hydrochloric
acid, diffusion provides one fifth of the hydrogen ions and four fifths chloride ions
needed.

Ions are transported through the bulk media by advection. Advection can occur naturally by thermal gradients or it can be forced by stirring. Advection is needed to transport ions near the surface where they can diffuse to the surface.

Most applications allow for simplification of the general mass balance expression. The migration term can be eliminated in the absence of an electrical field. Migration can also be neglected when the background electrolyte concentration is high relative to the active species concentration. Interfacial reactions often incorporate the effects of advection in a mass transport boundary layer and eliminate the convection term. The reaction term is often equated to the change in concentration with respect to time. If these assumptions and simplifications are appropriate, the equation can be simplified to:

$$\frac{\partial c_i}{\partial t} = \nabla(D\nabla c_i) \tag{4.53}$$

This equation is also known as Fick's Second Law. Other simplifications lead to other forms of the equation.

4.4.2 Diffusion for Fast Moving, Thin Reaction Zones

If the reaction occurs rapidly, the thickness of the reaction zone becomes very small and the reaction interface moves rapidly. In such cases the reaction position is moving at a rate similar to the diffusion rate. The appropriate expression for this application is Fick's Second Law, which for constant effective diffusivity can be written as [7]:

$$\frac{\partial c_i}{\partial t} = D_{\text{eff}} \nabla^2 c_i \tag{4.54}$$

where ∇^2 depends upon the coordinate system selected for the problem. For spherical coordinates and no angular dependance [8],

$$\nabla^2 = \frac{1}{r^2}\frac{\delta}{\delta r}\left(r^2\frac{\delta}{\delta r}\right) \tag{4.55}$$

Application of Eqs. (4.54) and (4.55) to rapid reactions in porous particles of porosity ε leads to:

Fig. 4.17 Schematic
illustration of a hydrated ion
in a beaker containing
solution that will participate
in a reaction at the metal
interface shown

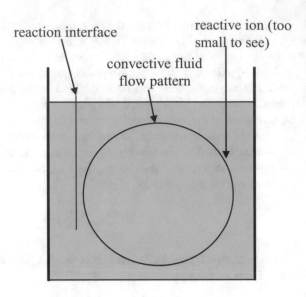

reaction interface

reactive ion (too
small to see)

convective fluid
flow pattern

$$\varepsilon \frac{\partial C_i}{\partial t} = \frac{D\varepsilon}{\tau} \left[\frac{\partial^2 C_i}{\partial r^2} + \left(\frac{2}{r} \right) \left(\frac{\partial C_i}{\partial r} \right) \right] \tag{4.56}$$

The reaction of a spherical particle can be represented with diffusion and reaction zones as shown in Fig. 4.17. The same approach can be used in ion exchange with the reacted portion treated as the loaded portion. In reactions involving pure metals or solid crystals, the entire particle is consumed. Consequently, only diffusion through the solution boundary layer and the reaction zone need to be considered for mass transport. Mass transport is combined with the reaction that occurs in the reaction zone.

In leaching of low-grade ore particles, the majority of the particle is unreactive. For such particles the diffusion through pores is more important than boundary layer diffusion. Thus, boundary layer diffusion is often neglected for ore particle leaching.

4.4.3 *Diffusion for Slow Moving or Stationary Reaction Zones*

Interfacial reactions require mass transport in the form of diffusion. If they do not involve migration and are nearly steady-state and slow moving, the flux is approximated by Fick's First Law:

$$J = \frac{dn}{dt}\frac{1}{A} = -D\frac{dc}{dx} \tag{4.57}$$

J is the flux or rate of mass transfer per unit area. A is the area. n is the number of moles transferred. t is the time. D is the diffusion coefficient. c is the concentration of the diffusing species. x is the thickness of the diffusion layer. x may be in terms of radius or other spatial coordinates depending upon the coordinate system. The negative sign indicates the direction of transport inward toward the particle center relative to the direction vector. This form of Fick's Law generally applies to reactions for which the reaction front can be considered as a stationary front relative to diffusion. It is often applicable to mineral leaching and metal corrosion. Application of this approach to particle leaching requires discrete particle sizes. The same solution method can be applied to corrosion through porous layers. It can also be used for other aqueous reactions involving slow moving reaction interfaces.

Fick's First Law for mass transport through simple diffusion for a reaction at a surface of area, A, can be rewritten as:

$$\frac{dn}{dt} = -DA\frac{(c_{\text{bulk}} - c_{\text{surface}})}{\delta} \tag{4.58}$$

The mass transport boundary layer thickness depends on flow rate, solution properties, and geometry.

The thickness of the diffusion layer or boundary layer thickness, δ, depends on the conditions. Reactions involving complete dissolution require diffusion through a solution boundary layer. This boundary layer is the layer of medium that is not very mobile. It is not very mobile because solvent molecules are bound to the surface and layers of solvent molecules near the surface tend to associate with surface and near surface molecules. As the distance from the surface increases the solvent molecules become more mobile. Solution in the bulk is usually moving due to natural or forced advection. Consequently, movement of bulk fluid transports ions to other parts of a system. However, bulk fluid flow cannot transport ions into regions where the solvent is immobile. Thus, the boundary or diffusion layer represents the distance from the surface at which an effective transition occurs between less mobile and more mobile solvent molecules. Figures 4.18 and 4.19 illustrate the relationship amongst the reaction interface, advection or convection, diffusion, and the diffusion or boundary layer.

The thickness of the mass transfer boundary layer, "δ", takes on different forms depending upon the geometry and flow as indicated by Eqs. 4.58–4.61 [9–11]:

The boundary layer thickness for laminar flow over a flat plate is:

$$\delta = 3l^{1/2}U_\infty^{-1/2}v^{1/6}D^{1/3} \tag{4.59}$$

l is distance from the leading edge. U_∞ is the bulk solution velocity. v is the kinematic viscosity. D is the diffusivity of the medium.

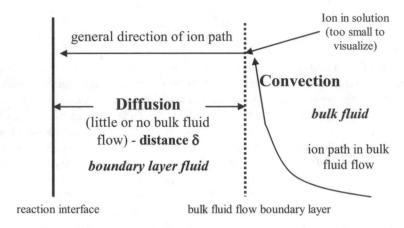

Fig. 4.18   Schematic illustration of fluid flow and ion transport near a reaction interface

Fig. 4.19 Comparison of rate
and the square root of
rotational velocity. The data
come from copper dissolution
in ferric sulfate aqueous
media

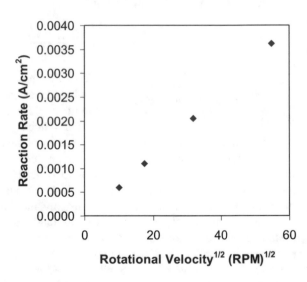

The boundary layer thickness for turbulent flow over a flat plate is:

$$\delta = l^{0.1}U_\infty^{-0.9}v^{17/30}D^{1/3} \tag{4.60}$$

The boundary layer thickness for laminar flow over a rotating disc is:

$$\delta = 1.61\omega^{-1/2}v^{1/6}D^{1/3} \tag{4.61}$$

where ω is the angular velocity.

Example 4.7 Calculate the boundary layer thickness for a disk rotating at 300 RPM if the kinematic viscosity of the liquid is 0.01 cm^2/s and the diffusivity of the fluid is 1×10^{-5} cm^2/s.

$$\delta = 1.61\omega^{-1/2}v^{1/6}D^{1/3}$$

$$\delta = 1.61\left(\frac{300\text{Rev}}{\text{min}}\frac{6.28}{\text{Rev}}\frac{1\text{ min}}{60\text{ sec}}\right)^{-1/2}\left(0.01\frac{\text{cm}^2}{\text{sec}}\right)^{1/6}\left(1 \times 10^{-5}\frac{\text{cm}^2}{\text{sec}}\right)^{1/3}$$

$$= 2.87 \times 10^{-3}\text{ cm}$$

Laminar flow around a spherical particle gives a boundary layer thickness: (based on the Frossling correlation [3]):

$$\delta = d_p\left[\frac{1}{2 + 0.6v^{-1/6}d_p^{1/2}U_\infty^{1/2}D^{-1/3}}\right] \tag{4.62}$$

d_p is the diameter of the particle.

Laminar and turbulent flow conditions must be determined prior to use of these equations. The evaluation is performed by first calculating the Reynolds number, $\mathfrak{R}$:

> *The Reynolds number can be calculated to determine if flow is laminar.*

$$\mathfrak{R} = \frac{\rho UL}{\mu} = \frac{UL}{v} \tag{4.63}$$

U is the characteristic velocity or generally the bulk velocity. L is the characteristic length. L is usually a tube or particle diameter or the length of a plate. v is the kinematic viscosity. The Reynolds number is used to determine the appropriate model. The Reynolds number is the ratio of inertial to viscous force. High Reynolds numbers indicate turbulent flow. Low Reynold's numbers are associated with laminar flow. The transition between laminar and turbulent flow in pipes occurs at $\mathfrak{R} \approx 2300$. For flow around particles the transition begins above a Reynolds number near 1.

Outer diffusion is insignificant compared to diffusion into a large particle. Inner diffusion occurs through reacted portions as well as microscopic pores. Inner diffusion also includes effects such as tortuosity, constriction, and porosity. These effects are often combined with diffusivity. The result is effective diffusivity D_{eff} [11]:

$$D_{\text{eff}} = \frac{D\sigma\varepsilon}{\tau} \tag{4.64}$$

σ is the constriction factor (constricted cross-sectional area of pore/normal cross sectional area of pore). ε is particle inner diffusion porosity. τ is the tortuosity (actual pore length from point A to B in the particle divided by the shortest distance from A to B in the particle).

> *Simplified mass transport equations are often used with mass transfer coefficients.*

Diffusion is sometimes expressed through a mass transfer coefficient. In some texts, Fick's first law is expressed in a rearranged form:

$$\frac{dn}{dt} = k_l A (C_b - C_s) \tag{4.65}$$

k_l is the mass transfer coefficient ($k_l = D/\delta$). Consequently, mass transfer coefficients are used in many applications. Use of mass transfer coefficients requires compliance with Fick's first law. Diffusivity and/or the boundary layer thickness values can be obtained from mass transfer coefficients.

The diffusivities of various species in water depend on conditions. Oxygen diffusivity i 2.25×10^{-9} m^2/s [1] in pure water. However, in more typical solutions, oxygen's diffusivity is 0.5×10^{-9} m^2/s [3]. The diffusivity for dissolved chlorine gas (in solutions with 0.12 m of dissolved salt) is 1.26×10^{-9} m^2/s. Diffusivity for chloride ion in sodium chloride (0.05 m) is 1.26×10^{-9} m^2/s [12] at room temperature. Diffusivity data are commonly found in many texts. The Stokes–Einstein Equation can be used to estimate diffusivity [7]:

$$D_i = \frac{kT}{4\mu\pi r_i} \tag{4.66}$$

μ is the solution viscosity (note that viscosity is a strong function of temperature also). T is the temperature. k is the Boltzman constant. r_i is the radius of species "i" that is diffusing. When using the Stokes–Einstein equation, some caution should be exercised. The radius of the ions, especially multivalent ions, may change with temperature. The radius change is often associated with hydration number. The Stokes–Einstein equation is often not accurate. However, it is effectively used to predict diffusivity changes due to changing viscosity and radii.

Another approach to determining diffusivity invokes molecular motion theory. Diffusion is a molecular motion process that is similar to a chemical reaction. Both involve internal energy and the probability of getting to the next step or position. In the case of reaction, the next step is usually interaction with another particle or atom. In the case of diffusion, the next step is displacement. A simplified expression for diffusion that utilizes molecular motion theory is:

$$D = D_o \exp\left(\frac{-E_D}{RT}\right) \tag{4.67}$$

In this expression D, is the diffusivity. D_o is the standard diffusivity. E_D is the activation energy for diffusion. R is the gas constant, and T is the temperature.

4.4.4 Identifying Mass Transport Control

Mass transport control can be identified by appropriate measurements and analysis. If increasing fluid velocity increases reaction rate, mass transport participates in kinetic control. If a fluid velocity increases and no change in rate is observed the reaction is not controlled by mass transport to the outer surface. It may still be controlled by inner diffusion if increased fluid flow does not impact the rate. Inner diffusion will be discussed later. If the increase in fluid velocity causes the rate to increase in proportion to mass transport control equations, the rate is mass transport controlled. Using Fick's first law, the rate of mass transport controlled reactions is inversely proportional to boundary layer thickness. Thus, if boundary layer thickness is decreased by 50%, the rate doubles. The boundary layer thickness is inversely proportional to the square root of rotational velocity for rotating disc experiments. Thus, for rotating disk experiments the rate should double if the rotational velocity is increased by four times if it is mass transport controlled. The data in Fig. 4.19 show a comparison of the square root of the rotational velocity to the reaction rate for a rotating disk experiment. The plot suggests the reaction is mass transport controlled because the rate increases in proportion to the square root of rotational velocity. A nonlinear relationship would suggest mass transport and reaction control. A horizontal line would indicate reaction or inner diffusion control. Similarly, the boundary layer thickness equations can be used to compare observed rate changes to the rates expected for mass transport control. If the rate change is not the same as expected, the rate may be controlled by both reaction and mass transport kinetics. Often, the rate is controlled by mass transport and reaction kinetics. In such cases a combined model must be used.

> *Mass transfer controlled reaction rates are often increased by increasing fluid flow rates, and the rates follow established mass transport equations. However, internal pore diffusion controlled reaction rates do not respond to increased fluid flow rates.*

4.5 Combined Mass Transport and Reaction Kinetics

The expression for a simple first order reaction at a surface of area A, is:

$$\frac{\mathrm{d}n}{\mathrm{d}t} = -AkC_s \tag{4.68}$$

Equating the first order reaction equation with the equation for diffusion from Fick's First Law gives:

$$-DA\frac{(C_b - C_s)}{\delta} = -AkC_s \tag{4.69}$$

Solving for C_s results in:

$$C_s = C_b\frac{1}{k\left(\frac{1}{k} + \frac{\delta}{D}\right)} \tag{4.70}$$

Substitution of this result into the equation for the reaction rate yields:

$$\frac{dn}{dt} = V\frac{dC_b}{dt} = \frac{-AC_b}{\left(\frac{1}{k} + \frac{\delta}{D}\right)} \tag{4.71}$$

Note that this equation indicates that when the rate constant is small, the inverse rate term becomes large and the overall rate low. Conversely, if the diffusion coefficient is small the diffusion term dominates. Both the diffusivity and rate constant must be large to have a large overall rate. Rearrangement of this equation, followed by integration for a batch system with constant area per unit of volume leads to:

$$\int_{C_o}^{C}\frac{dC_b}{C_b} = \ln\frac{C_o}{C_b} = \int_{o}^{t}\frac{-A}{V\left(\frac{1}{k} + \frac{\delta}{D}\right)}dt = \frac{-At}{V\left(\frac{1}{k} + \frac{\delta}{D}\right)} \tag{4.72}$$

> *Many applied reaction systems involve reaction and mass transport control.*

This equation can be used to predict concentration as a function of time for batch systems. The constants in this equation can be calculated based on measurements of concentration versus time.

4.5.1 Application of Mass Balancing to Different Reactor Types

One important reactor type for general chemical reactions is a plug flow reactor. Leaching in heaps is a very crude approximation of a very large plug flow reactor.

> *Some metallurgical processes behave like plug-flow reactors.*

Although heap leaching has complicating factors that limit its ability to be modeled as a plug flow reactor, some of the concepts are applicable. A plug flow type of reactor assumes that a reaction occurs in a tube. Reactant enters the tube, reacts inside, then exits partially or completely converted to product(s). The mass balance for a plug flow reactor leads to the equation: [11]

$$\frac{dCq}{dV} = r \tag{4.73}$$

q is the volumetric flow rate. If the reaction rate, r, is first order, the equation can be rearranged to:

$$\frac{dCq}{dV} = -kC \tag{4.74}$$

Rearrangement of this expression gives:

$$kdV = -q\frac{dC}{C} \tag{4.75}$$

Integration of both sides of the equation results in:

$$V = \frac{q}{k}\ln\frac{C_{init}}{C_{final}} \tag{4.76}$$

If both sides of the equation are divided by a unit area, such as one meter squared, the equation can be rewritten as:

$$l = \frac{q_{per.m^2}}{k}\ln\frac{C_{init}}{C_{final}} \tag{4.77}$$

This equation can be used to estimate the length of a reactor or height of a column for a given change in concentration.

Example 4.8 For a hypothetical leaching column with uniform particles of pure metal oxide mixed with inert rocks, estimate the initial column height needed to achieve an 80% reduction in the reactant concentration needed for leaching assuming a volumetric flow rate of 0.00286 L per meter squared per minute and a reaction constant of 9×10^{-9} s^{-1}. (Assume the size reduction due to leaching can be neglected for this approximation.)

$$l = \frac{q_{per.m^2}}{k}\ln\frac{C_{init}}{C_{final}} = \frac{\frac{0.00286\ \text{dm}^3}{\text{m}^2\ \text{min}}\frac{1\ \text{m}^3}{1000\ \text{dm}^3}}{\frac{60\ S}{1\ \text{min}}\frac{9\times10^{-9}}{\text{sec}}}\ln\frac{C_{init}}{0.2C_{init}} = 8.52\ \text{m}$$

Thus, the height of such a hypothetical heap leaching column would be 8.53 m.

Another type of reactor is a Continuous Stirred Tank Reactor (CSTR). This type of reactor is used in many hydrometallurgical applications, including leaching. A mole balance for this type of reactor for steady-state operation leads to:

$$V = \frac{q(C_{in} - C_{out})}{r} \tag{4.78}$$

> **Some metallurgical processes behave like CSTR reactors.**

Note that the concentration in a well-stirred CSTR is C_{out}. The concentration may refer to a product or a reactant. As an example, the rate of gold leaching is often based on the oxygen reactant concentration. The concentration of oxygen in a stirred tank will often be maintained at a constant level. Thus, the gold leaching rate may be constant for a given particle size and oxygen concentration.

Example 4.9 As a simplified example, calculate the CSTR volume and residence time needed to process 100 m³ per minute of leaching slurry containing 35 wt% solids (2.65 g/cm³ density for the solids) with an input concentration of precious metal of 1×10^{-6} m and an output concentration of 1×10^{-5} m if the concentration of dissolved oxygen is 0.0002 m and the effective precious metal leaching reaction rate constant is 8.75×10^{-6} s^{-1} (molar basis) and the observed metal leaching rate is first order with respect to oxygen. The reaction is effectively mass transport controlled but can be modeled using a simple first order chemical reaction rate equation in which k is effectively D/δ. (Assume the CSTR is well-stirred, and the particles are uniform in size and leaching characteristics. Assume 4 mol of metal are produced per mole of oxygen. Note this is an oversimplified view of many leaching scenarios.)

The liquid volume fraction can be calculated using the equation:

$$V_{fraction(liquid)} = \frac{1}{1 + \frac{\rho_l W_{fraction(solid)}}{\rho_s (1 - W_{fraction(solid)})}} = \frac{1}{1 + \frac{(1)(0.35)}{(2.65)(1-0.35)}} = 0.83$$

The volumetric flow rate of solution is:

$$q = q_{slurry}(FractionLiquid) = 100 \text{ m}^3/\text{min} \cdot (0.83) = 83 \text{ m}^3/\text{min}.$$

The equation for CSTR volume is:

$$V_{liquid} = \frac{q(C_{in} - C_{out})}{r} = \frac{q(C_{in} - C_{out})}{-kC_{O_2}}$$

$$V_{liquid} = \frac{83 \frac{\text{m}^3}{\text{min}} \frac{1 \text{ min}}{60 \text{ S}}(1 \times 10^{-6} \text{ m} - 1 \times 10^{-5} \text{ mMe})}{-\frac{4 \text{ m(Me)}}{1 \text{ m(O}_2)} \frac{8.75 \times 10^{-6}}{\text{sec}}(0.0002 \text{ mO}_2)} = 1.78 \times 10^3 \text{ m}^3$$

The reactor also needs to account for the solids by multiplying by 100/83 to give $2.14 \times 10^3 m^3$.

The residence time is calculated by dividing volume by flow rate.

$t = V/q = 2140 \text{ m}^3/100 \text{ m}^3/\text{min} = 21.4 \text{ min}.$

A more in-depth analysis of mixed mass transport and kinetics will be provided later.

4.6 Models for Reactions Involving Particles

4.6.1 Mixed-Control Kinetic Models Involving Particles or Surfaces

Overall reaction rates are often controlled by one or more of the individual rate processes. They can also be controlled by anomalous reactions such as precipitation. They may also be controlled by growth of a reaction product layer. Some product layers are tight and cause a rapid decrease in reaction rate. Such restrictive product layers are referred to as passive films. The process of passive film formation is called passivation. Passivation makes the analysis of kinetic data more challenging.

The previous models were developed for mass transport or reaction controlled kinetics. Kinetic control by multiple modes requires a combined approach. A rate balance is often the best method to solve such problems. Figure 4.20 illustrates a rate balance shown mathematically as Eq. 4.79. The rate of accumulation in the reaction system is the "in" – "out" plus reaction. It assumes that no precipitation or passivation occurs.

$$R_{accum."i"} = \sum R_{input"i"} - \sum R_{output"i"} + \sum R_{react."i"} \qquad (4.79)$$

This approach can be applied generally to any system. It can be adapted to multiple reactions and multiple inputs and outputs. For multiple reaction systems, the output from the first reactor becomes the input for the next reactor. Each species in a multiple species system requires a rate balance equation.

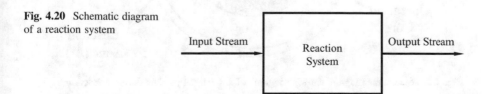

Fig. 4.20 Schematic diagram of a reaction system

4.6.2 Shrinking-Core Particle Kinetic Model (Inner Diffusion and Reaction Control)

> *The shrinking-core model is one of the most useful hydrometallurgical reaction models.*

Typical mineral oxidation or extraction reactions can be represented by Fig. 4.21. Inner diffusion is much slower than outer diffusion. Consequently, outer diffusion is generally neglected for extraction from ores. Thus, Fick's first law is used to represent the diffusion rate. Typical reaction rate kinetics is also considered. Oxygen is a common reactant. Thus, it will be used as the example for this derivation. A common reaction involving oxygen is $2FeS_2 + 7O_2 + 2H_2O \leftrightarrow 2Fe^{2+} + 4SO_4^{2-} + 4H^+$. Extraction of metals from minerals is generally a very slow process. Therefore, such systems are often assumed to be steady-state. The steady-state assumption allows for simplification of the rate balance equation. Steady-state reactions have no accumulation ($R_{accum} = 0$). Thus, the resulting rate balance for oxygen which is needed for this reaction yields:

$$R_{O_2 react.} = R_{O_2 diffusion} \tag{4.80}$$

This assumes there are no oxygen generating reactions. If it is further assumed that the reaction proceeds relatively slowly with a very negligible reaction zone, Fick's first law can be applied (Eq. 4.57), leading to:

$$R_{diffusion} = \frac{1}{A_{diff}} \frac{dn_{O_2}}{dt} = -D \frac{(C_{fin.} - C_{init.})}{\tau(x_{fin.} - x_{init.})} \tag{4.81}$$

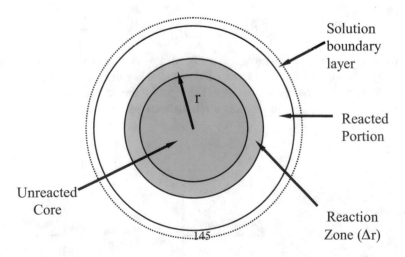

Fig. 4.21 Schematic diagram showing regions of a spherical particle during a reaction

which can be rearranged to:

$$\frac{dn_{O_2}}{dt} = -\frac{A_{\text{diff}}D(C_{O_2\text{bulk}} - C_{O_2\text{surf}})}{\tau(r_o - r)} \tag{4.82}$$

The corresponding rate equation for a first-order surface reaction is:

$$R_{O_2\text{react}} = \frac{1}{A_{\text{react}}}\frac{dn_{O_2}}{dt} = -k_f C_{O_2\text{surf}}\left(1 - \exp\left(\frac{\Delta G}{RT}\right)\right) \tag{4.83}$$

which can be rearranged to:

$$\frac{dn_{O_2}}{dt} = -A_{\text{react}}k_f C_{O_2\text{surf}}\left(1 - \exp\left(\frac{\Delta G}{RT}\right)\right) \tag{4.84}$$

Solving Eqs. 4.82 and 4.84 for the surface oxygen concentration ($C_{O2\text{surf}}$) leads to:

$$C_{O_2\text{surf}} = \frac{A_{\text{diff}}DC_{O_2\text{bulk}}}{\tau(r_o - r)k_f A_{\text{react}}\left(1 - \exp\left(\frac{\Delta G}{RT}\right)\right) + A_{\text{diff}}D} \tag{4.85}$$

Substituting Eq. 4.74 into 4.71 leads to:

$$\frac{dn_{O2}}{dt} = -\frac{A_{\text{diff}}DC_{O_2\text{bulk}}}{\tau(r_o - r)}\left(1 - \frac{A_{\text{diff}}D}{\tau(r_o - r)k_f A_{\text{react}}\left(\left(1 - \exp\frac{\Delta G}{RT}\right)A_{\text{diff}}D\right)}\right) \tag{4.86}$$

which after rearranging leads to:

$$\frac{dn_{O_2}}{dt} = \frac{A_{\text{diff}}DC_{O_2\text{bulk}}}{\tau(r_o - r)}\left[\frac{1}{\left(1 + \frac{\frac{A_{\text{diff}}D}{\tau(r_o-r)}}{k_f A_{\text{react}}\left(1 - \exp\frac{\Delta G}{RT}\right)}\right)}\right] \tag{4.87}$$

Notice that the driving force for the diffusion process

$$(C_{O_2\text{bulk}} - C_{O_2\text{surf}})$$

is replaced by:

$$C_{O_2\text{bulk}}\left[\frac{1}{\frac{R_{\text{diff}}}{R_{\text{diff}}} + \frac{R_{\text{diff}}}{R_{\text{react}}}}\right] \tag{4.88}$$

The change in the number of moles of reactant can be calculated by integration. However, it is usually more convenient to express the rate of reaction in terms of the particle fraction reacted. The number of moles reacted is not as practical. The conversion to fraction reacted begins with:

$$n_{\text{particle}} = \frac{\rho M_f 4\pi r^3}{3M_w} \tag{4.89}$$

M_f is the mass fraction of reacting material in the particle. M_w is the molecular weight of the reacting portion of the particle. ρ is the density of the reacting portion of the particle. The fraction of the particle that has reacted can be expressed as:

$$\alpha = \frac{\frac{4}{3}\pi r_o^3 - \frac{4}{3}\pi r^3}{\frac{4}{3}\pi r_o^3} = 1 - \frac{r^3}{r_o^3}, \quad [r = r_o(1-a)^{1/3}] \tag{4.90}$$

Substituting Eq. 4.80 into Eq. 4.79 leads to:

$$n_{\text{particle}} = \frac{\rho M_f 4\pi r_o^3(1-\alpha)}{3M_w} \tag{4.91}$$

which after differentiation becomes:

$$\frac{dn_{\text{particle}}}{dt} = -\frac{\rho M_f 4\pi r_o^3}{3M_w}\frac{d\alpha}{dt} \tag{4.92}$$

Reaction stoichiometry is used to link reactant and particle mole changes. The stoichiometry factor, S_f, is the number of moles of reactant per mole of consumable in the particles. Substitution of the stoichiometry factor into Eq. 4.81 leads to:

$$\frac{dn_{O_2}}{dt} = -\frac{\rho M_f S_f 4\pi r_o^3}{3M_w}\frac{d\alpha}{dt} \tag{4.93}$$

Equation 4.83 can be directly related to Eq. 4.76 and the resulting equation can be solved for the fraction reacted, α. Appropriate substitutions for variables such as areas are needed prior to solving.

The area terms for the shrinking-core scenario are given as:

$$A_{\text{react}} = \frac{A_f 4\pi r^2}{\Psi} = \frac{A_f 4\pi r_o^2(1-\alpha)^{2/3}}{\Psi} \tag{4.94}$$

$$A_{\text{innerdiff}} = \frac{\varepsilon\sigma M_f 4\pi r r_o}{\Psi} = \frac{\varepsilon\sigma M_f 4\pi r_o^2(1-\alpha)^{1/3}}{\Psi} \tag{4.95}$$

$$A_{\text{outerdiff}} = \frac{M_f 4\pi r_o^2}{\Psi} \quad \text{(not generally used)} \tag{4.96}$$

ε is the particle porosity. M_f is the mass fraction (approximately equal to the area fraction) of reacting material relative to the inert material. σ is the constriction factor (cross-sectional area of pore constrictions/average pore cross-sectional area). ψ is the sphericity factor (area of a sphere/actual particle surface area of equivalent diameter particle). Note that if δ is not small compared to r_o, then $A = 4\pi(r_o + \delta)^2$ for the outer diffusion area. Using Eqs. (4.87) and (4.83) along with the appropriate expressions for area leads to:

$$\frac{d\alpha}{dt} = \frac{3M_w}{\rho M_f S_f 4\pi r_o^3} \frac{A_{\text{diff}} D C_{O_2\text{bulk}}}{\tau(r_o - r)} \left[\frac{1}{\left(1 + \dfrac{\frac{A_{\text{diff}} D}{\tau(r_o - r)}}{k_f A_{\text{react}}\left(1 - \exp\frac{\Delta G}{RT}\right)}\right)}\right] \qquad (4.97)$$

Further rearranging for the conditions of no material reacted at time zero leads to:

$$\int_0^t \frac{3M_w}{\rho M_f S_f 4\pi r_o^3} dt = \int_0^\alpha \left[\frac{\tau(r_o - r)}{A_{\text{diff}} D C_{O_2\text{bulk}}} + \frac{1}{k_f A_{\text{react}} C_{O_2\text{bulk}}\left(1 - \exp\left(\frac{\Delta G}{RT}\right)\right)}\right] d\alpha \qquad (4.98)$$

After substitution for the areas and radii in terms of the fraction reacted, Eq. 4.88 becomes:

$$\int_0^t \frac{3M_w C_{O_2\text{bulk}}}{S_f \rho M_f r_o} dt = \int_0^\alpha \left[\frac{\Psi \tau r_o (1 - (1 - \alpha)^{1/3})}{\varepsilon \sigma M_f D (1 - \alpha)^{1/3}} + \frac{\Psi(1 - \alpha)^{-2/3}}{k_f M_f \left(1 - \exp\left(\frac{\Delta G}{RT}\right)\right)}\right] d\alpha \qquad (4.99)$$

The indefinite integral of $(a + bx)^n dx$ is equal to $(a + bx)^{n+1}/(n + 1)b + C$ (for $n \neq -1$). This indefinite integral can be used to solve the equation. Thus, application of the indefinite integral for constant reactant concentration, leads to:

$$\frac{3M_w C_{O_2 b.}}{S_f \rho r_o \Psi} t = \frac{3r_o}{2D_{\text{eff}}} \left[1 - \frac{2}{3}\alpha - (1 - \alpha)^{2/3}\right]$$
$$+ \frac{3}{k_f\left(1 - \exp\left(\frac{\Delta G}{RT}\right)\right)}\left[1 - (1 - \alpha)^{1/3}\right] \qquad (4.100)$$

which is one final form of the shrinking core mixed kinetic (diffusion and reaction control) leaching model (with $D_{\text{eff}} = D\varepsilon\sigma/\tau$). It is nearly the same as that given by Wadsworth [9] except for minor changes in nomenclature. Equation 4.100 shows that the reaction is r_o^2 dependent for diffusion-controlled systems. It shows r_o dependence for reaction controlled systems.

Note that there are three terms. The left term is a constant multiplied by the time. The middle term is a diffusion constant multiplied by a function of the fraction reacted. This middle term is the diffusion term. The right term is a reaction constant multiplied by a function of the fraction reacted. The right term is the reaction term. Assuming diffusion control would have eliminated the right term. Conversely, the assumption of reaction control would have removed the middle term. Data analysis can similarly reveal whether diffusion or reaction is rate controlling. If extraction time is proportional to $(1 - 2/3\ \alpha-(1 - \alpha)^{2/3})$, the rate is inner diffusion controlled. If the exraction time is proportional to $(1-(1 - \alpha)^{1/3})$, the rate is reaction controlled.

Example 4.10 A low-grade ore leaching test gsives the following results:

Fraction reacted	Time (min)
0.14	15
0.39	45
0.60	75

Determine if the process is likely diffusion or reaction controlled assuming that boundary layer diffusion is not rate limiting.

If the process is diffusion controlled, a plot of $(1 - 2/3\ \alpha-(1 - \alpha)^{2/3})$ versus time should be linear. As shown in Fig. 4.22, $(1 - 2/3\ \alpha-(1 - \alpha)^{2/3})$ versus time is not linear for the data in this example. Thus, the reaction is not likely diffusion controlled.

If the process is reaction controlled, a plot of time versus $(1-(1 - \alpha)^{1/3})$ should be linear. The plotting results presented in Fig. 4.23 show that the relationship between time and $(1-(1 - \alpha)^{1/3})$ is linear. Consequently, the rate is likely reaction controlled.

For many reactions, the initial portion of the reaction is reaction controlled. After the material near the surface is extracted, it often becomes diffusion controlled. Equation 4.100 shows this relationship as a function of the fraction reacted and other constants. If the surface reaction rate constant is high the diffusion term dominates. If the diffusion coefficient is large, the surface reaction becomes dominant.

Fig. 4.22 Plot of $1-2/3\alpha-(1 - \alpha)^{2/3}$ versus time for data in Example 4.10

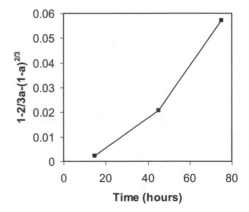

Fig. 4.23 Plot of $(1 - (1 - \alpha)^{1/3})$ versus time for the data given in Example 4.10

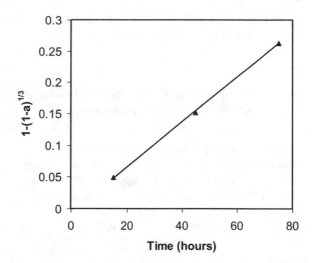

The same type of approach that was used to solve the shrinking core problem can be applied to a wide variety of other related problems. The key to applying such equations is to ensure correct assumptions. The shrinking core model is often applicable to mineral oxidation of low grade ores. The oxidation rate must be relatively slow. Such assumptions are commonly appropriate for commercial ore leaching operations.

An alternative method for determining whether a reaction follows mixed kinetics can be used. Division of all of the terms in Eq. 4.100 by $1 - (1 - \alpha)^{1/3}$ leads to a linear form. A subsequent plot of $t/(1 - (1 - \alpha)^{1/3})$ versus $(1 - 2\alpha/3 - (1 - \alpha)^{2/3})/(1 - (1 - \alpha)^{1/3})$ can be used for evaluation. If the relationship is linear, the rate is likely controlled by diffusion and reaction kinetics.

4.6.3   Inner Diffusion Controlled Rate Model

Inner diffusion rate control applies for extraction of disseminated valuable entities inside a nonreactive matrix. Alternatively, this model applies when a product layer forms during extraction from pure particles. It applies when inner diffusion distances increase with increasing reaction. The associated equation for reaction time is [13]:

$$t = \frac{S_f M_f \rho r_o^2}{2 D_e C} \left(1 - \frac{2}{3}\alpha - (1 - \alpha)^{2/3} \right) \tag{4.101}$$

t is time. M_f is mass fraction of the valuable entity within the matrix. ρ is molar density. r_o is the outer radius of the particle. S_f is the stoichiometric coefficient (moles of reactant consumed per mole of particle reacted). D_e is the effective diffusivity of the reactant ions. C is the molar concentration of the reactant ions. α is the fraction of total material reacted.

4.6.4 Outer (Boundary Layer) Diffusion Controlled Rate Model for Small Particles

Outer diffusion applies to systems involving pure *small* particles. Particles must be approximately smaller than a few micrometers in well-stirred solutions. Alternatively, this model can apply to larger particles under low flow conditions. Outer diffusion rate control can be expressed as [13]:

$$t = \frac{S_f \rho r_o^2}{2DC} (1 - (1 - \alpha)^{2/3}) \tag{4.102}$$

4.6.5 Outer (Boundary Layer) Diffusion Controlled Rate Model for Large Particles

This model is derived for reactions involving pure *large* particles approximately greater than 0.1 mm in well-stirred systems. The mathematical description is [13]:

$$t = \frac{k' r_o^{3/2}}{C} (1 - (1 - \alpha)^{1/2}) \tag{4.103}$$

k' is a proportionality constant.

4.6.6 Outer (Boundary Layer) Diffusion Controlled Rate Model

This model applies when outer diffusion controls heterogeneous reactions. Thus, it can apply to small ore particle leaching. It can also apply to small pure particle leaching when product layers grow in proportion to reaction. The associated equation is expressed as [13]:

$$t = \frac{S_f M_f \rho r_o}{2k_1 D_e C} \alpha \tag{4.104}$$

k_1 is the mass transfer coefficient.

4.6.7 Inner Diffusion with Pore Properties Variations

Change in pore characteristics as a function of initial size can be accommodated. Comminution leaves fewer pores per unit of volume in fine particles compared to large particles. Consequently, it is often helpful to accommodate this phenomenon using the equation:

$$t = \frac{r_o^{2-n_{ro}} \rho \psi M_{fY} S_f}{2kDC} \left(1 - \frac{2}{3}\alpha - (1-\alpha)^{2/3}\right) \qquad (4.105)$$

n_{ro} represents the power of the relationship between initial particle size and pore properties. If pore properties are not related to size, this equation reduces to the traditional inner diffusion controlled rate model. The value of n_{ro} can be found from a plot of the slopes of time versus the fraction reacted terms versus the logarithm of the initial particle size.

4.6.8 Surface Reaction Controlled Rates Involving Particles

This model is derived for reaction-controlled applications. It accommodates pure particles with or without product layers. It can also be appropriate for particles with pieces disseminated in a non-reactive matrix:

$$t = \frac{S_f M_f \rho r_o}{2k_s C} \left(1 - (1-\alpha)^{1/3}\right) \qquad (4.106)$$

k_s is the surface reaction rate constant.

4.6.9 Simple Diffusion of Solute from a Porous Sphere

This model accounts for diffusion of solute from a porous sphere. Applications may include copper diffusion from an agglomerate that has been "acid cured". This model is based on Fick's second law. The resulting equation is:

$$\alpha = 1 - \frac{6}{\pi} \sum_{n=1}^{\infty} \frac{1}{n^2} \exp\left(\frac{D_{\text{eff}} n^2 \pi^2 t}{r_o^2}\right) \qquad (4.107)$$

4.6.10   Combined Model Weighting for Ore Leaching Applications

> *Often, real-world applications require appropriate application of multiple kinetic models.*

In many leaching scenarios the particles have significant heterogeneity. They consist of sections of one mineral embedded within a matrix as shown in Fig. 4.24.

If a particle is a valuable mineral particle it is liberated. The probability of finding liberated particles can be estimated from particle distributions. The probability of finding a valuable mineral particle of size r_{vmp} multiplied by the probability that the valuable mineral particle is equal to or larger than host rock particles facilitates an estimate of the liberation. Mathematically, this is expressed in a discrete form as [14]:

$$P_{\text{Lib}} \cong \sum_{r_{hrp}} \sum_{r_{vmp}} f\left(r_{hrp} \pm \frac{\Delta r}{2}\right) \Delta r_{hrp} f\left(r_{vmp} \pm \frac{\Delta r}{2}\right) \Delta r_{vmp} \dots \quad \text{for} \quad (r_{vmp} \geq r_{hrp})$$

$$(4.108)$$

The subscripts "hrp" and "vmp" represent host rock particle and valuable mineral particle, respectively. A simplified form of Eq. (4.108) to estimate liberation is expressed as [15]:

$$P_{\text{Lib.Est.}} = \exp\left(\frac{-\sqrt{r^*_{vmp} r^*_{hrp}}}{r^*_{vmp}}\right)\left(1 - \exp\left(\frac{-\sqrt{r^*_{vmp} r^*_{hrp}}}{r^*_{hrp}}\right)\right); \dots for(r_{vmp} \geq r_{hrp})$$

$$(4.109)$$

The asterisk in Eq. (4.108) represents the characteristic particle size. This size usually corresponds to the 63.8% passing size or it can be approximated by the d_{80} size. Thus, the liberated fraction is found from the probability of finding an ore

Fig. 4.24 Illustration of relationships between valuable mineral particles and host rock particles in an ore

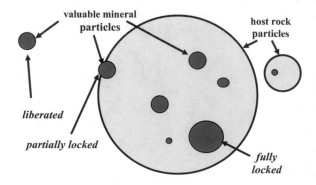

particle that is also a valuable mineral particle. The probability that a valuable mineral particle is either partially or fully locked within a host rock particle is determined from the nonliberated fraction $(1 - P_{Lib})$ or $(1 - P_{Lib.Est.})$. The fraction of partially locked valuable mineral particles can be estimated to be:

$$P_{\text{Partially-Locked.Est}} = [1 - P_{\text{Lib.Est}}] \left[1 - \left(\frac{r^*_{\text{hro}} - r^*_{\text{vmp}}}{r^*_{\text{hrp}}} \right)^3 \right] \dots (\text{for}\dots r^*_{\text{hrp}} > r^*_{\text{vmp}})$$

(4.110)

$$P_{\text{Partially-Locked.Est}} = [1 - P_{\text{Lib.Est.}}] \dots (\text{for}\dots r^*_{\text{hrp}} < r^*_{\text{vmp}})$$ (4.111)

The fraction of fully locked (nonexposure) valuable mineral particles can be estimated by [15]:

$$P_{\text{Fully-Locked.Est}} = [1 - P_{\text{Lib.Est.}}] \left[\left(\frac{r^*_{\text{hro}} - r^*_{\text{vmp}}}{r^*_{\text{hrp}}} \right)^3 \right] \dots \infty (\text{for}\dots r^*_{\text{hrp}} > r^*_{\text{vmp}})$$

(4.112)

Note that if the host rock particle size is less than the valuable mineral particle size, none of the material is fully locked.

The effect of host rock particle size reduction on valuable mineral particles size can be estimated. The location of the valuable mineral particle in a host rock particle during comminution determines the number and size of progeny (see Fig. 4.25 and Eq. 4.113). The sum of the progeny and their relative sizes in relation to the host rock size reduction are used for the estimation.

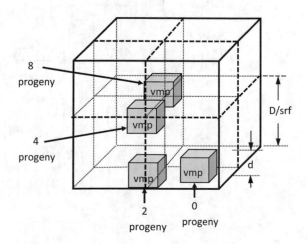

Fig. 4.25 Illustration of relationship possibilities amongst valuable mineral particles (vmp) in a host rock particle undergoing a size reduction with the relative vmp progeny outcomes

> *Calculated liberated, partially-locked, and fully-locked valuable mineral particle fractions can be used to appropriately weight kinetic models for more accurate overall leaching modeling.*

Using size distributions of valuable and host rock particles, the leaching characteristics can be determined. First, the fraction of locked, partially-locked, and liberated particles is calculated.

The resulting size of valuable mineral particles for a corresponding host rock particle size reduction can be calculated using progeny size and outcome probabilities as shown in Eq. 4.113 as follows:

$$
\begin{aligned}
d_{\text{new}} = & \frac{\left(\frac{D}{\text{srf}}\right)^3}{\left[\left(\frac{D}{\text{srf}}\right)+d\right]^3} d + \left(\frac{1}{n_{\max}}\right) \sum_{n=1}^{n_{\max}} \left[\left(\frac{d^3}{2}\left(\frac{n}{n_{\max}}\right) \right)^{1/3} \left(\frac{\frac{3d}{n_{\max}}\left(\frac{D}{\text{srf}}\right)^2}{\left(\frac{D}{\text{srf}}+d\right)^3} \right) \right. \\
& + \left(d^3\left(1-\frac{n}{2n_{\max}}\right) \right)^{1/3} \left(\frac{\frac{3d}{n_{\max}}\left(\frac{D}{\text{srf}}\right)^2}{\left(\frac{D}{\text{srf}}+d\right)^3} \right) \\
& + \left(d^3\left(\frac{n}{n_{\max}}\right)^2 \right)^{1/3} 12\left(\frac{\left(\frac{D}{\text{srf}}\right)d^2\left[\left(\frac{n^2-(n-1)^2}{n_{\max}^2}\right)\left(\frac{n^2-(n-1)^2}{n_{\max}^2}\right)\right]}{\left(\frac{D}{\text{srf}}+d\right)^3} \right) \\
& + \left(d^3\left(1-\frac{n}{n_{\max}}\right)\left(\frac{n}{n_{\max}}\right) \right)^{1/3} 24\left(\frac{\left(\frac{D}{\text{srf}}\right)d^2\left[\left(\frac{n^2-(n-1)^2}{n_{\max}^2}\right)\right]}{\left(\frac{D}{\text{srf}}+d\right)^3} \right) \\
& + \left(d^3\left(1-\frac{n}{n_{\max}}\right)^2 \right)^{1/3} 12\left(\frac{\left(\frac{D}{\text{srf}}\right)d^2\left[\frac{n^2-(n-1)^2}{n_{\max}^2}\right]}{\left(\frac{D}{\text{srf}}+d\right)^3} \right) \qquad (4.113) \\
& + \left(d^3\left(\frac{n^3}{n_{\max}^3}\right) \right)^{1/3} 8\left(\frac{8d^3\left(\frac{n^3-(n-1)^3}{n_{\max}^3}\right)}{\left(\frac{n^3-(n-1)^3}{n_{\max}^3}\right)^3} \right) \\
& + \left(d^3\left(1-\frac{n}{n_{\max}}\right)\left(\frac{n}{n_{\max}}\right)^2 \right)^{1/3} 24\left(\frac{8d^3\left(\frac{n^3-(n-1)^3}{n_{\max}^3}\right)}{\left(\frac{D}{\text{srf}}+d\right)^3} \right) \\
& + \left(d^3\left(1-\frac{n}{n_{\max}}\right)^2\left(\frac{n}{n_{\max}}\right) \right)^{1/3} 24\left(\frac{8d^3\left(\frac{n^3-(n-1)^3}{n_{\max}^3}\right)}{\left(\frac{D}{\text{srf}}+d\right)^3} \right) \\
& \left. + \left(d^3\left(1-\frac{n}{n_{\max}}\right)^3 \right)^{1/3} 8\left(\frac{8d^3\left(\frac{n^3-(n-1)^3}{n_{\max}^3}\right)}{\left(\frac{D}{\text{srf}}+d\right)^3} \right) \right]
\end{aligned}
$$

in which D is the diameter of the host rock particle, d is the diameter of the original valuable mineral particle, d_{new} is the new diameter of the valuable mineral particle after reducing the size of the host rock by the size reduction factor, srf, and n is the discretised number in the summation between 1 and the maximum number, n_{max}. The value of d_{new} becomes more accurate as the n_{max} increases.

The time for leaching is determined [15] using appropriate leaching models. The leaching model for liberated particle leaching results in [14]:

$$t(r_{vmp})_{lib} = k_{lib} r_{vmp} [1 - (1 - \alpha_{lib})^y] \tag{4.114}$$

k_{lib} is a reaction constant (sec/cm). α is the fraction reacted. y is 1/3, 1/2, or 2/3 for reaction control, rapid flow/fine particle, and slow flow/larger particle leaching conditions, respectively. A simplified estimate for partially-locked particle leaching is [15]:

$$t(r_{vmp})_{Part-Lock} \approx \left[\frac{4\pi}{\left(\frac{4}{3}\pi\right)^{2/3}}\right] k_{lib} r_{vmp} [1 - (1 - \alpha_{lib})^y] \tag{4.115}$$

Leaching kinetics for locked valuable mineral particles can be estimated using the shrinking-core or pore diffusion model.

Third, the overall leaching is determined by appropriate weighting of the various leaching models. The weighting is based on the fraction of material associated with the model. In other words, if 50% of the particles are partially-locked, the overall model will be 50% determined by the partially-locked leaching model.

4.7   Electrochemical Kinetics

A rigorous approach to electrochemical reaction kinetics would utilize the mass transport balance equation. This equation includes the effects of migration, convection, and diffusion as well as reaction. However, in most circumstances, migration currents of active species are small. Convective or advective flow can be coupled to diffusion using the boundary layer thickness. Consequently, the mass balance equation can be reduced to reaction and diffusion coupled to advection. The mass balance equation must be related to the rate of reaction. The rate of electrochemical reactions is modeled using the Butler-Volmer equation. Each equation is valid separately for the extreme cases of mass transport or electrochemical control. However, most electrochemical reactions are not controlled by mass transport or electrochemical kinetics. Instead, most electrochemical reaction rates are controlled by a combination of mass transport and electrochemical kinetics. Therefore, modeling combined kinetic control requires combined equations.

The Butler-Volmer equation assumes the concentration at the electrochemical interface is equivalent to the bulk concentration. If mass transport becomes a limiting factor, this assumption will no longer be valid. Thus, the bulk concentration, which is implicit in the i_o value (i_o is a direct function of bulk concentration), must be multiplied by the surface concentration divided by the bulk concentration. This procedure effectively replaces the bulk concentration term inherent in i_o with the desired surface concentration. The resulting Butler-Volmer equation is:

> **The Butler-Volmer Equation can be adapted to include the effect of mass transport.**

$$i = i_o \left[\frac{C_{sa}}{C_{ba}} \exp\left(\frac{\alpha_a F(E - E_{eq})}{RT} \right) - \frac{C_{sc}}{C_{bc}} \exp\left(\frac{-\alpha_c F(E - E_{eq})}{RT} \right) \right] \qquad (4.116)$$

where the "s" subscript denotes the surface concentration, "b" represents bulk concentration, "a" represents anodic, and "c" represents cathodic.

> **The limiting current density is the maximum electrochemical reaction current density based on diffusion control.**

If mass transport is controlling the current flow, Fick's law, combined with a conversion from molar rates to current densities leads to the expression for the mass transfer limiting current density, i_l:

$$i_l = \frac{nFDC_b}{\delta} \qquad (4.117)$$

In cases where the species that is mass transport limited is also a significant migration current carrier, a modified expression can be used for more accurate results:

$$i_l = \frac{nFDC_b}{\delta(1 - t_j)} \qquad (4.118)$$

In which t_j is the transference number for the species "j" under the test conditions.

Note that the boundary layer thickness can be calculated using the equations shown previously. However, when appropriate equipment is available it is often convenient to measure the limiting current density directly. Thus, it is often helpful to use the previous simple limiting current density expression in a different form:

$$\delta = \frac{nFDC_b}{i_l} \qquad (4.119)$$

Substitution of this equation into the expression for general Fick's first law yields:

$$i = \frac{nFD(C_b - C_s)}{\delta} \tag{4.120}$$

Substitution of the previous expression for the boundary layer thickness in terms of the limiting current density into this expression for current density leads to:

$$i = \frac{nFD(C_b - C_s)i_l}{nFDC_b} \tag{4.121}$$

This equation can be rearranged to give:

$$i = i_l\left(1 - \frac{C_s}{C_b}\right) \tag{4.122}$$

Additional rearrangement results in:

$$C_s = C_b\left(1 - \frac{i}{i_l}\right) \tag{4.123}$$

This expression can be used in the Butler-Volmer Eq. (4.116). The resulting substitution and rearrangement gives:

$$i = \frac{i_o\left[\exp\left(\frac{\phi_a F(E - E_{\text{nat.rxn}})}{RT}\right) - \exp\left(\frac{-\phi_c F(E - E_{\text{nat.rxn}})}{RT}\right)\right]}{1 + i_o\left[\left(\frac{1}{i_{la}}\right)\exp\left(\frac{\phi_a F(E - E_{\text{nat.rxn}})}{RT}\right) - \left(\frac{1}{i_{lc}}\right)\exp\left(\frac{-\phi_c F(E - E_{\text{nat.rxn}})}{RT}\right)\right]} \tag{4.124}$$

Although this form of the Butler-Volmer Equation is large, it is very useful for electrochemical reactions that are or could be partially diffusion limited.

In this equation i_{la} represents the limiting current density of the limiting anodic reactant. The term i_{lc} represents the limiting current density of the limiting cathodic reactant. Alternatively, the the previous expression for limiting current density can be used. This expression provides a realistic determination of current density when mass transport is partially or completely controlling kinetics. It assumes migration currents are negligible.

4.8   Crystallization Kinetics

Metal ions will combine with oppositely charged ions to form precipitate crystals if the concentrations of the ions exceed the solubility limits. The process of precipitation or crystallization is very important to many industries.

Most precipitates do not form immediately when the solubility product is exceeded. Usually, crystallization requires ion activities that exceed the solubility product by a significant margin. In cases where the solubility product is low, the ion activities generally must exceed the solubility product by a large margin. In contrast, when solubilities are high, the margin can be small. The margin above the solubility product is known as supersaturation, S.

$$S = \frac{C}{C_{solubility}} \qquad (4.125)$$

C is the concentration of the species that is being evaluated. $C_{solubility}$ is the solubility limit of the species being evaluated. The next

> **Crystallization kinetics are often related to supersaturation.**

requirement for crystallization after supersaturation, $S > 1$, is nucleation (if crystals are not already present). Usually, however, nucleation is delayed by a period of time referred to as an induction time. The induction time can be expressed using an equation such as [16]:

$$t = 10^{-10} \exp\left(\frac{32\pi\sigma^3 v}{15kK^2T^3}\right)(\ln(S))^{-2} \qquad (4.126)$$

t is the induction time. σ is the surface energy. v is a frequency term. k is Boltzman's constant. K is a reaction constant. T is absolute temperature. S is supersaturation. The rate of nucleation in terms of the number of crystals formed when diffusion is controlling nucleation is given as [16]:

$$N = 10^{22} \exp\left(\frac{48\pi\sigma^3 v}{15kK^2T^3}\right)(\ln(S))^{-2} \qquad (4.127)$$

It should be noted that the surface energy term drops significantly between nucleation in a pure solution to nucleation at a surface. Because the surface provides an energetic template that increases the surface energy, nucleation occurs more readily on high energy surfaces. High energy surfaces include glass and metal. Low energy surfaces include materials such as Teflon®.

Crystal growth rates are usually of concern to industrial personnel. Seed crystals are usually used in crystallizers to bypass the nucleation process. Nucleation that is more difficult to control than crystal growth. Various models are used to determine or predict the rate of crystal growth. Each model is specific to the conditions in the crystallizer. In many cases that involve adsorption-controlled crystallization, the rate of crystal growth tends to follow a linear rate law [16]:

$$R = \frac{DV_m C_{solubility}(S-1)}{\delta} \qquad (4.128)$$

D is the diffusivity. V_m is the molar volume of the precipitate. $C_{solubility}$ is the solubility limit concentration of the relevant species. S is the saturation. δ is the diffusion layer thickness that is usually on the order of 100 µm under stagnant conditions. Nonlinear cases can often be treated using equations of the form [16]:

$$R = kN_s(C - C_{solubility})^x \tag{4.129}$$

k is a reaction rate contstant. N_s is the number of seed crystals. C is the concentration of the desired species. $C_{solubility}$ is the solubility limit concentration of the relevant species. x is an adjustable factor that is often close to 2. The radius of a spherical growing crystal can be estimated from an equation derived by Turnbull that is based upon relevant diffusion equations [17]:

$$\frac{r^2}{2D} + \frac{r}{G} = mKt \tag{4.130}$$

r is the particle radius. D is the diffusion coefficient. G is the interface transfer coefficient. m is the concentration of precipitate in solution. K is a rate constant. t is time.

It should be noted that the size of crystals is dynamic even in cases where no net growth occurs. When small crystals are present in saturated solutions, some of the crystals will tend to grow at the expense of others shrinking. This phenomenon in which small crystals are consumed and large crystals grow is surface energy related. It is somewhat analogous to grain growth in metals at elevated temperatures. This phenomenon is called Otswald Ripening.

Another important aspect relating to crystal growth that should be remembered is the effect of impurities. Small changes in the levels and types of impurities can have enormous impacts on the type, shape, size, nucleation rate, and growth rate of the associated crystals. Impurities influence important rate limiting steps in the crystallization processes.

4.9   Overview of Surface Reaction Kinetics

Aqueous surface reactions require dehydration, reaction, and surface diffusion steps.

Nearly all of the kinetic processes in chemical metallurgy that occur in aqueous media involve reactions or events near interfaces. The importance of understanding various processes that occur near interfaces provides the motivation for including this last section in the chapter on kinetics. Any one of the steps that occurs between the time a reactive ion is in the bulk solution and the time a reaction takes place can become the rate limiting step.

Reactions at interfaces begin with a hydrated ion in solution. This ion needs to be transported to a surface to react. In this example a metal electrode with a negative applied potential will be considered. As shown in Fig. 4.14, the ion is extremely far from the surface in terms of an atomic scale. Consequently, convective mass transport is needed to enhance the ion mobility in its journey toward the reaction interface.

As the ion approaches the surface boundary layers of water provide significant drag on the moving bulk solution. The drag near the surface is such that the influence of moving bulk fluid is negligible within a certain distance of the surface. This distance is known as the boundary layer thickness. Even with turbulent flow in the bulk solution, the ions that are within a few micrometers of the surface are not influenced by the bulk flow. Consequently, the ions are forced to diffuse toward the surface between the boundary layer and the interface as shown in Fig. 4.15.

After diffusion through the boundary layer, ions must dehydrate. The dehydration process begins at the boundary of the electrical double layer as shown in Fig. 4.26. The electrical double layer consists of surface adsorbed water and ions that are close enough to have their water sheath touch the surface water sheath. The process of dehydration is also accompanied by charge transfer. However the ion must come very close to the surface for charge transfer to occur. Only part of the hydration sheath or covering is lost at this reaction stage. Precipitation events that do not involve charge transfer follow the same process without the charge transfer step.

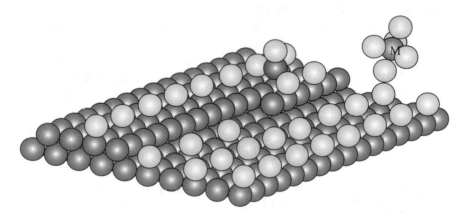

Fig. 4.26 Schematic diagram illustrating a hydration metal cation at the edge of the electrical double layer (only hydration water separates the charged metal surface from the metal cation). The dark spheres represent metal atoms. The lightest spheres represent water molecules

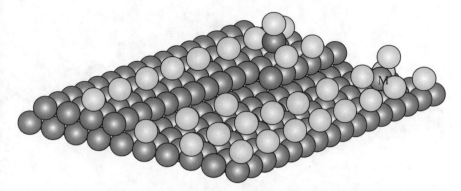

Fig. 4.27 Schematic diagram illustrating a partially hydrated, adsorbed metal atom that has been reduced in charge by the acquisition of an electron at the surface of the underlying metal. Note that in order for the metal atom to reach its place at the surface it needed to displace hydration water molecules, then become reduced in charge through electron acquisition. The dark spheres represent metal atoms. The lightest spheres represent water molecules

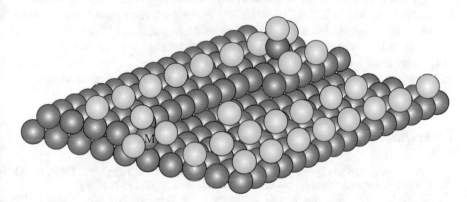

Fig. 4.28 Schematic diagram illustrating a further dehydrated, adsorbed metal atom that has moved along the surface by surface diffusion to the surface ledge as shown. At the ledge the metal atom has lost some hydration water in exchange for more association with metal atoms. The dark spheres represent metal atoms. The lightest spheres represent water molecules

After partial dehydration and charge transfer as illustrated in Fig. 4.27, additional steps must occur. As the ion diffuses along the surface it encounters surface steps or ledges. The presence of the ledge facilitates further dehydration. The ledge also provides additional association with surface atoms. Consequently, additional bonds are formed as illustrated in Fig. 4.28.

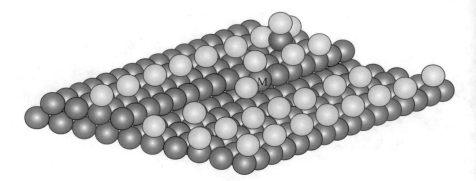

Fig. 4.29 Schematic diagram illustrating a further dehydrated, adsorbed metal atom that has moved along the surface edge by surface diffusion to the surface kink site as shown. At the kink site the metal atom has lost additional hydration water in exchange for more association with metal atoms. The dark spheres represent metal atoms. The lightest spheres represent water molecules

Edge associated atoms further diffuse until additional bonds can be formed at kink sites as shown in Fig. 4.29. Migration to the kink allows the metal atom to become further dehydrated and more completely integrated. Additional bonding is accomplished. This overall process continues with other ions as the surface grows. In contrast, the dissolution process follows the reverse order. Consequently, growth and dissolution are generally most prevalent around the edges of dislocations.

References

1. P.W. Atkins, *Physical Chemistry,* 3rd ed (W. H. Freeman and Co., New York, 1986), p. 692
2. L. Pauling, *General Chemistry* (Dover Publications, New York, 1970), p. 566
3. J.E. Bailey, D.F. Ollis, *Biochemical Engineering Fundamentals*, 2nd edn. (McGraw-Hill Publishing Company, New York, 1986), pp. 473–476
4. M.L. Free, *Bioleaching of a Sulfide Ore Concentrate - Distinguishing Between the Leaching Mechanisms of Attached and Nonattached Bacteria* (M.S. Thesis, University of Utah, 1992)
5. S. Nagpal, T. Oolman, M.L. Free, B. Palmer, D.A. Dahlstrom, Biooxidation of a refractory pyrite-arsenopyrite-gold ore concentrate, in *Mineral Bioprocessing*, ed. by R.W. Smith, M. Misra (TMS, Warrendale, 1991), p. 469
6. D.A. Jones, *Principles and Prevention of Corrosion* (Macmillan Publishing Company, New York, 1992)
7. R.B. Bird, W.E. Stewart, E.N. Lightfoot, *Transport Phenomena* (Wiley, New York, 1960)
8. M.C. Potter, J.F. Foss, *Fluid Mechanics* (Great Lakes Press Inc., Okemos, 1982)
9. M.E. Wadsworth, Principles of leaching, in *Rate Processes of Extractive Metallurgy*, ed. by H.Y. Sohn, M.E. Wadsworth (New York, Plenum Press, 1979), pp. 133–197
10. J.D. Miller, Cementation, in *Rate Processes of Extractive Metallurgy*, ed. by H.Y. Sohn, M.E. Wadsworth (Plenum Press, New York, 1979), pp. 197–244
11. H.S. Fogler, *Elements of Chemical Reaction Engineering* (Prentice-Hall, Englewood Cliffs, 1986)

12. R.E. Treybal, *Mass Transfer Operations*, 3rd ed. (McGraw-Hill Publishing Company, New York), p. 198
13. O. Levenspiel, *Chemical Reaction Engineering*, 2nd edn. (Wiley, New York, 1972), pp. 357–373
14. M.L. Free, Modeling of heterogeneous material processing performance. Can. Metall. Q. **47** (3), 277–284 (2008)
15. M.L. Free, Predicting leaching solution acid consumption as a function of pH in copper ore leaching, in *7th International Copper 2010-Cobre 2010 Conference, Proceedings of Copper 2010*, Vol. 7, pp. 2711–2719 (2010)
16. M.L. Free, Modelling heap leaching PLS from valuable mineral associations and leaching parameters, in *Proceedings of Heap Leach Solutions*, 2015, Reno, Nevada, ed. by M. Evatz, M.E. Smith, D. van Zyl (Infomine, Vancouver, BC, Canada, 2015), pp. 395–402
17. Sarig, Fundamentals of aqueous solution growth, in *Handbook of Crystal Growth*, Chap. 19, ed. by D.T.J. Hurle (Elsevier, 1994), pp. 1167–1216
18. R.W. Bartlett, *Solution Mining* (Gordon and Breach, 1992), p. 240

Problems

(4-1) Calculate the boundary layer thickness when a 0.1 cm (diameter) sphere is stirred in a solution where the relative fluid velocity past the sphere is 0.1 cm/sec, the kinematic viscosity is 0.01 cm^2/sec, and the diffusivity is 1×10^{-5} cm^2/sec.

(4-2) Calculate the rate of mass transfer per unit area (mol/sec cm^2) of an ion in solution when the stationary mass transfer boundary layer thickness is 0.005 cm, the difference between the bulk and surface concentration is 0.15 mol/liter, and the diffusivity of the ion is 3×10^{-5} cm^2/s. (answer: 9×10^{-7} mol/sec cm^2)

(4-3) Derive an expression for leaching time as a function of fraction reacted for a solid particle of mineral X that is gradually and completely leached away by compound Y. Assume that both diffusion and reaction may be rate controlling. Note that for particles in this problem, it can be assumed that the fluid boundary layer thickness is approximately proportional to $r^{1/2}$.

(4-4) A pure particle of solid malachite (CuCO$_3 \cdot$ Cu(OH)$_2$) 0.3 mm in diameter dissolves completely in a well stirred acid bath according to the reaction:
CuCO$_3 \cdot$ Cu(OH)$_2$ + 4H$^+$ $\leftrightarrow$ 2Cu^{2+} + 3H$_2$O + CO$_2$
Assuming acid depletion is negligible and that the rate of reaction is completely controlled by diffusion, calculate (a) the time required for a particle to be 95% reacted if the required time for 50% reaction is 5000 s, and (b) the time required to leach 98% of a particle that is three times the size of the original particle for which leaching data is available. (Assume the particle is relatively large by kinetic model standards.)
(answer for 95% reaction of the original particle is 13,252 s).

(4-5) Using the accompanying chalcopyrite leaching data for 12 μm particles, determine the activation energy for the chalcopyrite reaction given assuming the reaction is not controlled by diffusion. Assume the leaching agent concentration is 1 molal. Assume the particle surface area is 1 m^2.

Temp (°C)	Time (min.)	Rate (1/min)
60	410	0.000141
75	386	0.000301
90	420	0.000380

(4-6) Using the following data (from Roman et al. 1974) from column leaching tests, determine if the reaction is diffusion controlled.

Time (days)	Fraction reacted
1.5	0.12
3.0	0.21
3.0	0.27
5.6	0.36
9.8	0.45
16.5	0.60
27.2	0.71
35.6	0.75

(4-7) Using the data from 4 to 6 and knowing that the concentration of leaching agent is 0.5 M, the valuable mineral density is 4.0 g/cm^3, the mass fraction of valuable mineral in the host rock matrix is 0.019, the sphericity is 0.8, and the molecular weight of the desired mineral is 221.1, determine the diffusion coefficient for the material with an approximate average diameter of 3 cm. (Assume the reaction is: $CuCO_3 \cdot Cu(OH)_2 + 4H^+ \leftrightarrow 2Cu^{2+} + 3H_2O + CO_2$)

(4-8) What will the current density be for the reaction $M^{2+} + 2e^- = M$ if the equilibrium exchange current density for the reaction is 1×10^{-7} A/cm^2, the mixed potential is –0.339 V and the equilibrium potential is –0.311 V. Assume the symmetry factor is 0.5.

(4-9) Calculate the rate of bacterial oxidation for a reaction that is controlled by Michaelis–Menten enzyme kinetics if the concentration of cells is 2 g/l, the maximum specific growth rate is 0.036 per hour, the yield coefficient is 2, the Michaelis constant is 0.015 g/l, and the concentration of the substrate that is oxidized by the bacteria is 0.05 g/l. (10 points)

Chapter 5
Metal Extraction

> *Commercial metal extraction requires chemical reaction, reagents, and practical application of engineering principles.*

Key Learning Objectives and Outcomes:
Understand basic metal extraction principles and terminology
Know specific reactions relevant to mineral leaching applications
Understand different types of leaching
Understand the role of bacteria in relevant leaching scenarios
Understand the importance of percolation and flow in heap leaching
Identify general conditions needed for different ore leaching scenarios

5.1   General Principles and Terminology

Aqueous metal extraction or leaching is the process of extracting metal or a valuable entity found within an object. Consider the copper minerals shown in Fig. 5.1. Each of these mineral samples contains a high concentration of copper. Most are nearly pure mineral specimens. The copper in these minerals can be extracted using a variety of processes. Metal extraction is typically performed using ore particles. Ore particles contain small pieces of the pure metal-bearing mineral. These small pieces are often finely disseminated in the host rock as shown in Fig. 5.2. Figure 5.2 shows a 3 cm diameter rock piece that contains approximately 1% chalcopyrite. The chalcopyrite is found as finely disseminated dark pieces that are more easily observed when magnified as shown in Fig. 5.3.

© The Minerals, Metals & Materials Society 2022
M. L. Free, *Hydrometallurgy*, The Minerals, Metals & Materials Series,
https://doi.org/10.1007/978-3-030-88087-3_5

Fig. 5.1 Copper-bearing
minerals (Chrysocolla,
azurite, malachite, bornite,
chalcocite, and chalcopyrite)

Fig. 5.2 Photograph of a 3
cm diameter chalcopyrite ore
sample. The tiny dark
blemishes are small grains of
chalcopyrite disseminated
within the host rock

Fig. 5.3 Magnified (7×)
view of a chalcopyrite ore
sample. The small dark
sections are pieces of
chalcopyrite disseminated
within the host rock matrix

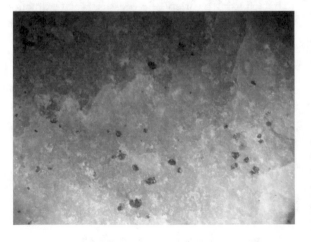

In other cases the distribution of valuable mineral involves larger valuable mineral particles and a less random distribution. Figure 5.4 shows a copper ore particle that has been partially leached. Note that the removal of copper around the outer edge of the rock is uniform in some sections and nonuniform in other areas. It is important to note, however, that the leaching solution penetrated the ore particle, which is often presumed to be "solid rock". However, rock particles have pores and cracks that can allow solutions to penetrate and extract the valuable entity. An example of the pore network after leaching is shown in Fig. 5.4. Figure 5.4 shows a leached ore particle or rock that was exposed to a blue dye under vacuum. The blue dye shows where leaching occurred. The blue dye reveals a combination of channels and pores as well as zones where the valuable mineral was removed. It is

> *Ore particles contain natural pores through which reagents can very slowly penetrate.*

important to note that leaching occurred in isolated areas within the rock particle. Thus, it is clear that there are microscopic pores that allow solution to penetrate into the interior of the rock particle. In other words, the desired mineral can be leached even though it may be locked inside of the host rock.

Figure 5.5 shows the cross-section of a copper oxide ore particle that has been leached for approximately 48 h in a sulfuric acid leaching solution at pH 1.5. The copper oxide minerals near the outer rim as well as acid consuming gangue minerals near the rim have been dissolved. It is also worth noting that the penetration depth is relatively constant except for one region in which the penetration is significantly greater due to local porosity or more reactive minerals.

Fig. 5.4 View of cross-sectioned copper oxide mineral rock particle that was leached, then vacuum impregnated with a blue dye to identify pore areas and leached sections of the particle. The horizontal length of the image is approximately 1 cm

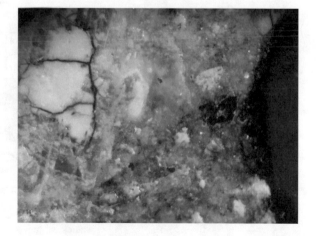

Fig. 5.5 Cross-section of a
copper oxide ore particle after
24 h of leaching. The
horizontal length of the image
is approximately 1 cm

Leaching of minerals and metallic compounds has become increasingly important in nonferrous metal production. The conditions that are needed for leaching to
occur are determined using thermodynamics.
Consequently, phase diagrams are often used to
ascertain the solution environment that will be
necessary for leaching. Figure 5.6 shows a simplified copper phase diagram that includes several
copper compounds. Other species have been
neglected to maintain simplicity. The combina-

> *Hydrometallurgical
> extraction is only possible
> under appropriate
> solution conditions.*

tion of thermodynamics and practical application are discussed in this chapter as
well as many other sources [1–37].

Fig. 5.6 Simplified copper
phase diagram showing phase
stability regions. Leaching
occurs in shaded regions
where soluble species are
stable

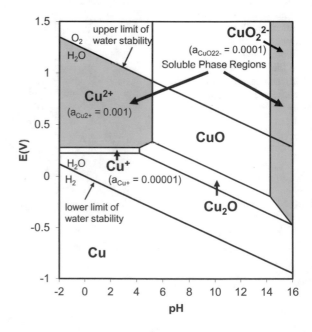

According to Fig. 5.6 the copper oxides are stable at high pH and potential values. The low pH and high potential regions of the diagram favor dissolved copper species. High pH and intermediate potentials also favor dissolved species. Because leaching implies solubilization, this diagram shows the conditions needed for extraction. Extraction through leaching is possible in regions with soluble species. It is also important to note that if the potential is above the water stability line, water will break down to form oxygen gas. If the potential is below the lower water stability line, water will decompose into hydrogen gas and hydroxide ions. Thus, only narrow potential regions are practical for most aqueous extraction.

> *Most metals are soluble, and therefore extractable, at very low pH and high potentials or at very high pH and moderate potentials.*

Phase diagrams and thermodynamics are important tools in assessing the viability of leaching under a prescribed set of conditions. However, distinguishing between what is possible in the context of leaching and what is practical from an engineering standpoint requires an evaluation of the applicable rates. Chemical, electrochemical, and mass transport kinetics were discussed in Chap. 4. Other related issues and principles also need to be discussed as they relate to leaching.

The main types of leaching are in-situ, dump, heap, atmospheric pressure tank, concentrate, and pressure leaching. Leaching under ambient conditions in heaps and dumps is often accelerated by bacteria. In-situ and pressure leaching are generally accelerated by the enhanced activity of oxygen at higher pressures.

> *Metal extraction is often referred to as leaching. Leaching types include in-situ, dump, heap, atmospheric tank, pressure and concentrate leaching.*

Often, leaching involves oxygen as an oxidant. Many leaching processes are limited by the supply of oxygen. Heap and dump leaching is performed using particles with varying sizes that are placed in piles. Heap and dump leaching rates are often influenced significantly by the permeability of the leaching material. Rates of extraction in heaps and dumps are often controlled by reagent availability.

5.1.1 Permeability and Fluid Flow Through Particle Beds

Permeability is the ability of a medium to conduct a fluid. As a fluid travels through a bed of particles, there is tremendous drag due to the surfaces over which the fluid flows. A schematic diagram of flow through a bed of particles is shown in Fig. 5.7. When solution is applied to a bed of particles, some regions become saturated with fluid.

> *Commercial metal extraction often requires leaching from beds of particles that must have adequate permeability to accommodate reasonable flows of solution.*

Fig. 5.7 Schematic diagram
illustrating solution
application and flow through
a bed of particles

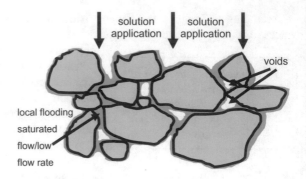

Other regions have voids as illustrated in Fig. 5.7. Solution flows more rapidly
through regions that have good permeability.

From a theoretical perspective, the flow through the pores that form between
particles in a bed can be described in a similar fashion to flow through tubes as
illustrated in Fig. 5.8.

Poiseuille described laminar flow through a single tube as [1]:

$$\frac{dV}{dt} = \frac{\Delta P \pi r^4}{8 \mu L} \tag{5.1}$$

where V is the volume of fluid, t is the time, P is the pressure drop, r is the radius
of the tube or pore, μ is the viscosity, and L is the thickness or height of the bed of
particles. This equation, known as the Poiseuille Equation, states that the flow rate
is directly proportional to the radius to the fourth power, thus the flow rate is

Fig. 5.8 Schematic diagram
illustrating solution
application and flow through
an array of tubes that
simulates flow through a bed
of particles

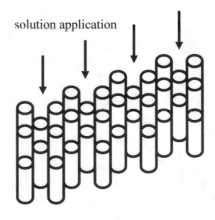

extremely sensitive to variations in
pore size which is a function of particle
size and particle size distribution. Thus
as particle size decreases, there is a

> **Pore or tube diameter is a critical parameter in determining flow.**

tremendous decrease in the flow rate through a bed of particles. Using the Poiseuille
Equation, Kozeny transformed the relationship in terms of bed porosity (not to be
confused with particle porosity), particle surface area, and flow area using a
characteristic constant [2], resulting in:

$$\frac{dV}{A_{bed}dt} = \frac{\varepsilon_{bed}^3 \Delta P}{K_K \mu (1 - \varepsilon_{bed})^2 S_p^2 L} \tag{5.2}$$

A_{bed} is the area of the bed through which fluid is flowing. ε_{bed} is the bed porosity or
fraction of the particle bed that contains voids. K_K is the Kozeny constant. S_p is the
specific surface area of the particles. Equation (5.2) is often referred to as the
Kozeny or the Kozeny-Carmen Equation due to contributions made by Carmen.
This equation reveals the importance of porosity in the bed. If the bed becomes
compacted significantly by heavy equipment, its ability to transport leaching
solutions and oxygen will be greatly diminished.

Example 5.1 Determine the new maximum fluid flow rate into a heap leaching
operation relative to the old flow rate if the old porosity is 0.40 and the new porosity
is 0.38 if all other factors are constant.

$$Q = \frac{dV}{dt} = \frac{A_{bed} \varepsilon_{bed}^3 \Delta P}{K_K \mu (1 - \varepsilon_{bed})^2 S_p^2 L} = \frac{k \varepsilon_{bed}^3}{(1 - \varepsilon_{bed})^2}$$

$$\frac{Q_{new}}{Q_{old}} = \frac{(1 - \varepsilon_{bed(old)})^2 \varepsilon_{bed(new)}^3}{\varepsilon_{bed(old)}^3 (1 - \varepsilon_{bed(new)})^2} = \frac{(1 - 0.4)^2 0.38^3}{0.4^3 (1 - 0.38)^2} = 0.803$$

Finally, on a more empirical and much less informative approach, there is
Darcy's law for fluid flow (Note the similarity to Eq. 5.2.) [3]:

$$\frac{dV}{A_{bed}dt} = \frac{k_i \Delta P}{\mu L} \tag{5.3}$$

k_i is the intrinsic permeability in units of
Darcys (Note that for water at 20 °C, the
hydraulic conductivity $k_{hydraulic}$ (m/s) or
coefficient of permeability or seepage coefficient is equal to 9.68×10^{-6} times k_i

> **The Darcy Equation is a simple equation that relates flow rate to pressure gradient and viscosity.**

(Darcy) where 1 Darcy $= 9.87 \times 10^{-13}$ m^2 [1]). The Darcy Equation is commonly applied in filtration and leaching. However, the equation gives no indication of the effect of the bed porosity or particle size on fluid flow.

Hydraulic conductivity can be mathematically expressed for saturated flow through beds as:

$$k_{\text{hydraulic}} = \frac{L}{\frac{\mu L^2}{k_i(\rho g L)}} = \frac{k_i \rho g}{\mu} \tag{5.4}$$

If the intrinsic permeability is 1 Darcy, the hydraulic conductivity for water at 20 °C is 9.68×10^{-6} m/s. The effective flow of water through such a particle bed would occur at that superficial solution velocity or 9.68×10^{-6} m/s. Note that the superficial velocity is equivalent to a volumetric flow rate per unit of area (9.68×10^{-6} m/s = 0.00968 l/m^2/s). Consequently, a heap cannot transmit a typical 0.003 l/m^2/s leaching solution application flow rate if the permeability of the heap is less than 3×10^{-6} m/s. Moreover, if the heap is operated near the permeability limit, instabilities are likely to arise and oxygen penetration will likely be very limited. Thus, most operations tend to ensure that the permeability is higher than 1 Darcy with a hydraulic conductivity of more than 9.68×10^{-6} m/s.

Effective clay liners for a leaching pad must have low hydraulic conductivity. The hydraulic conductivity of the clay portion of the liner is often restricted between 1×10^{-9} m/s and 1×10^{-8} m/s [3].

Another useful tool for evaluating flow through beds is the residence time. The residence time is the average time the fluid is in residence in the bed of particles. The residence time for saturated flow through a particle bed is defined as:

$$t_{\text{residence}} = \frac{\text{Volume}}{\text{Flow Rate}} = \frac{AL}{\frac{dV}{dt}} = \frac{AL}{\frac{Ak_i(\Delta P)}{\mu L}} = \frac{AL}{\frac{Ak_i(\rho g L)}{\mu L}} = \frac{\mu L}{k_i \rho g} \tag{5.5}$$

Flow through most heaps is not saturated with respect to water flow. Consequently, the pressure gradient is not the hydrostatic head pressure from the top to the bottom of the heap. Instead, a very small pressure gradient that represents an effective hydrostatic head can be used.

Many heaps in the copper industry have air injection. Thus, flow may be saturated with respect to air. Therefore these equations can find use in industry as written for air injection/infiltration. They can also be modified with respect to head pressure for solution flow applications. In practice the hydrostatic head pressure depends on flow patterns. Head pressure is also related to the drainage system.

These equations have specific application to evaluating a leaching operation. If too much fluid is applied, the bed of particles will fill with leaching fluid. This condition is known as flooding. It is also referred to as saturated flow. Under saturated conditions gas penetration is minimized. Gas penetration through the liquid phase is not sufficient for most scenarios. Thus, it is imperative that solution flow rates are low enough to prevent flooding. Low flow rates lead to solution percolation and some gas void space. The open space in particle beds allows needed convective gas transport. In many systems forced gas advection through perforated pipes is also needed.

Another factor that can influence leaching performance is capillarity. Capillary forces are particularly important when aqueous solutions are confined in narrow spaces or tubes. The capillary force draws fluid up in glass tubes and creates the meniscus on the walls of glass vessels. The height the fluid is drawn up is expressed as:

$$h_{capillary} = \frac{2\gamma \cos \theta}{\rho_l g r} \tag{5.6}$$

γ is the surface tension (72 mN/m for a typical aqueous fluid without surfactant). θ is the contact angle of solution on the particle or vessel surface (θ is often close to 0° for typical mineral particles). ρ_l is the liquid density. g is the gravitational acceleration. r is the radius of the tube or pore through which the liquid is drawn upward ($r \approx r_{particle}$ (void fr/solid fr)$^{1/3}$). Typically, the capillary rise or height is less than one meter. It is often only of concern near the bottom due to fine particle accumulation. Thus, for very deep beds capillary rise is not very significant. However, it can be significant for heap leaching if agglomeration is not utilized. In extreme cases, capillary rise can create flooding near the bottom of leaching operations. It may also make air injection and more difficult.

> *Capillary rise contributes to solution retention and can contribute to flow problems when fines accumulate near the bottom of heaps.*

Many leaching operations allow the material to "rest". The rest period facilitates solution management. It allows for control of concentrations of dissolved metal in the effluent stream. It allows for utilization of less expensive pumping infrastructure. Most operations actively leach or irrigate for a period of time. During the leaching time the solution is applied at a rate accommodated by the pumping infrastructure. Many irrigation systems operate at one pumping speed. An appropriate period of resting follows to regulate the average application rate. During the rest period no solution is applied. The term "irrigation rate" is often used to give an average rate of leachate flow through a particle bed. The irrigation rate is defined as [3]:

$$\text{Irrigation Rate} = \text{Application Rate}\left(\frac{\text{Leach Time}}{\text{Leach Time} + \text{Rest Time}}\right) \tag{5.7}$$

5.1.2   In-Situ Leaching

In-situ leaching involves mineral oxidation without material removal. Figure 5.9 illustrates this technique. In-situ extraction utilizes material with appropriate porosity. The porosity is either natural or induced by means such as explosives. In-situ leaching can be performed safely if the water tables and geological formations are appropriate. It is commonly applied to soluble salt extraction. Solutions must be properly contained and recovered as indicated in Fig. 5.9. In some cases, the effective water table can be drawn down through additional pumping as shown in Fig. 5.9. Appropriate pumping can prevent seepage from the in-situ leaching zone and the underlying aquifer.

In-situ leaching reduces costly excavation and reclamation costs. For deep deposits, in-situ leaching may be the only viable method of extraction. Many small scale in-situ leaching projects have been accomplished. However, due to uncertainties with regard to solution control, in-situ leaching has not been practiced on a large scale.

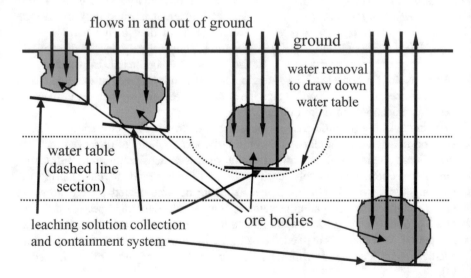

Fig. 5.9 Schematic diagram of in-situ leaching scenarios. Adapted from course notes from Milton E. Wadsworth

5.1.3   Dump Leaching

Dump leaching is characterized by very low-grade ore leaching. It involves run-of-mine material that has not undergone significant crushing or processing. Dump leaching practice generally consists of dumping the low-grade run-of-mine material. The material is dumped over the edges of mining terrain under which a liner has been placed. It is sometimes placed on lined leaching pads. Leaching solutions are applied at the top of the dump and allowed to percolate through the dump. A schematic diagram of a typical dump is presented in Fig. 5.10.

> *Dump leaching is a very significant commercial metal extraction method that is likely to continue to increase in importance due to low cost and improving technologies.*

The action of the leaching solution may be amplified many times by bacterial activity in ores containing sulfides such as pyrite. The bacteria that are naturally found on the ore particles can make a process economically feasible. In copper dump leaching the solutions generally contain acid. In gold dump leaching, the solutions generally contain cyanide and hydroxide. The leaching solutions are then passed through a stripping process to concentrate the metal. The leaching solutions are combined with makeup leaching solution before being reapplied at the top of the dump. Some dumps are hundreds of feet deep. Dumps may take decades to leach.

One of the most common ores that is leached in dumps is copper ore. Copper ore generally contains various copper oxides. In some locations copper sulfide ores are leached. In most dump leaching operations, oxygen penetration is critical to the

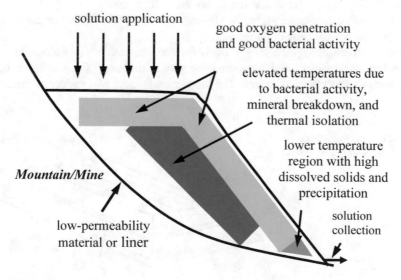

Fig. 5.10 Schematic diagram of a typical dump leaching scenario

leaching process. The oxygen is utilized by bacteria to drive their metabolism. The action of bacteria is covered in more detail in Chap. 4. Oxygen is also used for chemical reactions.

5.1.4   Heap Leaching

Heap leaching is leaching performed on low to medium grade ores. The ores must have enough valuable material to justify additional size reduction. Most heap leaching operations are carried out on a much shorter time scale than dump leaching. Often the leaching time for ore such as simple oxides can be 45 days or less due to the high permeabilities, small particles, and shallow depths. However, for sulfide ores or larger, run-of-mine material placed on leaching pads, leaching times may approach 600 days.

Often, each lift or vertical section is less than 10 m high. These conditions allow much greater oxygen penetration than is commonly found in dump leaching. Shorter lifts often reduce compaction, reduce precipitation, and speed extraction.

Heap leaching is generally practiced using lifts or sections that are 3-10 meters high and cover large areas.

A general view of a heap leaching system is shown in Fig. 5.11. Actual heap leaching operations are presented in Figs. 5.12 and 5.13. In operations requiring large volumes of oxygen, air is injected at the bottom of the heap.

The heap leaching lifts are constructed on top of pads. The pads often consist of foundation, underliner, liner, and overliner. Some pads contain additional layers as shown in Fig. 5.14. The additional layers are a liner and a drainage layer. Leak detection sensors are often placed in the drainage layer. Leak detection monitoring is often required for project permitting.

Pad construction begins with a properly formed foundation. The foundation must be smooth, compacted, and contoured for proper drainage. Contouring is often performed to direct fluid to a sump. The foundation is made of a variety of materials.

Fig. 5.11 Schematic diagram of a typical heap with two lifts

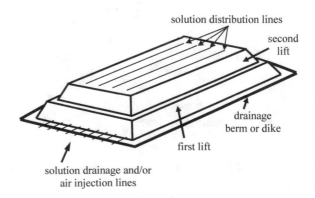

solution distribution lines

second lift

drainage berm or dike

first lift

solution drainage and/or air injection lines

Fig. 5.12 Examples of heap leaching operations

Fig. 5.13 Examples of heap leaching operations

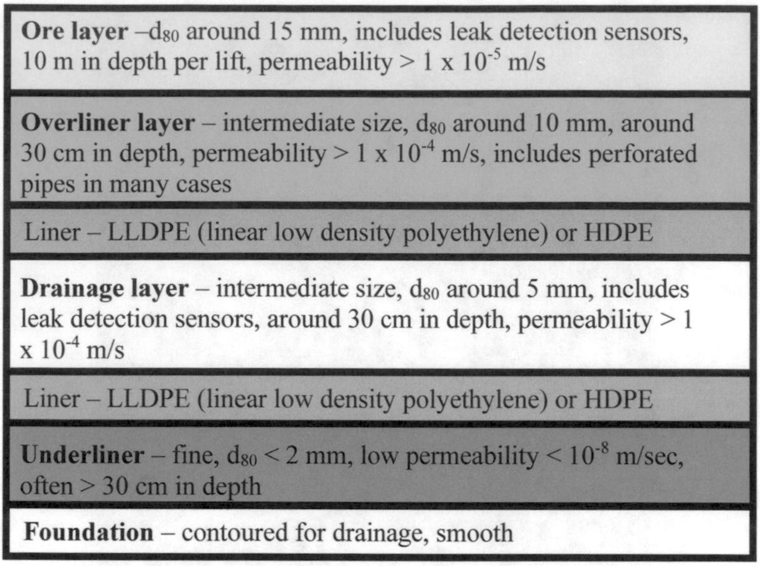

| Ore layer $-d_{80}$ around 15 mm, includes leak detection sensors, 10 m in depth per lift, permeability $> 1 \times 10^{-5}$ m/s |
| Overliner layer – intermediate size, d_{80} around 10 mm, around 30 cm in depth, permeability $> 1 \times 10^{-4}$ m/s, includes perforated pipes in many cases |
| Liner – LLDPE (linear low density polyethylene) or HDPE |
| Drainage layer – intermediate size, d_{80} around 5 mm, includes leak detection sensors, around 30 cm in depth, permeability $> 1 \times 10^{-4}$ m/s |
| Liner – LLDPE (linear low density polyethylene) or HDPE |
| Underliner – fine, $d_{80} < 2$ mm, low permeability $< 10^{-8}$ m/sec, often > 30 cm in depth |
| Foundation – contoured for drainage, smooth |

Fig. 5.14 Illustration of liner layers for heap and dump leaching

The underliner is made with crushed particles less than 30 mm in diameter. A typical d_{80} size for underliner material is 2 mm. The underliner material generally contains fine material. The underliner has a saturated hydraulic conductivity less than 1×10^{-8} m/s. The underliner layer is often 30 cm or greater in depth.

The liner is usually a geomembrane material. Geomembranes are often made of linear, low-density polyethylene (LLDPE) or high-density polyethylene (HDPE). However, some are made of polyvinyl chloride (PVC). The average thickness of the liners is around 2 mm.

> *Heap and dump leaching take place on carefully prepared pads that are lined to prevent leakage.*

The overliner is typically made of crushed material with a d_{80} of around 10 mm. The top size of the overliner material is around 30 mm. Pipes are often inserted into the overliner to solution drainage and/or gas injection. The overliner protects the liner from haulage equipment and facilitates solution collection. The overliner layer is often 30 cm or greater in depth.

The drainage layer may consist of material sized between that of the overliner and underliner. The drainage layer facilitates the removal of leaked solution. It also minimizes hydrostatic head pressure above the secondary liner. Reduced head pressure decreases leakage into the underliner.

Ore is stacked directly onto the overliner. The ore is typically crushed to a d_{80} size between 25 and 8 mm with the most common size near 12.5 mm. Appropriate size distribution is critical to leaching performance Particles that are too large do not allow appropriate distribution. Ore that is too fine can result in fines migration. Additionally, particle transport can lead to local structural instability. Consequently, crushed ore is often agglomerated before leaching. Particle transport can lead to void formation and channeling. Fine particles can also plug channels. Generally, ores with more than 20% clay are difficult to utilize in heaps and are often blended with ores with fewer fines.

The permeability of the stacked ore is targeted to be at least 10 times greater than the application rate. Figures 5.15, 5.16 and 5.17 show the effects of low, moderate, and excessive permeability on leaching using point source application. Note that excessive permeability results in channeling and nonuniform solution distribution when point source application is made.

Agglomeration is a technique used to increase permeability and to distribute leaching reagent. It is accomplished by mixing ore with water and a binder. The ore is mixed with water and binder in a mill or by conveyance and stacking. In alkaline leaching, the binder is often cement. In acidic leaching, acid is commonly used to facilitate binding and to maintain low pH levels throughout the heap when acid consuming minerals are present. The use of acid during agglomeration or ore mixing is commonly used with oxide and secondary sulfide ore deposits. The agglomerated ore is stacked on the pads by stackers or by haul trucks. The preferred placement is made by retreat stacking.

> *Agglomeration is often performed to enhance permeability and distribute leaching reagents.*

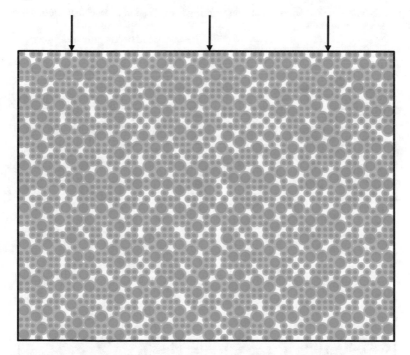

Fig. 5.15 Schematic diagram of point source (arrows) application of leaching solution using an ore with low permeability

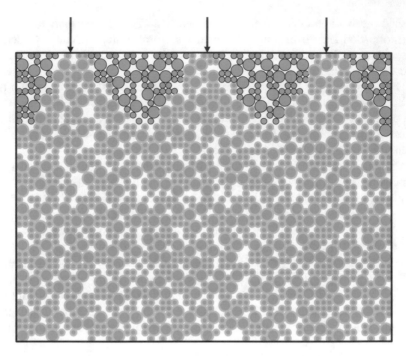

Fig. 5.16 Schematic diagram of point source (arrows) application of leaching solution using ore
with moderate permeability

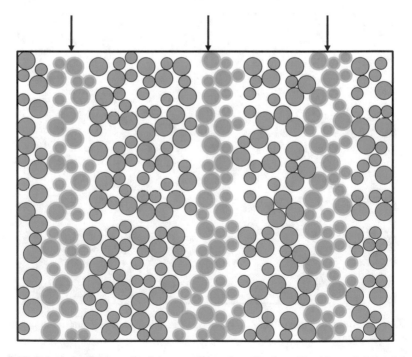

Fig. 5.17 Schematic diagram of point source (arrows) application of leaching solution using ore
with excessive permeability

Retreat stacking is accomplished by conveyors that lift the ore to the edge of a stack. The material then cascades down the stack. The stacker retreats as the material accumulates to the desired height. The stack height is the height of one layer known as a lift.

> *Optimum permeability allows for desired fluid and gas flow as well as solution distribution.*

After stacking is complete, solution distribution lines are commonly placed on top of the heap with minimal compaction by associated light weight equipment. In some cases cable systems are used to pull solution lines across newly stacked material without the need to drive vehicles on most of the area that is to be leached. In some cases, often with run-of-mine ore, where haul trucks place the ore on the pads, a bulldozer with a few large prongs behind is driven to "rip" the surface to facilitate greater solution penetration through the pad area compacted by haul trucks.

In cases where acid has been added during agglomeration, the acid soaked ore is allowed to "cure" or leach for a few days prior to solution application. This acid curing process allows for the relatively concentrated acid to leach the valuable entity prior to solution application to the heap.

Solution application is performed after solution distribution lines are placed on the heap. Leaching solution is applied to the top of the heap. The application rate is on the order of 0.002 d m^3 s^{-1} m^{-2} (equivalent to 2×10^{-6} m/s) or in the range of 0.005–0.01 m^3 h^{-1} m^{-2}.

A variety of solution distribution methods are used. A common distribution system utilizes drip emmiters. After recovery reaches an acceptable level, another lift is constructed on top of the first lift. A new solution application system is then installed and utilized.

When sulfide minerals are present in significant quantities, bacterial activity makes acid-based heap leaching more feasible. When sulfides are present, oxygen is generally provided using perforated pipes installed at the bottom of the heap. Air application rates are often in the range of 0.005–0.01 m^3 min^{-1} m^{-2} or about 60 times higher than the solution application rates. However, the amount of air needed is often based on the quantity of sulfides present.

The overall connection of heap leaching and metal recovery is shown in Fig. 5.18. The solution that is applied to the heap extracts the desired metal into the pregnant leaching (or product laden) solution (PLS).

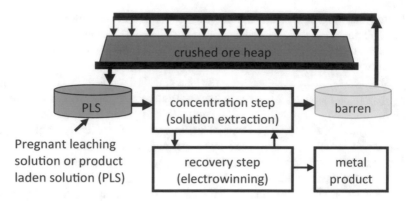

Fig. 5.18 Simplified flow sheet diagram of heap leaching and its integration with overall metal production

The PLS is concentrated, usually by solvent extraction. The depleted leaching solution leaving the solvent extraction process is called the raffinate or barren solution. The raffinate is generally returned to the leaching process. The concentrated solution is often called the rich electrolyte. Metal is generally recovered from the concentrated solution by electrowinning.

The metal content in the electrowinning cells is replenished by returning the depleted, lean electrolyte back to the solvent extraction

> *Metal-bearing solution leaving a leaching operation is often called the pregnant leaching solution or PLS.*

stripping unit. The process of solvent extraction will be discussed in the next chapter.

5.1.5   Modeling Heap and Dump Leaching Performance

Heap and dump leaching can be modeled using the fluid flow and kinetic equations previously discussed. Equations such as the shrinking-core model are commonly applied. Such models are initially applied to individual particle sizes. Equations are combined for all relevant ore sizes for an overall model. In addition, localized mass balances must be utilized. As leaching solution percolates into the heap or dump, compositions change. The effects of such changes must be incorporated into the model. These changes are generally vertical. Consequently, heaps and dumps are often evaluated as tall plug-flow reactors. Calculations are often facilitated by considering an array of voxels or cubes. An illustration of this approach is shown in Fig. 5.19.

Fig. 5.19 View of heap and dump leaching modeling approach with sequential voxels or unit boxes of particles treated separately from 1 to N

feed solution

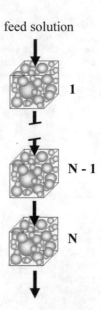

The effects of heat generation or consumption are included using thermodynamics and mass balancing. Fluid flow effects such as channeling are accommodated using fluid dynamics. Leaching of ore is generally modeled based on the shrinking core kinetics model.

Agglomerated ores have two additional issues to be addressed in modeling. Often the agglomeration is performed using a concentrated form of the leaching solution. Thus, leaching often begins when the material is agglomerated. Agglomerated ore is often exposed to this more concentrated solution for a day or more before solution lines are set up and leaching solution is applied. This period of delay and concentrated solution exposure is sometimes called the cure time.

Consequently, when the leaching solution is applied to the top of a heap, the solution collects the previously dissolved entities in the agglomerates and rock particles. Therefore, the initial PLS is high.

The initial solution concentration in the PLS is dependent on the rate of diffusion out of agglomerates and pores inside of rock particles such as those depicted in Figs. 5.20 and 5.21, respectively. It is also dependent on the application flow rate, heap height, particles size, and other factors.

Leaching in general is further complicated by the heterogeneous nature of the rock particles. Rock particles often contain a variety of minerals. The rock particle in Fig. 5.22 shows several minerals in an ore particle. Some minerals dissolve readily. Some minerals form passive films. Other minerals can form

> *Modeling leaching performance is challenging due to the complex nature of ores.*

as precipitates during leaching. Other minerals are inert. Often, bacteria are involved. Thus, leaching modeling is generally quite complex and results can vary widely. However, despite the complexity, leaching modeling can be reasonably successful.

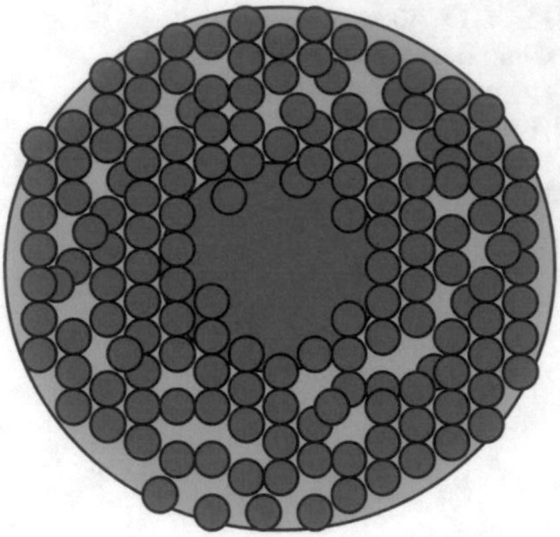

Fig. 5.20 Illustrating agglomerated ore that has been exposed to concentrated leaching solution during agglomeration, resulting in concentrated product

Fig. 5.21 Illustrating a network of pores in a solid particle

Fig. 5.22 Schematic diagram of ore particle with different minerals contained in the interior and at the outer surface. These mineral constituents interact with leaching solution by dissolution, alteration, precipitation, or inertness

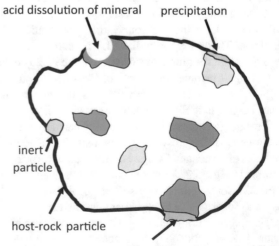

acid dissolution of mineral   precipitation

inert particle

host-rock particle

mineral alteration product layer

5.1.6 Solution Application Techniques

> *Most heap and dump leaching solutions are applied by sprinklers or drip emitters.*

In heap and dump leaching operations solutions are generally applied by trickling/dripping and sprinkling, although in rare cases ponding, or injecting are utilized. Trickling or dripping is performed using perforated pipes or tubes that allow for continual dripping across heaps or dumps and is restricted to relatively low flow rates. Often, drip emitters, which are placed at the ends of small diameter tubes that are connected to a central pipe, are used. Drip emitters are becoming increasingly popular—especially in arid climates due to water conservation. Sprinkling is a common method of solution application involving common types of sprinklers that provide a relatively high rate of fluid delivery in a relatively narrow range of flow rates. Ponding is carried out by simply retaining the solution in a pond at the top of dumps. Ponding is usually the least effective method of application due to channeling/short circuiting problems and reduced oxygen penetration. Ponding is rare. Injecting is another method of solution application involving direct pumping of the solution into the heap or dump through injection pipes, although it is rarely practiced.

5.1.7 Commercial Heap Leaching Applications

5.1.7.1 Copper

Phase diagrams for typical copper oxide and sulfide scenarios are presented in Figs. 5.23 and 5.24 as well as Fig. 5.6, shown previously. These diagrams show that an oxidizing, acidic environment is generally needed for metal extraction.

Copper extraction is often performed using dump or heap leaching. These processes were described previously. A variety of metals can be extracted by these processes. Copper is the metal most commonly extracted by these processes. Consequently, copper will be used as the general example. Copper is dump and heap leached primarily from oxide ores, although sulfides are also leached.

Copper oxide ores often include malachite ($CuCO_3 \cdot Cu(OH)_2$), azurite ($2CuCO_3 \cdot Cu(OH)_2$), and chrysocolla ($CuO \cdot SiO_2 \cdot 2H_2O$). Malachite is a deep-green monoclinic crystal, and azurite is a dark purple-blue monoclinic crystal. Chrysocoola is hydrated silicate with CuO associated in the lattice. Leaching often occurs according to the reactions:

$$4H^+ + CuCO_3 \cdot Cu(OH)_2 \leftrightarrow 2Cu^{2+} + CO_2 + 3H_2O \qquad (5.8)$$

$$6H^+ + 2CuCO_3 \cdot Cu(OH)_2 \leftrightarrow 3Cu^{2+} + 2CO_2 + 4H_2O \qquad (5.9)$$

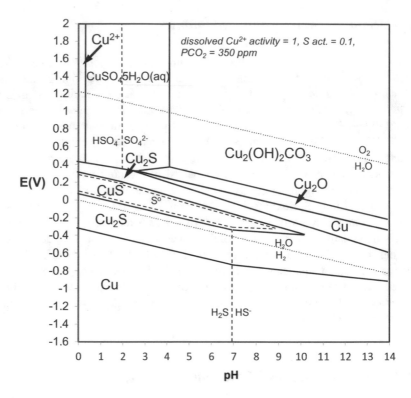

Fig. 5.23 Copper Eh/pH diagram based on data in Ref. [25] along with the addition of copper sulfate heptahydrate

$$H^+ + CuO \cdot SiO_2 \cdot 2H_2O \leftrightarrow Cu^{2+} + SiO_2 \cdot nH_2O + (3-n)H_2O \qquad (5.10)$$

In each reaction, two moles of hydrogen are consumed for each mole of copper released. One mole of sulfuric acid would be required for each mole of copper leached. However, acid consumption is usually much higher due to acid consuming minerals in the ore. An example of an acid consuming gangue mineral is calcite. Calcite consumes acid according to the reaction:

$$2H^+ + CaCO_3 \leftrightarrow Ca^{2+} + CO_2 + H_2O \qquad (5.11)$$

Acid consuming minerals such as calcite increase acid costs. In addition, acid consuming minerals contribute to precipitate formation, which can be detrimental.

Reactions that are commonly encountered in copper dump or heap leaching of sulfide ores include [11]:

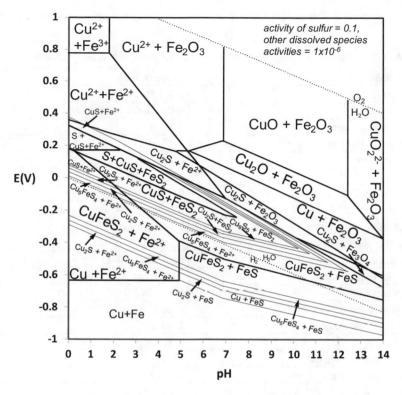

Fig. 5.24 Eh–pH diagram for copper, iron, sulfur, and water. Developed based on data in Ref. [25]

$$Cu_2S + 2Fe^{3+} \leftrightarrow Cu^{2+} + 2Fe^{2+} + CuS \tag{5.12}$$

$$CuS + 2Fe^{3+} \leftrightarrow Cu^{2+} + 2Fe^{2+} + S^{\circ} \tag{5.13}$$

$$Cu_5FeS_4 + 12Fe^{3+} \leftrightarrow 5Cu^{2+} + 13Fe^{2+} + 4S^{\circ} \tag{5.14}$$

$$CuFeS_2 + 4Fe^{3+} \leftrightarrow Cu^{2+} + 5Fe^{2+} + 2S^{\circ} \tag{5.15}$$

The last reaction (5.15) is generally too slow for typical commercial heap or dump leaching.

In addition to Eqs. 5.15–5.18, it is important to realize that other sulfides such as pyrite are often present in significant quantities. Pyrite can have an important influence on the overall consumption of oxygen. Also, it should be noted that the ferric ions are generated from ferrous ions by bacteria. Elemental sulfur is usually oxidized by the bacteria as indicated in Eq. 5.13.

Some general parameters for copper heap leaching are presented in Table 5.1.

Table. 5.1 Typical copper heap leaching parameters for oxide and secondary sulfide ores

Parameter	Range or method
Ore grade	0.3–1.5% copper
Size (area)	500,000–5,500,000 m^2
Overall heap height	7–60 m
Lift (individual layer) height	3–10 m
Stacking method	Conveyors/truck dumping
Liner	1.5–2 mm HDPE
Solution application	0.005–0.01 m^3/h/m^2
Air injection	0–0.01 m^3/min/m^2
Application line spacing	0.3–1.0 m
Emmitter spacing on line	0.3–1.0 m
Ore size (crushed)	80% minus 10–15 mm
Ore size (run of mine)	80% minus 80–120 mm
Agglomeration	Common (60–90 s. if used)
Agglomeration acid addition	5–20 kg/tonne of ore
Rest time before leaching	2–20 days
Application solution acid	5–15 kg/m^3
Application solution temp	15–25 °C
Total leaching time (oxides)	50–180 days
Total leaching time (sulfides)	150–600 days
PLS (Preg. Leach. Soln.) Cu	1–6 kg/m^3
PLS (Preg. Leach. Soln.) Acid	5–10 kg/m^3
PLS (Preg. Leach. Soln.) Temp	15–30 °C

> *Pyrite is important to bacterial oxidation, oxygen consumption, acid production, and iron control in sulfide ore leaching.*

5.1.7.2   Gold

Gold is more commonly leached in tanks than in heaps. Consequently, most gold extraction will be discussed in a subsequent section with precious metals. However, it is useful to indicate that heap leaching of low-grade gold ore is commonly practiced using 50–500 ppm sodium cyanide solution at pH 10.5–11.5. Heap design and flow rates are similar to those used for copper heap leaching.

5.1.7.3   Iron

Although iron is not leached as a desired mineral, it is often extracted during the extraction of other metals. A pH-Eh diagram for iron and water is presented in Fig. 5.25. This diagram shows that although iron must be oxidized to extract it from oxide minerals, the potential for extraction is relatively low. Correpondingly, it is relatively easy to corrode iron or steel in water. At low and high pH levels iron species are soluble. At moderate pH levels iron forms oxides, which are not structurally very stable from a corrosion passivation perspective. The relative effect of sulfur and iron in water is presented in the copper section.

5.2   Bioleaching/Biooxidation

Bioleaching is commonly applied to heap leaching as well as concentrate leaching, so it will be introduced in this section.

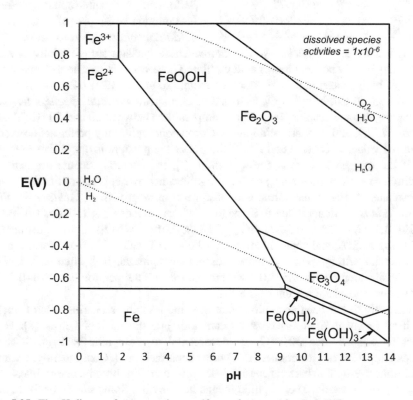

Fig. 5.25 Eh–pH diagram for copper, iron, sulfur, and water. After Ref. [35]

Bioleaching has been practiced for many centuries, yet bacteria that performed most dump leaching were not identified until the 1940s [4]. Three classes of bacteria are capable of mineral oxidation or leaching. These clasess are autotrophs, mixotrophs and heterotrophs. Autotrophic bacteria, the most common mineral oxidizing bacteria, derive their energy from the oxidation of inorganic compounds. Their energy comes from ions such as ferrous ions. They obtain their cell carbon from dissolved carbon dioxide [5]. Autotrophs leach minerals indirectly by providing oxidized inorganic species. Alternatively they secrete enzymes that oxidize minerals. Mixotrophic

> *Bacteria can enhance metal extraction by regenerating needed mineral oxidants such as ferric ions.*

bacteria obtain their energy from inorganic compound oxidation. Mixotrophs obtain at least some of their cell carbon from organic carbon sources [5]. Mixotrophs can leach essentially the same way as autotrophic bacteria. Heterotrophic bacteria obtain their energy and cell carbon from organic carbon sources such as sucrose. Heterotrophs leach minerals by providing acids such as citric, formic, acetic, oxalic, lactic, pyruvic, and succinic acids that are produced as by-products of organic carbon metabolism [5].

> *Microorganisms need appropriate nutrients and conditions to thrive.*

Most sulfide mineral bioleaching is performed by autotrophic bacteria. A common example of bacteria is *Acidithiobacillus ferrooxidans*, formerly known as *Thiobacillus ferrooxidans*. These bacteria are are relatively slow growing, but do not require organic carbon. These bacteria can be extremely tolerant of high concentrations of dissolved metal (20 g/l As, 120 g/l Zn, 72 g/l Ni, 30 g/l Co, 55 g/l Cu, 12 g/l U_3O_8 [4, 6]). Bacteria such as *acidithiobacillus ferrooxidans* are also mesophilic. They prefer temperatures between 20 and 50 °C (ideally 33 to 37 °C), and they are acidophilic. Correspondingly, they prefer acidic media with pH values between 1 and 2.5. If the pH or temperature is out of these ranges, or if the oxygen level falls below 1 ppm [7], the bacteria become dormant. In addition to these requirements, bacteria require other nutrients such as nitrogen and phosphate. A traditional formulation for maximum growth of *Acidithiobacillus ferrooxidans* nutrients is the 9 K medium which includes 3 g $(NH_4)SO_4/l$, 0.1 g KCl/l, 0.5 g K_2HPO_4/l, 0.5 g $MgSO_4 \cdot 7H_2O/l$, 0.01 g $Ca(NO_3)_2/l$, pH adjustment to 2.3 using H_2SO_4, and approximately 45 g $FeSO_4 \cdot 7H_2O$ /l [4]. However, in many commercial settings where growth rates are not optimal, it is often sufficient to supply air along with 0.5–1.0 kg ammonium sulfate per ton and 0.1–0.2 kg potassium phosphate per ton of ore [7].

Recently, it has been determined that in some leaching scenarios, other bacteria such as *Leptospirillum ferrooxidans* can dominate the microbial flora [8]. High leaching temperatures (>50 °C) are detrimental to mesophilic bacteria. In some heaps temperatures greater than 70 °C have been measured. Consequently, bacteria other than mesophilic bacteria are needed. Thus, there has been increased interest in thermophilic bacteria. Thermophilic means heat loving. Some successes for higher temperature applications have been reported using archae and bacteria, some of which have been found in natural hot springs [9, 35].

Microorganisms need an environment without excessive toxins. Most bacteria can tolerate or adapt to some levels of toxins. Often, bacteria can adapt to relatively high levels of some toxins if allowed to slowly acclimate to increasing levels of toxins. Some observed tolerance limits of Thiobacillus ferrooxidans for specific ions are presented in Table 5.2.

> *Microorganisms can slowly adapt to their environment. In some cases microorganisms can slowly adapt to very toxic environments.*

The metabolic pathways through which the bacteria oxidize ferrous ions to ferric ions are complex. Likewise, the bacterial oxidation of sulfur to sulfate is not simple. *Acidithiobacillus ferrooxidans* are capable of oxidizing sulfur and ferrous ions. Other bacteria such as *Acidithiobacillus thiooxidans* are also capable of sulur oxidation. Both bacteria are commonly found together in ores. The final step in the ferrous ion oxidation involves the formation of water:

$$O_2 + 4H^+ + 4e^- \leftrightarrow 2H_2O \tag{5.16}$$

which has a standard potential of 0.82 V inside the cell where the pH is approximately 7. In contrast, the potential of the ferrous/ferric half-cell reaction:

$$Fe^{3+} + e^- \leftrightarrow Fe^{2+} \tag{5.17}$$

is 0.770 V under standard conditions. The potential is slightly higher or lower under actual leaching conditions. Thus, the formation of water within the bacteria drives the oxidation of ferrous ions to ferric ions. The overall reaction is:

$$O_2 + 4H^+ + 4Fe^{2+} \leftrightarrow 4Fe^{3+} + 2H_2O \tag{5.18}$$

Table. 5.2 Thiobacillus ferrooxidans tolerance limits for selected metal ions [6]	Metal	Tolerance limit (mg/l)
	As	>20,000
	Sb	80–300
	Fe	>50,000
	Zn	3000–80,000
	Cu	>100
	Pb	20
	Se	80
	Ca	>200
	Mg	>200

The process of biooxidation in an actual leaching situation is shown schematically in Fig. 5.8. As shown in Fig. 5.8, the bacteria, which are typically one micrometer in size, can be attached to mineral particles or suspended in the solution. Most bacteria tend to attach to mineral particles. The bacteria that are attached to the particles perform most of the oxidation of ferrous ions. Some types of attached bacteria are also believed to oxidize the sulfur to sulfate without the use of ferric ions from the bulk solution [10].

For leaching of a typical sulfide mineral such as pyrite (FeS_2), the following reactions are likely to occur when bacteria are present:

$$2Fe^{2+} + 0.5O_2 + 2H^+ \leftrightarrow 2Fe^{3+} + H_2O \quad \text{(biotic)} \tag{5.19}$$

$$FeS_2 + 2Fe^{3+} \leftrightarrow 3Fe^{2+} + 2S^{\circ} \quad \text{(abiotic)} \tag{5.20}$$

$$2S^{\circ} + 3O_2 + 2H_2O \leftrightarrow 2SO_4^{2-} + 4H^+ \quad \text{(biotic)} \tag{5.21}$$

with the overall reaction given as (Fig. 5.26):

$$FeS_2 + 3.5O_2 + H_2O \leftrightarrow Fe^{2+} + 2SO_4^{2-} + 2H^+ \quad \text{(overall)} \tag{5.22}$$

Alternatively, other abiotic reactions ($14Fe^{3+} + FeS_2 + 8H_2O \rightarrow 15Fe^{2+} + 16H^+ + 2SO_4^{2-}$) can be important. As with many reactions, looking at the overall reaction does not give a very informative view of the process of getting there. With most

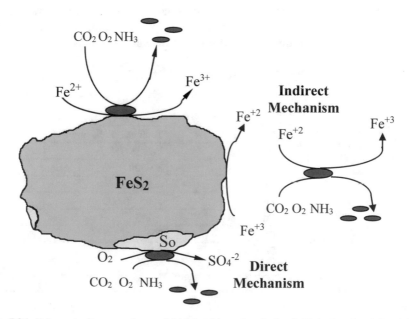

Fig. 5.26 Schematic diagram of potential bioleaching of pyrite by *Acidithiobacillus ferrooxidans*

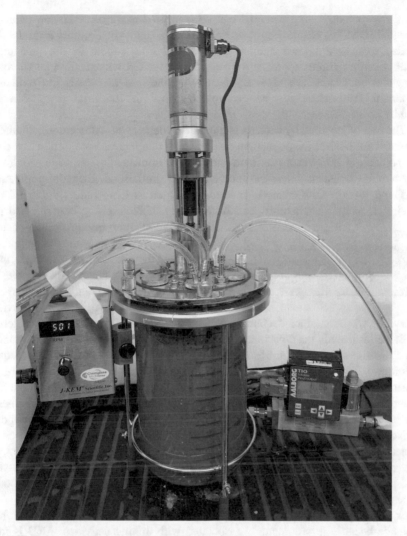

Fig. 5.27 Laboratory bioreactor with associated air flow, feed flow, and temperature control tubing

sulfide oxidation, the bacteria play a very important role in oxidation or leaching. In fact, the rate of leaching with bacteria is nearly always orders of magnitude greater than without bacteria present. A laboratory bioreactor is shown in Fig. 5.27.

One important factor that complicates the evaluation of bacterial leaching kinetics is the dynamic nature of bacterial populations and the diversity of types of bacteria within the population. When bioleaching operations are initiated, bacterial populations are low.

Because bioreactions such as Eq. 5.19 are electrochemical, rates can be monitored electrochemically. In practice the rate can be measured from the rate of change in potential. The potential can be measured using an oxidation reduction potential (ORP) electrode. Ferrous oxidation rates can be determined from the time needed oxidize a known amount of added ferrous sulfate to the preaddition ORP. A more rigorous approach for determining the ferrous oxidation rate involves principles in Chaps. 2 and 4. By measuring the rate of ferrous ion oxidation, important information can be determined about the bacterial population and activity.

The rate of leaching by bacteria is population dependent. After a core population becomes established the population and leaching rate increase proportionally. Once the population achieves a maximum sustainable number, the rate of leaching usually remains constant unless the food supply is diminished. Bacterial populations also tend to shift with changes in conditions such as temperature, dissolved iron level, pH, oxygen availability, etc. Additionally, if bioleaching is performed in a continuous flow reactor setting, bacteria can be removed with the solids, to which they are often attached, more quickly than they are capable of replenishing the population. This condition of excessive bacteria removal is known as washout, since eventually such a condition will wash out the bacteria.

> *Commercial biooxidation facilities utilize a consortium or group of different microorganisms.*

Bacterial leaching can become a problem where ground is disturbed. Ground is often disturbed for roads, infrastructure, and mining. Often these activities require blasting or heavy equipment use. Blasting reduces the particle size and increases mineral exposure area. Mineral exposure is particularly problematic for fine tailing material. Fine tailing material has a high surface area per unit of volume. Minerals that are at or near the surface are exposed to oxygen and water. Newly exposed surfaces often release needed nutrients for bacteria. As a result of the exposure, bacterial activity can be accelerated. In areas where sulfide minerals are present, the bacterial activity may lead to acid production. Equation 5.21 shows one possible acid producing sulfide reaction. Acid produced by such reactions can leach other minerals and release a variety of dissolved metals. This scenario is called Acid Rock Drainage (ARD). This natural process is sometimes referred to as acid mine drainage because it is commonly associated with mining activities. ARD is most common for sulfide tailings due to enhanced sulfide mineral exposure.

Prevention of ARD is dependent on reducing mineral exposure and bacterial activity. Reductions in water and air exposure are important to ARD prevention. Consequently, prevention schemes often include capping exposed material with low permeability materials. Materials such as clay can be used for capping. ARD can also be controlled by mixing mined material with limestone to prevent acidification. The addition of biocides that kill bacteria has also been tested.

5.2.1 Commercial Bioleaching Applications

The BIOX® Process, originally developed by Gencor (now BHP Billiton Process Research) and marketed through Gold Fields, is used in several commercial operations throughout the world. Locations of operations include Ashanti, Ghana, Fairview South Africa, Sao Bento, Brazil, Jinfeng, China, and Wiluna, Australia. BIOX has been commercially available for more than 15 years. One operation processes nearly 1000 tonnes of sulfide concentrate each day. Additional plans to build a BIOX plant for the Dachang project in Qinghai, China [11]. The BIOX® Process uses a mixture or consortium of three bacteria: *Thiobacillus ferrooxidans*, *Thiobacillus thiooxidans*, and *Leptospirillum ferroxidans*. The process utilizes temperatures between 40 and 45 °C at pH levels between 1.2 and 1.6 with 20% solids. Dissolved oxygen levels are maintained above 2 ppm. The total residence time is 4–6 days. Air is injected to provide both oxygen and carbon dioxide that are necessary for bacterial growth and mineral oxidation to take place. Excess process heat generated by sulfide oxidation is removed. Additional nutrients and carbonate can be added in some cases to optimize performance. The capital cost for a 720 tonnes per day plant was estimated to be $25 million in 1994. Operating costs have been estimated at $17/tonne [12]. Adaptation of BIOX mesophilic cultures to nickel sulfide concentrates has been referred to as BioNIC [13].

> *The BIOX Process is used commercially at several plants.*

The GEOCOAT process consists of coating concentrate onto gangue particles that are stacked on heaps and leached [14]. The associated heap is inoculated with sulfide-oxidizing bacteria that may include moderate thermophiles [14].

The BioCOP process consists of thermophilic culture based tank leaching. Extracted metal is concentrated by subsequent solvent extraction and recovered by electrowinning [15]. It was developed by Billiton and demonstrated by Billiton and Codelco in 2000. The demonstration facility was eventually expanded to include a 20,000 tons per annum plant at a Chuquicamata site by 2003 [15].

The BACOX process utilizes naturally occurring bacteria. This process uses higher temperatures (50 °C) than traditional bioleaching. The BACOX process also includes environment optimization for improved performance [11].

5.3 Precious Metal Leaching Applications

Leaching of ores containing precious metals can be more complex than nonprecious-metal ore leaching. Complexity arises from the difficulty of extracting precious metals from "refractory" ores. The term "refractory" refers to ores for which extraction by direct leaching yields poor recoveries or excessive leaching times. The common reason for refractory ores is gold locked inside a host mineral matrix. Most

> *Refractory gold ores require additional processing steps.*

often the matrix is sulfidic. However, the recoveries from refractory ores can be enhanced greatly by pretreating the ore. Pretreatment often consists of roasting, pressure leaching, or bioleaching to decompose the host matrix. Thus, with refractory ores there are often two entirely different leaching operations. The first operation is pretreatment. The second operation is metal extraction by leaching. General flow sheets for typical aqueous precious metal extraction will be presented in Chap. 10.

5.3.1 Cyanide Leaching

The most common precious metal extractant is cyanide. Cyanide is one of the few compounds that can dissolve gold in oxidizing media. It is believed that the cyanide first adsorbs and forms an intermediate [16]:

$$AuCN_{ads} + e^- \leftrightarrow Au + CN^- \quad \text{(occurs in opposite direction)} \tag{5.23}$$

forming a passive $AuCN_{ads}$ film, which occurs as an anodic reaction. This anodic reaction is followed by additional reaction of cyanide to form the aurous cyanide complex [16]:

$$AuCN_{ads} + CN^- \leftrightarrow Au(CN)_2^- \tag{5.24}$$

The Pourbaix diagram for gold and cyanide is presented in Fig. 5.28.

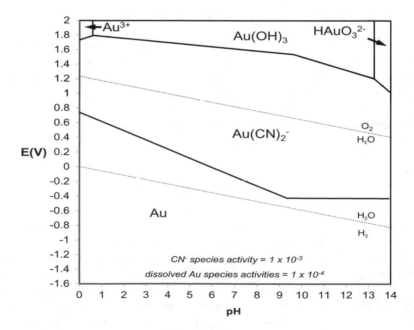

Fig. 5.28 Pourbaix diagram for gold and cyanide based on data in Ref. [17]

The addition of cyanide to water reduces the gold oxidation potential by approximately 1 Volt due to the strong Au(CN)⁻ complex.

The related overall half-cell reaction for silver and cyanide is:

$$Ag(CN)_2^- + e^- \leftrightarrow Ag + 2CN^-$$

The associated Pourbaix diagram for silver and cyanide is shown in Fig. 5.29. The reaction of silver and cyanide results in an insoluble silver cyanide compound at pH levels below about 3.5.

Equation 5.23, requires a counter reaction that consumes electrons. The electron consuming reaction is water or hydrogen peroxide formation. Hydrogen peroxide decomposes into oxygen and hydrogen ions or oxygen and water. It is believed that the most important electron consumption reaction in gold leaching with cyanide is [18]:

$$O_2 + 2H^+ + 2e^- \leftrightarrow H_2O_2 \quad (\text{or } O_2 + 4H^+ + 4e^- \leftrightarrow 2H_2O) \tag{5.25}$$

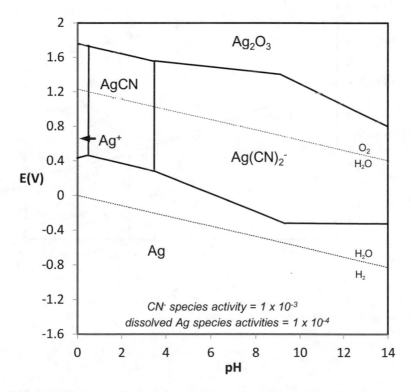

Fig. 5.29 Pourbaix diagram for silver and cyanide based on data in Refs. [19, 20]

> *Oxygen availability is generally crucial to gold leaching because it is the key oxidant.*

Thus, oxygen plays a critical role in gold extraction. After the hydrogen peroxide forms, it breaks down into water. Thus, the overall reaction for gold/cyanide leaching in alkaline solutions:

$$4Au + 8CN^- + O_2 + 2H_2O \leftrightarrow 4Au(CN)_2^- + 4OH^- \qquad (5.26)$$

The reaction is always carried out in alkaline media (pH >10—usually 10.5–11). Cyanide reacts to form hydrogen cyanide gas ($H^+ + CN^- \leftrightarrow HCN(g)$) at lower pH levels. The Pourbaix diagram for cyanide is shown in Fig. 5.30.

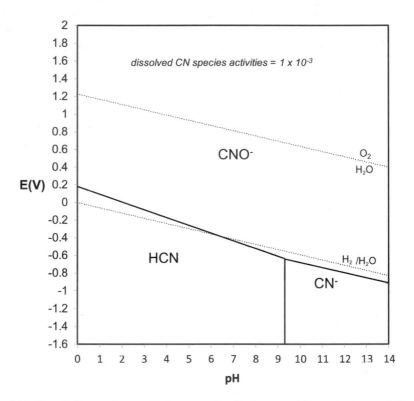

Fig. 5.30 Eh–pH diagram for cyanide in water. The dissolved cyanide species have activities of 0.001. Based on thermodynamic data in Appendix D

Hydrogen cyanide gas is toxic. Consequently, it is critical to perform cyanide leaching above approximately pH 10.5. Most cyanide leaching takes place with 0.1 to 1 kg NaCN per metric ton of ore. Concentrations often range between 50 and 1000 ppm cyanide. Heap and pad leaching operate at the lower end of this range. Concentrate leaching is performed at higher cyanide levels than heap and dump leaching. If cyanicides are present cyanide requirements may be greater. Cyanicides are compounds such as iron and copper that complex with cyanide and increase cyanide consumption. A small amount of dissolved lead is generally beneficial to leaching.

> *Very alkaline conditions (above pH 10.5) are needed to prevent hydrogen cyanide gas formation from cyanide solutions.*

Natural carbonaceous material in gold ore can reduce recovery. Carbonaceous material can adsorb solubilized gold. Gold adsorbed on carbonaceous material in the ore remains with the ore. Gold is difficult to recover from natural carbonaceous material. Consequently, carbonaceous material effectively "robs" the precious-metal-bearing (pregnant) solution of its gold. The resulting undesirable stripping of the pregnant solution by materials in the ore is called "preg robbing". Carbonaceous material can often be consumed by combustion in roasting pre-treatment operations.

The most common method of gold leaching is agitated tank or vat leaching. Approximately 50% of the gold produced in 2004 was extracted by agitated tank or vat leaching [21]. However, heap leaching is growing in popularity. In 2004 10% of worldwide gold production came from heap leaching [21].

5.3.2 Other Precious Metal Extractants

Generally, halides such as chlorine, bromine, and iodine as well as sulfur compounds such as thiocyanate (SCN-), thiourea (NH_2CSNH_2), and thiosulfate ($S_2O_3^{2-}$) are good lixiviants (extractants) for gold in aqueous media. Typically, the reaction with chloride and other halides (X) is:

$$Au(X)_4^- + e^- \leftrightarrow Au + 2(X)^- + (X)_2 \tag{5.27}$$

> *Cyanide toxicity has driven research to evaluate alternative, less toxic gold lixiviants such as thiosulfate, thicyanate, and chloride.*

Gold extraction with halides is usually much faster than cyanide. Halide leaching is done in solutions that are acidic and corrosive. Thus, the cost of halide-based extraction is higher than cyanide-based extraction. Therefore, halides are rarely used for gold extraction.

Typical reactions involving thiocyanate, thiourea, and thiosulfate are [22]:

$$Au(SCN)_2^- + e^- \leftrightarrow Au + 2SCN^- \tag{5.28}$$

$$Au(NH_2CSNH_2)_2^+ + e^- \leftrightarrow Au + 2(NH_2CSNH_2) \tag{5.29}$$

$$4Au + 8S_2O_3^{2-} + O_2 + 2H_2O \leftrightarrow 4Au(S_2O_3)_2^{3-} + 4OH^- \tag{5.30}$$

However, it should be noted that the gold oxidation reaction in thiosulfate solution involves copper ions $(4Cu(S_2O_3)_3^{5-} + 16NH_3 + O_2 + 2H_2O \leftrightarrow 4Cu(NH_3)_4^{2+} + 8S_2O_3^{2-} + 4OH^-)$ [22].

Associated Pourbaix diagrams for thiocyanate and thiosulfate are presented in Figs. 5.31 and 5.32, respectively. Both thiocyanate and thiourea leaching are performed in acidic media. In contrast, leaching with thiosulfate is generally performed in alkaline media. Leaching with thiocyanate and thiosulfate requires the use of an added oxidant. Ferric and cupric ions are often preferred for oxidation due to the

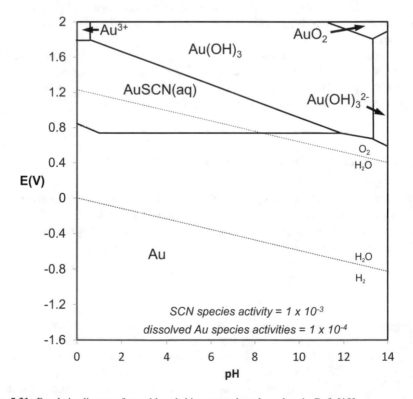

Fig. 5.31 Pourbaix diagram for gold and thiocyanate based on data in Ref. [19]

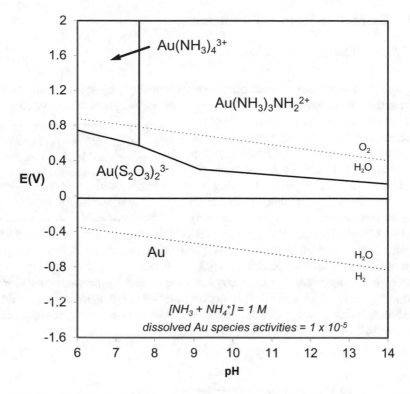

Fig. 5.32 Pourbaix diagram for gold and thiosulfate in ammonia/ammonium solution and 0.1 M Na_2SO_3 based on data in Ref. [23]

slow oxidation using oxygen. Copper is more common for thiosulfate applications. The use of such oxidants generally requires the use of stabilizers such as oxalate for iron and ammonia for copper. The stabilizers keep the ions from precipitating in moderately alkaline media. The use of ammonia also results in an ammonia-gold complex $(Au(NH_3)_2^+)$. Thiosulfate advantages include lower toxicity and low consumption by sulfide minerals relative to cyanide. However, concentrations of sulfur based lixiviants are much higher than for cyanide. Each of these sulfur compounds has potential for application as a precious metal lixiviant. However, at present they are not widely used in industry. High reagent costs, process control complexity, and low industry acceptance have limited their use.

5.4   Extraction from Concentrates

5.4.1   Concentrate Leaching

High-grade concentrated ores can often be economically leached in leaching vessels such as the one depicted in Fig. 5.33. Concentrate leaching is not the most common method of leaching. It requires milling and concentration as preceeding steps. However, large tonnages of calcined zinc concentrates are leached each year. In recent years the application of concentrate bioleaching has increased. Several bioleaching plants leach concentrates as discussed in the bioleaching

> *Metal sulfide minerals, concentrated by flotation, are commonly leached in vessels at elevated temperature, pressure, and oxygen.*

section. Many commercial operations utilize high pressure autoclaves to oxidize sulfide minerals as pretreatments to gold extraction from refractory ores. A typical high pressure autoclave leaching (HPAL) vessel is depicted in Fig. 5.34. Chalcopyrite concentrate leaching has been performed by pressure oxidation at Freeport's Morenci operation. This concentrate leaching operation has utilized ultrafine grinding to a d_{80} size less than 10 µm. More details regarding pressure leaching will be discussed in a subsequent section.

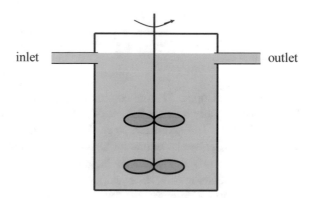

inlet   outlet

Fig. 5.33 Schematic diagram of a leaching vessel with impellars for mixing

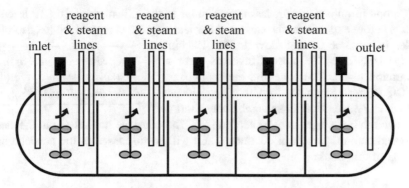

Fig. 5.34 Schematic diagram of a high pressure autoclave leaching (HPAL) vessel

5.4.2 Ambient Pressure Concentrate Leaching

In the zinc industry, zinc sulfides are often roasted and then leached as a concentrate to remove iron:

$$ZnO \cdot Fe_2O_3 + 4H_2SO_4 \leftrightarrow ZnSO_4 + Fe_2(SO_4)_3 + 4H_2O \tag{5.31}$$

The resulting ferric sulfate is then separated by precipitation processes that will be discussed later.

5.4.3 Agitated Leaching of Gold Ores and Concentrates

Gold ores and in some cases concentrates are commonly leached in tanks. The process and reactions for leaching are described in previous sections. Agitated leaching is performed with 40–50% by weight solids in the slurry. Air or oxygen is often added due to its necessity for gold leaching. Leaching is often combined with concentration using activated carbon. Because this is a concentration process it will be discussed in more detail later. However, because it is used in leaching tanks, some discussion is included here. The use of activated carbon particles (1–5 mm in diameter) in leaching (CIL) facilitates low concentrations of dissolved gold in leaching solutions. In CIL circuits the gold that dissolves during leaching is rapidly adsorbed onto carbon in the same solution. The CIL process allows for higher recoveries of gold when carbonaceous, "preg-robbing" material is present. However, CIL processing results in significant attrition of carbon and associated losses of fine particles of gold-loaded carbon that report to the tailings. Leaching is generally performed for 24–48 h.

> *Non-refractory gold ores are commonly leached in large tanks under near ambient conditions.*

Carbon in pulp (CIP) processing is similar to the CIL method. In CIP leaching occurs before exposure to the carbon. After leaching the carbon is fed to adsorption tanks in counter current flow to the leaching slurry. Thus, the CIP process involves a set of adsorption tanks. Consequently, the capital cost for CIP is greater than that for CIL. The residence time in CIP adsorption tanks is relatively short (1 h), although 6–8 stages may be needed [24]. The other method related to the use of activated carbon, carbon in columns (CIC) will be discussed in the concentration section of this book.

> *Activated carbon is used extensively to adsorb and concentrate gold.*

5.4.4   Miscellaneous Atmospheric Pressure Leaching Methods

5.4.4.1   Nickel

Nickel sulfide compounds can be broken down sequentially using chlorine gas:

$$Ni_3S_2 + Cl_2 \leftrightarrow Ni^{2+} + 2NiS + 2Cl^- \tag{5.32}$$

$$2NiS + Cl_2 \leftrightarrow Ni^{2+} + NiS_2 + 2Cl^- \tag{5.33}$$

$$NiS_2 + Cl_2 \leftrightarrow Ni^{2+} + 2S^\circ + 2Cl^- \tag{5.34}$$

$$2S^\circ + 6Cl_2 + 8H_2O \leftrightarrow 2SO_4^{2-} + 16H^+ + 12Cl^- \tag{5.35}$$

An Eh–pH diagram for nickel, sulfur, and water is presented in Fig. 5.35. Other reactions using chlorides such as ferric or cupric chloride such as:

$$2FeCl_3 + CuS \leftrightarrow Cu^{2+} + 2Fe^{2+} + S^\circ + 3Cl^- \tag{5.36}$$

are also used. One distinct advantage to leaching with chlorine, ferric ions, and cupric ions is that the oxidizing agent can be regenerated during electrowinning, although this is seldom done in practice.

5.4.4.2   Zinc

Zinc is commonly leached as a concentrated zinc sulfide that has been calcined to form zinc oxide. The associated Pourbaix diagram (Fig. 5.36) shows leaching is feasible in acidic or alkaline media at low oxidation potentials.

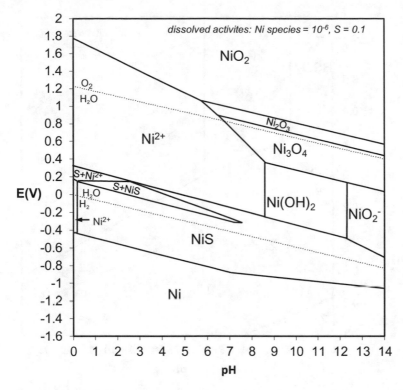

Fig. 5.35 Eh–pH diagram for nickel and sulfur in water based on data in Ref. [25]. The dissolved nickel species have an activity of 0.000001 and the sulfur activity is 0.1

Many ambient pressure leaching processes have been developed for copper sulfides including Arbiter, BHAS, Bromide, CANMET, CENIM-LINETI, CLEAR, Cuprex, Cymet, Dextec, Ecochem, Electroslurry, Elkem, GALVANOX, Intec, Minemet, Nenatech, Nitric Acid, and USBM. A general overview of associated chemistries of many of these processes is summarized in the literature [27]. However, none of these processes has been practiced commercially on a large scale for sufficient time to provide needed data and confidence for widespread utilization. An example of a commercial ambient pressure leaching tank is shown in Fig. 5.37.

Many other possible processes can be used at ambient pressure for metal extraction from minerals. As discussed in the bioleaching section, several processes have been used successfully on a commercial scale for sulfide concentrate leaching.

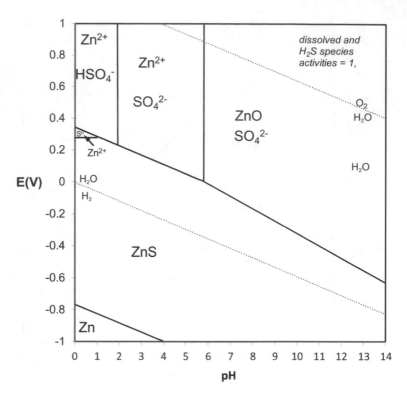

Fig. 5.36 Eh–pH diagram for zinc and sulfur in water. All dissolved species activities are unity. Created based on data in Ref. [26]

5.4.4.3 Processing of Titanium Dioxide

Titanium is often recovered as TiO_2 from illmenite ores which contain $FeTiO_3$. One common approach is to leach the illmenite ore in sulfuric acid:

$$FeTiO_3 + 2H_2SO_4 \leftrightarrow TiOSO_4 + FeSO_4 + 2H_2O \qquad (5.37)$$

The resulting solution of titanium oxide and ferrous sulfate is chilled to precipitate ferrous iron as ferrous sulfate heptahydrate. The solution is then heated and diluted with water to form $TiO(OH)_2$. Seed crystals of TiO_2 are added as nuclei for precipitate formation. Some titanium sulfate is also added to reduce remaining ferric iron and prevent it from coprecipitating as ferric hydroxide with the titanium oxide hydroxide. The $TiO(OH)_2$ is calcined to form TiO_2.

Rare earth elements, which include the lanthanide (Ln) elements, are leached in either sulfuric acid or sodium hydroxide from bastnasite or monazite minerals.

Fig. 5.37 View of
commercial leaching tank

5.4.5   *Pressure Leaching of High Grade Ores*
and Concentrates

Pressure leaching is the process of aqueous metal extraction under high pressure
conditions. Pressure leaching is performed in an autoclave. Pressure leaching occurs
naturally in-situ (using hydrostatic head pressure). Pressure leaching is usually
oxidative. Therefore, it is commonly pressure oxidation (POX). Most often pressure
oxidation is performed using high pressures of oxygen gas (5–50 atm). Elevated
temperatures, usually close to 200 °C, are also used. As a result of the elevated
temperature and pressure, the rate of metal extraction is rapid. The average residence time in pressure oxidation circuits is approximately 2 h. Pressure oxidation is
used to extract some metals such as zinc from zinc calcines. Pressure oxidation is
also used as a pretreatment for precious metal extraction operations. Pressure
leaching has also been used to control various precipitation reactions.

Many mineral decomposition reactions involve oxidation and reduction. Often the cathodic portion is supplied by oxygen under low or high pH conditions:

$$4H^+ + 4e^- + O_2 \leftrightarrow 2H_2O \quad \text{(low pH)} \tag{5.38}$$

$$2H_2O + 4e^- + O_2 \leftrightarrow 4OH^- \quad \text{(high pH)} \tag{5.39}$$

Increasing oxygen pressure increases the rate of these reactions. Thus, mineral decomposition is often increased by increased oxygen pressure. Under ambient conditions, the partial pressure of oxygen is 0.21 atm. Increasing the partial pressure to 10 atm in an autoclave leads to a 50-fold increase in the oxygen availability. In addition, with a typical autoclave temperature of 150 °C, the corresponding rate increase, assuming an activation energy of 25 kJ/mole, would be 20 times the rate at 25 °C. The kinetics advantage of autoclave leaching is obviously impressive with typical kinetics enhancements that can easily approach a factor of 1000. In addition, HPAL offers some potential advantages for iron removal, arsenic stabilization, and reduced oxygen consumption that are generally not available under ambient conditions.

5.4.6 Elevated Pressure Commercial Aluminum Ore Leaching

Pressure leaching has been applied to bauxite ore processing through the Bayer process since late in the nineteenth century [28]. In the typical Bayer process both gibbsite $(Al(OH)_3)$ and diaspore or boehmite $[AlO(OH)]$ are dissolved from bauxite ore. The related Eh–pH diagram for aluminum and water is shown in Fig. 5.38. The Eh–pH diagram indicates that aluminum oxide can be dissolved

> *Activated carbon is used extensively to adsorb and concentrate gold.*

at either very acidic or highly alkaline conditions. The diagram also shows that dissolution is possible without reduction or oxidation.

The Bayer process occurs at elevated temperature and pressure to expedite the dissolution process. This process takes place at about 150–200 °C in 3–12 M NaOH at a pressure of 4–8 atm [29] as shown below:

$$Al(OH)_3 + OH^- \leftrightarrow AlO_2^- + 2H_2O \tag{5.40}$$

$$AlO(OH) + OH^- \leftrightarrow AlO_2^- + H_2O \tag{5.41}$$

Gibbsite is more easily dissolved at lower temperature and hydroxide concentration than boehmite.

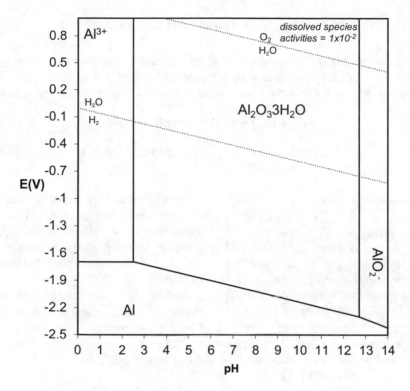

Fig. 5.38 Eh–pH diagram for aluminum in water based on data in Ref. [37]. The dissolved species activities are 0.01

The next process step is water addition to form gibbsite:

$$AlO_2^- + 2H_2O \leftrightarrow Al(OH)_3 + OH^- \tag{5.42}$$

The resulting gibbsite precipitate is then calcined to alumina. The alumina is electrolytically reduced to aluminum metal in molten salt bath electrowinning cells.

Sodium hydroxide is regenerated and oxalate is removed by the addition of lime (CaO). Associated reactions are:

$$Na_2CO_3 + CaO + H_2O \leftrightarrow 2NaOH + CaCO_3 \tag{5.43}$$

$$H_2COO + CaO \leftrightarrow CaCOO + H_2O \tag{5.44}$$

Some calcium aluminate also forms.

5.4.7 Sulfide Mineral Concentrate Pressure Oxidation

Pretreatment of refractory precious metal ores as well as leaching of some copper sulfide concentrates is performed by pressure leaching on an industrial scale [30]. Typical reactions for refractory gold ore, pressure oxidation pretreatment include:

$$FeS_2 + H_2O + 3.5O_2 \leftrightarrow FeSO_4 + H_2SO_4 \tag{5.45}$$

$$FeAsS + 1.5H_2O + 3.25O_2 \leftrightarrow FeSO_4 + H_3AsO_4 \tag{5.46}$$

Pressure oxidation of sulfides is often performed to extract copper or other metals. It is also performed to break down the sulfide matrix for improved subsequent extraction of gold by cyanide leaching.

These reactions are usually carried out in acidic media due to the acid forming tendencies of the sulfidic minerals. The resulting leaching slurry is then converted to alkaline media prior to cyanide leaching.

Metals such as copper are also extracted using pressure oxidation. Chalcopyrite reactions are sensitive to temperature. Proper control of temperature allows for control of products such as elemental sulfur and acid production. At low temperatures of 100–160 °C the common reactions are [31]:

$$2CuFeS_2 + 2.5O_2 + 5H_2SO_4 \rightarrow 2CuSO_4 + Fe_2(SO_4)_3 + 4S^\circ + 5H_2O \tag{5.47}$$

$$CuFeS_2 + 2Fe_2(SO_4)_3 \rightarrow CuSO_4 + 5FeSO_4 + 2S^\circ \tag{5.48}$$

The iron generally precipitates as goethite [31]:

$$Fe_2(SO_4)_3 + 4H_2O \rightarrow 2FeOOH + 3H_2SO_4 \tag{5.49}$$

At higher temperatures (180–230 °C) the corresponding reactions include [31]:

$$2CuFeS_2 + 8.5O_2 + H_2SO_4 \rightarrow 2CuSO_4 + Fe_2(SO_4)_3 + H_2O \tag{5.50}$$

$$2CuFeS_2 + 16Fe_2(SO_4)_3 + 16H_2O \rightarrow 2CuSO_4 + 34FeSO_4 + 16H_2SO_4 \tag{5.51}$$

$$2FeSO_4 + 0.5O_2 + H_2SO_4 \rightarrow Fe_2(SO_4)_3 + H_2O \tag{5.52}$$

$$Fe_2(SO_4)_3 + 3H_2O \rightarrow Fe_2O_3 + 3H_2SO_4 \tag{5.53}$$

Thus, at low temperatures the sulfur product is elemental sulfur. At higher temperatures, sulfuric acid is the sulfur product.

Another useful chemistry for sulfide oxidation is the use of chloride and acid at 220 °C with oxygen as practiced in the Platsol Process [32]. The addition of the choride with acid and oxygen allows base metals to dissolve as sulfates along with precious metals that are dissolved as chloride complexes.

Other chemistries such as those involving nitric acid have been successful in niche markets [33]. The process includes regeneration of the nitric acid.

5.4.8 Commercial Scale Nickel Concentrate Leaching

Another important use of pressure leaching is the Sherrit-Gordon process. In this process nickel is extracted from a pentlandite, (Fe, Ni)S, concentrate. This process takes place at 70–90 °C at 7–10 atm using ammonia and oxygen [16]. The reactions can be expressed as:

$$NiS + 6NH_3 + 2O_2 \leftrightarrow Ni(NH_3)_6^{2+} + SO_4^{2-} \tag{5.54}$$

$$4FeS + 9O_2 + 8NH_3 + 4H_2O \leftrightarrow 2Fe_2O_3 + 8NH_4^+ + 4SO_4^{2-} \tag{5.55}$$

This process dissolves the nickel as an ammonia complex; iron oxide forms as a precipitate.

Much more detailed nickel leaching information is presented in Chap. 10 in connection with nickel ore processing and associated flow sheets.

5.4.9 Uranium Ore Pressure Leaching

Uranium ores can be leached under pressure using sulfuric acid in autoclaves. Sample pressure leaching reactions for uranium oxide minerals are [30]:

$$UO_2HCO_3 + e^- \leftrightarrow UO_2 + HCO_3^- \text{ (occurs anodically in practice)} \tag{5.56}$$

$$UO_2CO_3 + H_2O + e^- \leftrightarrow UO_2HCO_3 + OH^- \text{ (occurs anodically)} \tag{5.57}$$

$$UO_2CO_3 + 2CO_3^{2-} \leftrightarrow UO(CO_3)_3^{4-} \tag{5.58}$$

5.4.10 Opportunities for Hydrometallurgical Recycling of Metals

Metals are one of the most recycled materials. Steel is usually rated as the most recycled material. However, much of the current recycling of metals is performed pyrometallurgically. Steel is ideal for recycling pyrometallurgically, rather than hydrometallurgically.

Fig. 5.39 View of e-waste shred—coarse (left), fine (right)

There are, however, opportunities for future hydrometallurgical recycling of metals such as metals from e-waste on a large scale. Figure 5.39 shows e-waste after general shredding and additional shredding. These types of materials can be leached in a variety of scenarios. Often, they can be processed with high recoveries and low costs.

5.4.11 Extraction of By-Products

Metals such as silver, tellurium, and gold are commonly recovered as by products. In copper porphyry ores, these elements end up in process waste. This waste is leached and the desired elements are separated and recovered [38]. The first leaching step in an autoclave dissolves copper and tellurium [38]. The gold is leached from the residue using chlorine as an oxidant and complexing agent [38]. Silver is leached using ammonia [38].

5.4.12 Rare Earth Elements (REEs) Leaching

Rare earth elements can be leached using sulfuric acid, hydrochloric acid, and nitric acid. As shown in Fig. 5.40, REEs such as lanthanum have a wide range of leaching conditions. However, it should be noted that some REE processing utilizes acid baking. Details about processing REEs, which often involve complex flow sheets, are presented in Chap. 10.

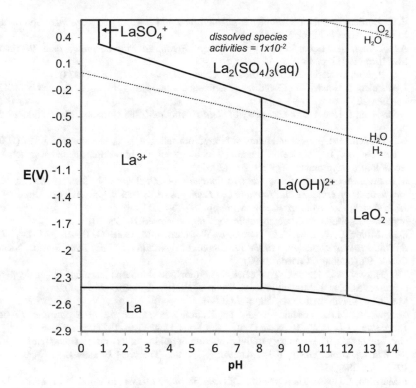

Fig. 5.40 Eh–pH diagram for Lanthanum and sulfur at 25 °C based on data in Ref. [39]

5.4.13 Acid Baking

Sulfuric acid has a high boiling point of 337 °C. Consequently, it is feasible to leach ores in sulfuric acid at elevated temperatures at atmospheric pressure. Some processes "acid bake" ores at temperatures between 200 and 300 °C. Acid baking can modify crystal structures while simultaneously leaching. The elevated temperature accelerates kinetics, and the concentrated acid enhances solubilization. In order to avoid excess acid consumption, acid baking is often performed with a high solids to liquid ratio. Acid baking is often performed for 0.5–2 h. The residue from acid baking is often suspended in water for further processing.

References

1. N. De Nevers, in *Fluid Mechanics* (Addison-Wesley Publishing Co., Reading, 1970)
2. R.J. Akers, A.S. Ward, Liquid filtration theory and filtration pretreatment, in *Filtration: Principles and Practice*, ed. C. Orr (Marcel Dekker, Inc., New York, 1977), pp. 169–250

3. R.W. Bartlett, in *Solution Mining: Leaching and Fluid Recovery of Materials* (Gordon and Breach Science Publishers, 1992)
4. A.E. Torma, New trends in biohydrometallurgy, in *Mineral Bioprocessing*, eds. R.W. Smith, M. Misra (1991), p. 43
5. C.L. Brierly, Bacterial leaching. *CRC Crit. Rev. Microbiol.*, **6**(3) 207 (1978)
6. J. Marsden, I. House, *The Chemistry of Gold Extraction* (Ellis Horwood, New York, 1993), pp. 199–229
7. J. Marsden, I. House, *The Chemistry of Gold Extraction* (Ellis Horwood, New York, 1993), p. 230
8. H.L. Ehrlich, Past, present and future of biohydrometallurgy. Hydrometallurgy **59**, 127 (2001)
9. D.B. Johnson, Importance of microbial ecology in the development of new mineral technologies. Hydrometallurgy **59**, 147 (2001)
10. M.L. Free, *Bioleaching of a Sulfide Ore Concentrate - Distinguishing Between the Leaching Mechanisms of Attached and Nonattached Bacteria.* M.S. Thesis, University of Utah, 1992
11. J. Chadwick, Golden horizons. Int. Min., pp. 74–75 (2011)
12. http://www.goldfields.co.za/com_technology.php, accessed 10 Oct 2012
13. D.M. Miller, D.W. Dew, A.E. Norton, M.W. Johns, P.M. Cole, G. Benetis, M. Dry, *The BioNIC Process; Description of the Process and Presentation of Pilot Plant Results*, Nickel/Cobalt, 97 (Sudbury, Canada, 1997)
14. T.J. Harvey, N. Holder, T. Stanek, Thermophilic bioheap leaching of chalcopyrite concentrates. Eur. J. Mineral Process. Environ. Protect. **2**(3), 253–263 (2002)
15. M.E. Clark, Hydrometallurgy **83**, 3–9 (2006)
16. Wadsworth, M.E., Leaching-metals applications, in *Handbook of Separation Process Technology*, ed., R.W. Rousseau (Wiley and Sons, New York, 1987), pp. 500–539
17. N.P. Finkelstein, The chemistry of the extraction of gold from its ores, in *Gold Metallurgy on the Witwatersrand.* ed. by R.J. Adamson (Cape and Transvaal Printers Ltd., Cape Town, 1972), pp. 284–351
18. J. Marsden, I. House, *The Chemistry of Gold Extraction* (Ellis Horwood, New York, 1993), pp. 265–266
19. K. Osseo-Asare, T. Xue, V.S.T. Ciminelli, Solution chemistry of cyanide leaching systms, in *Precious Metals: Mining, Extraction and Processing*, ed. V. Kudryk, D.A. Corrigan, W.W. Liang (1984), pp. 173–197
20. J. Marsden, I. House, in *The Chemistry of Gold Extraction* (Ellis Horwood, New York, 1993), p. 253
21. J. Marsden, I. House, in *The Chemistry of Gold Extraction* (Ellis Horwood, New York, 1993), p. 505
22. J. Marsden, I. House, in *The Chemistry of Gold Extraction* (Ellis Horwood, New York, 1993), pp. 299–305
23. W. Stange, The process design of gold leaching andcarbon-in-pulp circuits. J. South Afr. Inst. Min. Metallurgy 13–26 (1999)
24. G. Senayake, W.N. Perera, M.J. Nicol, Thermodynamic studies of the gold(III)/(I)/(0) redox system in ammonia-thiosulfate solutions at 25°C, in *Hydrometallurgy 2003: Proceedings 5th International Symposium Honoring Prof. I. M. Ritchie*, ed. C. A. Young, A. Alfantazi, C. Anderson, A. James, D. Dreisinger, B. Harris (TMS, Warrendale, 2003), pp 155–168
25. R.M. Garrels, C.L. Christ, *Solutions, Minerals, and Equilibria* (Jones and Bartlett Publishers, Boston, 1990), p. 245
26. P. Hayes, in *Process Principles in Minerals and Mateials Production*, 3rd edn (Hayes Publishing Co., Brisbane, 2003), p. 238
27. M.L. Free, Electrochemical coupling of metal extraction and electrowinning, in *Electrometallurgy 2001*, ed. J.A. Gonzales, J. Dutrizac (2001), pp. 235–260
28. E. Jackson, *Hydrometallurgical Extraction and Reclamation* (Ellis Horwood Limited, Chichester, 1986), pp. 57–58
29. E. Jackson, *Hydrometallurgical Extraction and Reclamation* (Ellis Horwood Limited, Chichester, 1986), pp. 61–62

30. D. Matthews Jr., Getchell mine pressure oxidation circuit four years after start up. Min. Eng. **46**(2), 115 (1994)
31. R.G. McDonald, D.M. Muir, Pressure oxidation leaching of chalcopyrite. Part I. Comparison of high and low temperature reaction kinetics and products. Hydrometallurgy **86**, 195–205 (2007)
32. C.J. Ferron, C.A. Fleming, P.T. O'Kane, D. Dreisinger, Pilot plant demonstration of the platsol process for the treatment of the NorthMet Copper-Nickel-PGM Deposit, SME Transactions, vol. 312 (2002)
33. C.G. Anderson, S.M. Nordwick,The application of sunshine nitrous-sulfuric acid pressure leaching to sulfide materials containing platinum group metals, in *Precious Metals 1994, Proceedings of the 18th Annual IPMI Conference* (Vancouver, B. C., June 1994), pp. 223–234
34. M.E. Wadsworth, Leaching-metals applications, in *Handbook of Separation Process Technology*, ed. R.W. Rousseau (Wiley and Sons, New York, 1987), pp. 500–539
35. D.C. Silverman, Corrosion **38**(8), 453–455 (1982)
36. D. Mikkelsen, U. Kappler, R.I. Webb, R. Rasch, A.G. McEwan, L.I. Sly, Visualisation of pyrite leaching by selected thermophilic archaea: nature of microorganism–ore interactions during bioleaching. Hydrometallurgy **88**(1–4), 143–153 (2007)
37. D.A. Jones, *Principles and Prevention of Corrosion* (Macmillan Publishing Company, New York, 1992)
38. C. Nexhip, R. Crossman, M. Rockandel, By-products recovery via integrated copper operations at Rio Tinto Kennecott Kennecott, in *Conference: Exchange of Good Practices on Metal By-Products Recovery—Technology and Policy Challenges* (European Union Commission, Brussels, November, 2015)
39. F.N. Yahya, l.N. M. Suli, W. hanisah, W. Ibrahim, R.A. Rasid, Thermodynamic evaluation of the aqueous stability of rare earth elements in sulfuric acid leach of monazite through pourbaix diagram. Mater. Res. Today Proc. 19, 1647–1656 (2019)

Problems

(5-1) Using the Poiseuille equation estimate the new flow rate of leaching solution as well as the new rate of oxygen flow into a dump leaching operation that results from a reduction in the average particle diameter from 0.5 mm to 0.4 mm. The current maximum (prior to flooding) leach solution flow rate is 0.01 l/m^2 s and the oxygen flow rate is 0.008 l/m^2 s. (Note that the pore radius is directly proportional to particle radius.) To keep the problem simple, assume that the number of effective pores per unit area remains constant for each particle size.

(5-2) For the same change in particle diameter (0.5–0.4 mm) mentioned in problem 5.1, what would the corresponding new leaching time (for 50% reaction) be according to the shrinking-core model if the current time is 3 years. Assume all other factors remain constant and that the reaction is diffusion-controlled. (answer: 1.92 years)

(5-3) If the porosity in a heap leaching operation decreases from 0.3500 to 0.3325 due to excessive compaction by haulage equipment, what will the corresponding decrease in solution flow rates be according to the Kozeny equation if the pressure gradient ($\Delta P/L$) remains constant.

(5-4) Calculate the height of solution hold-up at the bottom of a dump leaching operation due to capillary forces if the diameter of the critical particles is 0.2 mm. Assume typical values for the surface tension, contact angle, and solution density. Also assume that the void fraction is 0.35 and the solid fraction is 0.65.

(5-5) Using typical values for bacterial leaching kinetics, estimate the rate of pyrite oxidation (in terms of iron dissolution) if all of the sulfur is converted to elemental sulfur by ferric ions in a continuous flow operation. If the ferrous concentration were cut in half, what would the rate of iron dissolution be? (Refer to Chap. 4 for relevant kinetics information.)

(5-6) The rate of oxygen consumption in a gold heap leaching operation is 6.67×10^{-6} l/m^2 s at STP, and 3% of the oxygen is consumed by the gold dissolution reaction involving cyanide. Estimate the rate of gold dissolution by cyanide leaching if the 15 m high heap contained 1500 kg/m^3 of material assaying at 2 g gold/ton if the rate limiting step is oxygen penetration. How long would it take to extract all of the gold at that rate? How long would complete extraction take if only 0.45% of the oxygen were utilized for the cyanide-gold reaction? (answer: 541 days)

(5-7) Calculate the rate of irrigation for a heap leaching facility if the equipment can only handle one flow rate of liquid (0.007 l/min/m^2) and the system is in operation 60% of the time.

(5-8) Calculate the average air residence time in a large heap leaching operation when air is injected at a pressure of 0.1 atm (relative to the air pressure above the heap) at the bottom of the heap that has an intrinsic permeability of 1.2 Darcys and a height of 50 m. Assume a viscosity for air of 0.0000183 kg s^{-1} m^{-1} (answer: 3.8×10^6 s or 1060 h).

Chapter 6
Separation of Dissolved Metals

Separating metal ions is the processing bridge that links extraction and recovery of pure metals.

Key Learning Objectives and Outcomes:
Understand basic metal concentration principles and terminology
Understand how solvent extraction is performed
Know how to evaluate solvent extraction processing
Understand how ion exchange is performed
Know how to evaluate ion exchange processing
Understand how carbon adsorption is performed
Understand how precipitation is performed
Understand how ultrafiltration is performed

Metals dissolved by extraction or leaching need to be separated in order to obtain purified products. Separation of metal ions is based on differences in thermodynamic properties of each metal as discussed in Chaps. 2 and 3. Dissolved metals are commonly separated using solvent extraction, ion exchange, carbon adsorption, precipitation, and ultrafiltration. This chapter discusses the fundamentals and applications of these methods.

6.1 Liquid–Liquid or Solvent Extraction

Liquid–liquid or solvent extraction is a common process for selectively concentrating metals. Liquid–liquid extraction is performed using an organic extractant that is dissolved in an organic phase. The organic phase is allowed to contact an aqueous phase containing the dissolved metal or metal ion complex. Thus, two liquids are used, hence the term, liquid–liquid extraction. The aqueous and organic phases are immiscible in each other. However, there is some loss of organic in the aqueous phase that is often less than 15 ppm [1].

The organic phase contains an extractant and a diluent. The diluent effectively dilutes the extractant. The diluent commonly consists of paraffins, naphthenes, and

© The Minerals, Metals & Materials Society 2022
M. L. Free, *Hydrometallurgy*, The Minerals, Metals & Materials Series,
https://doi.org/10.1007/978-3-030-88087-3_6

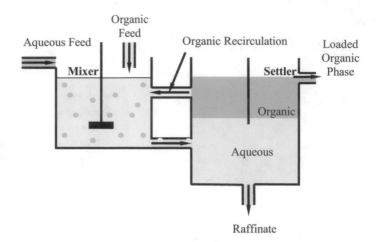

Fig. 6.1 Schematic diagram of a laboratory solvent extraction mixer and settler in the loading stage

alkyl aromatics. Diluents are needed to facilitate pumping, processing, and settling of the extractant that is often viscous and difficult to manage without a diluent. The diluent also helps to distribute the extractant more effectively in organic phase droplets. The diluent effectively extends the presence of extractant at the droplet interface. Thus, the diluent is also referred to as the extender.

Solvent extraction is carried out using a mixer. The mixer disperses the organic phase in the aqueous phase as small droplets. Small droplets enhance the extraction kinetics. A schematic diagram of small scale and industrial scale solvent extraction systems are presented in Figs. 6.1 and 6.2.

The mixing stage is interfaced with a settling stage as illustrated in Figs. 6.1 and 6.2. The settling stage allows for separation of the phases or phase disengagement. After loading, the organic phase is scrubbed to remove unwanted metal ions.

> *Solvent extraction loading involves intimate mixing of organic and aqueous phases during which metal ions are selectively absorbed into the organic medium.*

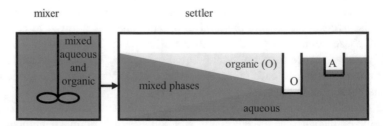

Fig. 6.2 Schematic diagram of an industrial solvent extraction mixer and settler in the stripping stage

Loaded organic phase is then stripped into a concentrated aqueous solution.

The aqueous phase often requires some conditioning. Conditioning steps often include clarification to remove particulate matter. Residual solvent is also removed from the aqueous phase when necessary. Crud, which typically consists of a mixture of aqueous, organic, and solid matter, must also be removed. Crud removal can become an important maintenance issue.

6.1.1   Types of Liquid–Liquid or Solvent Extractants

Two main events are required for liquid–liquid metal extraction. One event is dehydration. The other event is charge neutralization. Because the extractants are organic, they do not have a high tolerance for hydrated ions. The charge on the ions cannot be accommodated by the nonpolar nature of most organic molecules. Since oil and water do not mix, it is evident that ions that are surrounded by water molecules do not mix in oil. The organic phase is essentially oil and does not easily accommodate water. Thus, dehydration is an important liquid–liquid extraction step.

The three basic types of solvent extractants are used. Each type facilitates dehydration and charge neutralization. These types are ion exchange, solvation, and coordination extractants. Coordination extractants typically undergo ion exchange as part of the overall extraction process. Solvation extractants do not formally exchange ions.

6.1.1.1   Ion Exchange Extractants

Ion exchange extractants include both basic and acidic compounds. The basic extractants often have a net excess of hydrogen ions. These ions are attracted to hydroxyl or other anions in solution. Some of these extractants used in the basic or alkaline pH range. Acidic extractants have a net excess of negative charge that attracts cations. Acidic extractants are generally used in the acidic pH range. In acid, hydrogen ions occupy active sites until exchanged with a metal cation. Some common acidic extractants are carboxylates, sulfonates, and phosphates. The basic extractants are nearly always primary, secondary, tertiary, or quaternary amines.

6.1.1.2   Solvating Extractants

Solvating extractants are effectively neutral. Consequently, they are sometimes referred to as neutral extractants. Solvating extractants remove metal ions and complexes by first replacing solvating water molecules. The replacement of water molecules with organic solvent molecules facilitates organic solubility. Next, ion associations (usually through protonation) are made. These associations effectively neutralize the overall charge. Most solvating extractants contain polar oxygen atoms that accept hydrogen ions. The accepted hydrogen ions associate with negatively charged metal complexes. Many metals form anionic complexes with chloride ions. The most common functional groups in solvating extractants include ketones, ethers, esters, and alcohols. Examples of solvating extractants are tri-n-butyl phosphate (TBP) and trioctyl phosphine oxide (TOPO).

6.1.1.3 Coordination Extractants

Coordination extractants act by spatially coordinating metal-bearing ions. The most common type of coordination extractants is chelating extractants. Chelating extractants utilize ion dissociation and association to form a coordinated complex. Coordinating extractants often have excess electron pairs on nitrogen or oxygen atoms. These pairs are separated from each other at the ends of heterocyclic organic rings. The proximity of the nitrogen and oxygen atoms on the ends of the rings aids in complex formation. Other factors that assist in extraction include and easily removed hydrogen atom. The hydrogen void creates a stable negative charge. The special arrangement between nitrogen, neutral oxygen, and anionic oxygen is designed for specific metals. The resulting structure is very selective. Typically, two of the extractant molecules coordinate and bond with one divalent metal ion in a very specific manner. Coordinating extractants for copper solvent extraction are often hydroxyoximes. Other general extractants such as carboxylates are also effective. Examples of coordination extractants are presented in Figs. 6.3 and 6.4. These figures illustrate the importance of structural or steric interactions.

> *Coordination extractants are often very selective for specific ions due to steric (atomic position constraints within a molecule) effects in the extractant that limit extraction to ions within a very narrow range of size.*

6.1.2 General Principles of Solvent Extraction

Most extractants involve ion exchange. Hydrometallurgical extractions, typically utilize ion exchange extractants. Often, extraction takes place in acidic media. Thus, acidic extractants are commonly used. A typical extraction reaction can be expressed by:

$$M^{n+} + nRH \leftrightarrow MR_n + nH^+ \tag{6.1}$$

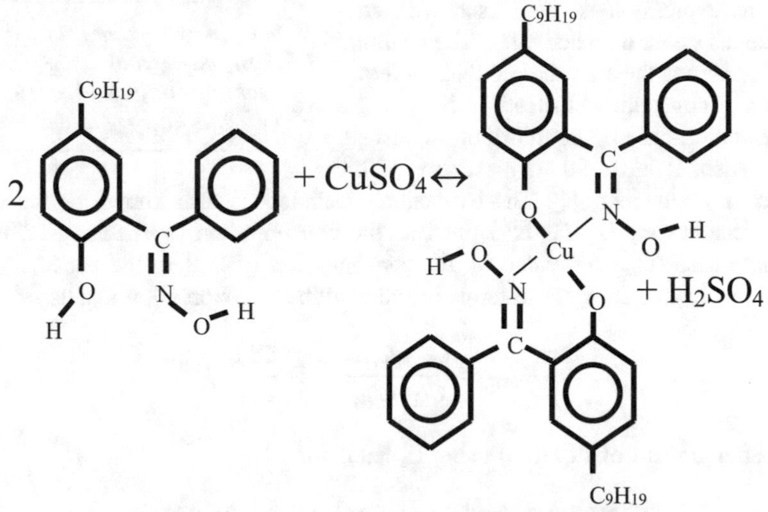

Fig. 6.3 Schematic representation of copper extraction by LIX-65 N, a hydroxyoxime

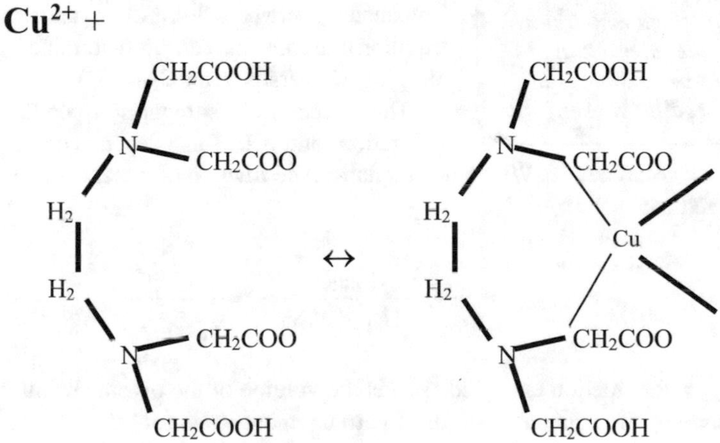

Fig. 6.4 Schematic representation of copper extraction by Ethylenediaminetetra-acetic acid or EDTA. Although it is not generally used for solvent extraction, it is important in many commercial products

The corresponding equilibrium constant can be expressed as:

$$K = \frac{a_{H^+}^n \, a_{MR_n}}{a_{M^{n+}} \, a_{RH}^n} \tag{6.2}$$

This expression is often used with concentrations rather than activities. The resulting constant is not the true thermodynamic equilibrium constant. Instead, it is a concentration-based equilibrium constant, K_{conc}. Also, it is useful to make use of the

> *Solvent extraction is based on the chemical reaction or exchange of ions between the aqueous and organic phases.*

extraction coefficient, E_C. The extraction coefficient is also known as the distribution coefficient, D_c. E_C is defined as the concentration of species M^{n+} in the organic phase (C_{MR_n}) divided by the concentration of M^{n+} in the aqueous phase (C_{Mn+}). Substitution of these terms in the equilibrium expression results in:

$$K_{conc} = \frac{C_{R_nM}C_{H^+}^n}{C_{M^+}C_{RH}^n} = E_C\frac{C_{H^+}^n}{C_{RH}^n} \tag{6.3}$$

Rearrangement of Eq. (6.3) using E_c leads to:

$$\log(E_C) = \log(K_{conc}) + n(\log C_{RH} - \log C_{H^+}) \tag{6.4}$$

> *Maximum extraction is based on thermodynamic equilibrium constraints for the interaction of ions in aqueous and organic phases.*

It is usually assumed that $-\log C_{H^+}$ is equal to the pH. K_{conc} effectively compensates for discrepancies between $-\log C_{H^+}$ and pH. The equilibrium constant can be determined by plotting $\log E_C$ versus $n(\log C_{RH} + pH)$.

The success of extraction depends on concentrations and pH. It also depends on the organic to aqueous volume ratio. With a mass balance, the fraction of metal extracted can be determined as:

$$F_{extracted} = \frac{V_{org}C_{MRn}}{V_{org}C_{MRn} + V_{aq}C_{Mn+}} = \frac{\frac{V_{org}}{V_{aq}}E_C}{\frac{V_{org}}{V_{aq}}E_C + 1} \tag{6.5}$$

$F_{extracted}$ is the fraction extracted. V_{org} is the volume of the organic solution. V_{aq} is the volume of the aqueous solution. Rearrangement leads to:

$$E_C = \frac{F_{extracted}}{\frac{V_{org}}{V_{aq}}(1 - F_{extracted})} \tag{6.6}$$

Further rearrangement leads to:

$$\frac{1}{F_{extracted}} = 1 + \frac{1}{E_C}\frac{V_{aq}}{V_{org}} \tag{6.7}$$

Equation (6.7) can be used as a tool for determining the extraction coefficient. The parameters in Eq. (6.7) are generally known. Thus, a plot of $1/F_{extracted}$ versus the aqueous to organic solutions should reveal a slope of $1/E_C$ and an intercept of 1. Corresponding volumetric flow rates, Q_o and Q_{aq} can be used in place of the respective volumes. Clearly, the volumetric ratio is important to the extraction effectiveness.

Loading is often visually apparent as shown in Fig. 6.5. As the solvent extractant loads the dissolved copper its color changes and it becomes more opaque. Copper bearing solutions also change their color as shown in Fig. 6.6.

Loading characteristics are important in liquid–liquid extraction. A typical liquid–liquid distribution curve is based on the interaction of the metal with organic and aqueous phases. The general format is the concentration of metal in the organic phase is equal to a function of the aqueous metal concentration. Most initial testing

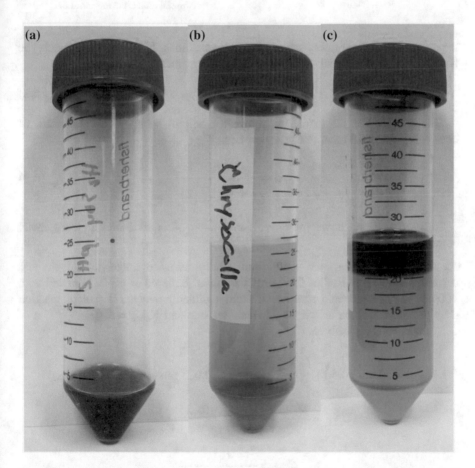

Fig. 6.5 **a** solvent extractant before loading, **b** chrysocolla leaching solution with dissolved copper, and **c** mixture of solvent extractant (top phase) and leaching solution (bottom phase) after loading by mixing.

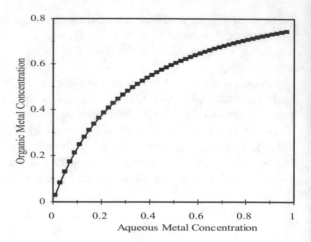

Fig. 6.6 Solvent extraction distribution isotherm for single-charged ions ($n = 1$)

is performed on a laboratory scale in batch tests. A mass balance is often used. The balance is most commonly used for the organic extractant:

$$C_{Rtot} = nC_{R_nM} + C_{RH} \qquad (6.8)$$

The balance for the extractant allows for helpful substitution. The result of combining Eqs. (6.3) and (6.8) for $n = 1$ is:

$$C_{RM} = \frac{C_{Rtot}}{1 + \left(\frac{C_{H^+}}{K_{conc}C_{M^+}}\right)} \qquad (6.9)$$

This expression allows for direct comparison of organic and aqueous metal concentrations.

Example 6.1 Calculate the equilibrium concentrations of metal in aqueous and organic phases if 50 cm^3 of 0.5 M MCl solution, buffered at pH 2, is mixed with 100 cm^3 of pure organic phase (extractant + diluent) that has 1 mol of reaction sites per liter. Assume activity coefficients are one and $K_{conc} = 6$.

A mass balance leads to:

$$M_{total} = 50 \text{ cm}^3 (0.5 \text{ M}) \frac{\frac{\text{mole}}{l}}{M} \frac{l}{1000 \text{ cm}^3} = 0.025 \text{ mole} = C_{M^+} V_{aq} + C_{RM} V_o$$

$$C_{M^+} = \frac{M_{total} - C_{RM} V_o}{V_{aq}} = \frac{0.025 - C_{RM} V_o}{V_{aq}}$$

$$C_{RM} = \frac{C_{Rtot}}{1 + \left(\frac{C_{H^+}}{K_{conc}C_{M^+}}\right)}$$

substitution leads to:

$$C_{RM} = \cfrac{C_{Rtot}}{1 + \left(\cfrac{C_{H^+}}{K_{conc}\frac{M_{total}-C_{RM}V_o}{V_{aq}}}\right)} = \cfrac{0.1}{1 + \left(\cfrac{0.01}{6^{\frac{0.025-C_{RM}(0.1)}{(0.05)}}}\right)}$$

Solve by iteration to find the value of C_{RM}, then use the mass balance to determine C_{M^+}

The value of C_{RM} is 0.0995.

$$C_{M^+} = \frac{M_{total} - C_{RM}V_o}{V_{aq}} = \frac{0.025 - 0.0995(0.1)}{0.05} = 0.301$$

Extraction tests are often initially performed in separatory funnels. The volume of extractant and aqueous solutions are varied. The funnels are vigorously agitated for a reasonable time period. The resulting aqueous phase is then analyzed. C_{RM} versus C_{M^+} data can be used to create a distribution curve. Extraction is generally performed at a specific temperature. Consequently, distribution curves are often called distribution isotherms. A typical distribution isotherm as shown in Fig. 6.6 for $n = 1$. A distribution isotherm for $n = 2$ is shown in Fig. 6.7. Equation (6.9) can be rearranged to:

$$\frac{1}{C_{RM}} = \frac{1}{C_{Rtot}} + \frac{C_{H^+}}{K_{conc}C_{M^+}C_{Rtot}} \tag{6.10}$$

Fig. 6.7 Solvent extraction distribution isotherm for double-charged ions ($n = 2$)

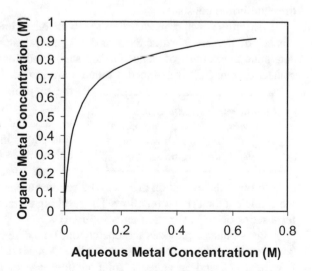

Fig. 6.8 Sample inverse concentrations plot used to determine the equilibrium constant

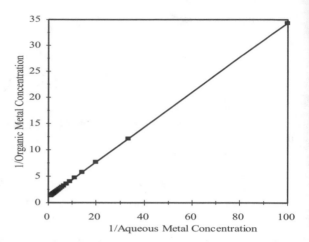

Plotting $1/C_{RM}$ versus $1/C_{M^+}$ from appropriate data will lead to a slope of $(C_{H^+})/(C_{Rtot}K_{conc})$. The associated intercept is $1/C_{Rtot}$ for $n = 1$. A representative plot is shown in Fig. 6.8. This approach assumes that C_{H^+} is constant. If n is not equal to one, an appropriate equation can be derived.

The appropriate extraction equation when $n = 2$ is:

$$C_{R_2M} = \frac{4C_{Rtot}K_{conc} + \frac{C_{H^+}^2}{C_{M^{2+}}} \pm \sqrt{(-4C_{Rtot}K_{conc} - \frac{C_{H^+}^2}{C_{M^{2+}}})^2 - 16K_{conc}^2 C_{Rtot}^2}}{8K_{conc}} \tag{6.11}$$

Both distribution isotherms shown in Figs 6.5 and 6.7 require knowledge about the equilibrium constant.

Distribution isotherms can be used with other information to determine the number of necessary extraction stages. A general mass balance is performed to determine the number of stages. Steady-state conditions are assumed. In addition, counter-current flow is assumed. Counter-current flow is the flow of one stream in the opposite direction of another interacting stream. In liquid–liquid extraction the aqueous feed enters the last extraction stage. In contrast, the organic feed enters the initial stage. The aqueous and organic streams flow in opposite directions through the extraction process. The counter-current flow makes the extraction more efficient than concurrent flow. The resulting extraction flow sheet is illustrated for a series of "*N*" extraction units in Fig. 6.9.

> *Commercial solvent extraction is constrained by a mass balance of the flows and concentrations in aqueous and organic phases.*

A mass balance provides the operational boundaries. The resulting operational boundary line is known as the operating line. A mass balance with an organic phase flow rate, Q_o, and an aqueous solution flow rate of Q_{aq} for the overall system containing the species "*y*" leads to:

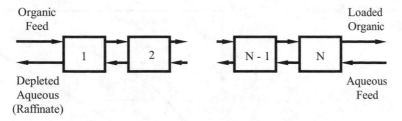

Organic
Feed

Loaded
Organic

Depleted
Aqueous
(Raffinate)

Aqueous
Feed

Fig. 6.9 Schematic diagram of counter-current solvent extraction at steady-state

$$C_{(N)Oy}Q_o + C_{(1)aqy}Q_{aq} = C_{(0)Oy}Q_o + C_{(N+1)aqy}Q_{aq} \qquad (6.12)$$

The stage of extraction is given in parentheses. The extraction stage for the aqueous feed is the previous stage $(N + 1)$. The extraction stage for the incoming organic feed is the "0" stage because it precedes stage 1. Rearrangement of the mass balance equation results in:

$$C_{(N)Oy} = \frac{Q_{aq}}{Q_o}\left[C_{(N+1)aqy} - C_{(1)aqy}\right] + C_{(0)Oy} \qquad (6.13)$$

> *McCabe-Thiele diagrams combine the constraints of thermodynamic equilibrium with the constraints of flows and concentrations to estimate performance for each loading or stripping stage or unit.*

Equation (6.13) is known as the operating line. It represents the possible conditions for operation. It is related to the equilibrium extraction isotherm in plant practice. The equilibrium extraction isotherm represents the thermodynamic limit for operation. The operating line represents the steady-state mass balance limit for operation. These two limits are often plotted together as shown in Fig. 6.10. Using the extraction distribution isotherm shown in Fig. 6.5, the operating line, which is Eq. (6.13), is presented with the distribution isotherm in Fig. 6.11. This diagram, which also shows the various stages in terms of the concentrations entering and leaving the stage, is often referred to as a McCabe–Thiele diagram.

Example 6.2 Calculate the concentration of metal in the organic phase in stage 2 of a series of two extraction units. The flow rate of the organic phase is 500 l/min. The flow rate of aqueous feed is 1000 l/min. The concentration of metal in feed organic is 0.2 g/l. The raffinate concentration is 0.5 g/l. The aqueous feed concentration is 4 g/l.

$$C_{(2)Oy} = \frac{Q_{aq}}{Q_o}\left[C_{(3)aqy} - C_{(1)aqy}\right] + C_{(0)Oy} = \frac{1000}{500}[4 - 0.5] + 0.2 = 7.2 \text{ g/l}$$

Fig. 6.10 McCabe–thiele
diagram

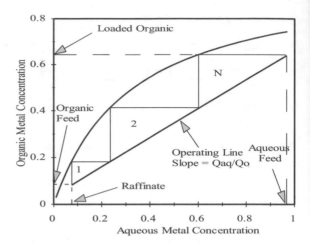

The importance of the upper extraction equilibrium line is intuitive from a chemical perspective. It represents the thermodynamic limit for the associated chemistry. The operating line represents the mass balance limits for plant operations. If the organic solution flow rate is low and the aqueous solution flow rate is high, the amount of the extracted solute in the organic phase would be low. Conversely, if the organic flow rate was high and the aqueous flow rate was low, more of the solute could be extracted. Thus, the operating line illustrates the important operating constraints imposed by mass balance.

Loaded organic is often washed to remove entrained aqueous phase and/or scrubbed to remove unwanted metal ions. The washing process utilizes relatively pure water and bleed electrolyte to ensure sufficient acidity to facilitate effective phase separation. Washing is particularly important in solutions with high impurity levels, such as chloride, that could be detrimental in electrowinning if carried over as entrained impurities. Scrubbing is usually performed with water, weak acids or bases, and/or metal salts. The scrubbed organic phase is sent for stripping. In contrast, the aqueous scrub raffinate is usually recycled to a leaching operation. Crud, which is a solid-like mixure of fine particles, vegetation, mold, organic degradation precipitates and some organic material, is also removed, often by selective pumping [2].

The stripping of the loaded organic phase is performed in acid for acidic extractants. Often the pH during the stripping is near zero, allowing the equilibrium in Eq. (6.1) to favor metal dissociation from the organic lixiviant. The dissociated metal is recovered in the aqueous stripping solution. The aqueous stripping solution has a higher metal content than the aqueous loading solution. The stripping solution also is generally quite pure in terms of other dissolved metal ions. Its purity is related to its selectivity in the loading and scrubbing operations. The stripping process is also represented in

> *Stripping is performed in concentrated solutions that force the equilibrium to shift from the loaded state toward the stripped state.*

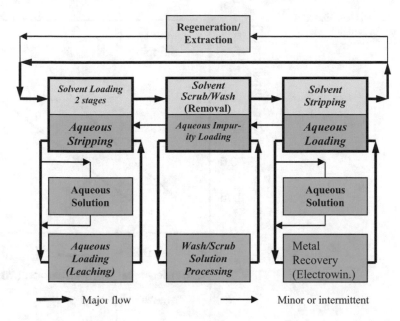

Fig. 6.11 Typical solvent extraction process flow sheet

McCabe–Thiele diagrams. The stripping McCabe–Thiele diagrams are shown with reversed axes. Figure 6.11 presents an example of a McCabe–Thiele diagram for stripping. The stripping diagram looks similar to the loading diagram. However, the axes are reversed and the equilibrium line is correspondingly inverted. In addition, the stripping diagram includes a narrow range of aqueous concentrations. The overall process of loading, stripping, scrubbing, etc. for a typical process is shown in Fig. 6.12.

The type of extractant as well as the solution conditions determines the selectivity. As seen from Eq. (6.1), most extraction involves hydrogen ions. Consequently, liquid–liquid extraction is pH dependent. At low pH values, the extraction is poor. Correspondingly, the extraction coefficient is low at low pH. The pH at which the extraction coefficient is one is termed pH_{50} or $pH_{1/2}$. Figure 6.13 illustrates the relationship between extraction coefficient, pH, and pH_{50}. At pH values below pH_{50} stripping is favored. Above pH_{50}, loading is favored.

Differences in affinities for ions in specific extractants allows for effective separations. In a two metal system metals often effectively compete against hydrogen ions for adsorption sites in the organic phase. Consequently, differences in ion affinities relative to pH can be observed in appropriate plots. Figure 6.14 illustrates metal loading for two metals as a function of pH. As illustrated in Fig. 6.14, the extractant has a higher affinity for Metal 1. At high pH values both metals will be extracted. At low pH values only Metal 1 will be extracted. If the pH is maintained at 3, very little of the impurity Metal 2 will load into the organic phase. At pH 3 a significant amount of Metal 1 will load.

Fig. 6.12 McCabe–theile
diagram for liquid–liquid
stripping

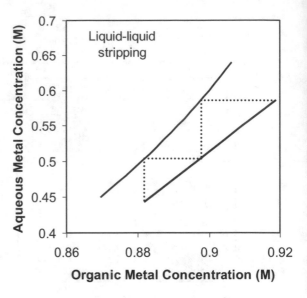

Fig. 6.13 Comparison of
extraction coefficient and pH
as it relates to the pH_{50} or
$pH_{1/2}$. ($n = 1$). Note that the
pH_{50} is equal to half of the
maximum loading
concentration only if the
aqueous concentration is the
same, which is rarely the case

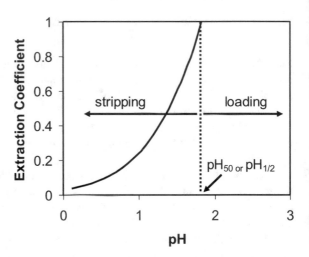

The efficiency of the extraction of Metal 1 would not be optimal at pH 3.
However, the separation of Metal 1 from Metal 2 would be effective at pH 3.
Conversely, the metals could be separated during the stripping process by stripping
at a pH between 5 and 6.5. Such a separation during stripping would not likely be
very efficient.

> *Efficiency is greatest when the pH*
> *favors only the desired metal.*

Metals with low pH hydrolysisPH are
usually extracted at low pH values. In
other words the tendency for hydrolysis
(e.g. $2H_2O + M^{2+} \leftrightarrow M(OH)_2 + 2H^+$)

Fig. 6.14 Typical plot of organic metal concentration versus pH during solvent extraction ($n = 1$)

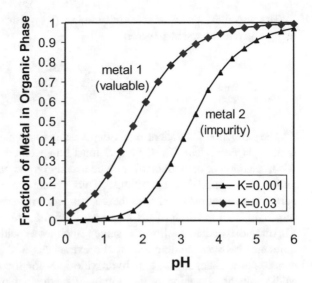

is often related to extraction (e.g. $2RH + M^{2+} \leftrightarrow R_2M + 2H^+$). A list of pKa values for common ions is presented in Table 6.1. As an example, ferric ions hydrolyze strongly and have a pKa value of 2.2. Zinc ions hydrolyze more weakly and have a pKa of 8.8. Thus, ferric ions tend to be extracted at lower pH values than zinc ions. However, with chelating extractants this hydrolysis/extraction trend is not as apparent.

Table 6.1 Selected Metal pKa values (after Ref. [3])

Ion	pKa
Fe^{3+}	2.2
Cr^{3+}	3.8
Al^{3+}	5.1
Cu^{2+}	6.8
Pb^{2+}	7.8
Zn^{2+}	8.8
Co^{2+}	8.9
Fe^{2+}	9.5
Ni^{2+}	10.6
Mg^{2+}	11.4
Ca^{2+}	12.6

The selectivity factor or selectivity index, $(S_{A:B})$ for specie "A" over specie "B" using a given extractant is given as:

$$S_{A:B} = \frac{E_A}{E_B} = \frac{\dfrac{K_A(C_{RH_A})^n}{C_{H^+}^{n_A}}}{\dfrac{K_B(C_{RH_B})^n}{C_{H^+}^{n_B}}} \tag{6.14}$$

The extraction coefficient is pH dependent. Consequently, the selectivity factor is also pH dependent. Thus, the pH must be specified with the selectivity factor. Also, selection of an extractant is made usually on the basis of the selectivity factor under plant operating conditions. Thus, conditions are driven by operating parameters rather than optimum theoretical performance parameters. An example of selectivity as a function of pH is shown in Fig. 6.15.

Anionic extractants follow the same principles as acidic extractants. The primary difference between anionic and acidic extractants is the ions exchanged. Anionic extractants exchange halide or hydroxide ions for the desired anions. Generally, acidic solvent extraction is the dominant method of metal extraction due to the positive charge found on dissolved metals. However, some metals, such as gold, can be extracted as anionic complexes.

Example 6.3 Calculate the selectivity factor of copper to iron for a system containing dissolved copper and iron at pH 2 if the extraction coefficient of copper is 36 and the extraction coefficient for iron is 4.

$$S_{Cu:Fe,pH2} = \frac{E_{Cu,pH2}}{E_{Fe,pH2}} = \frac{36}{4} = 9$$

Commercial Solvent Extraction

Fig. 6.15 Example comparison of selectivity index versus pH

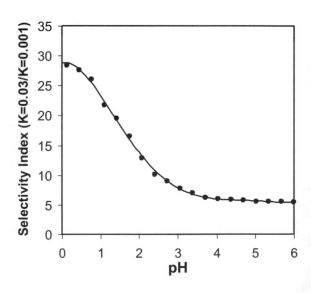

6.1.2.1 Copper

Copper solvent extraction is typically performed using LIX extractants that are mixed with diluents in ratio that is often near 1 part extractant with 7–10 parts dilutent. Diluents are often compounds made up of iso-alkanes.

Copper solvent extraction circuits generally consist of two extraction stages and one stripping stage [4]. The extraction stages are commonly found in series, although it is not uncommon to see series–parallel configurations that allow for flexibility in the solvent extraction circuit.

Incoming pregnant leaching solutions commonly contain around 3 g of dissolved copper per liter. pH levels are often around 2. Iron is also usually present at levels similar to copper. Extraction from the PLS is often near 90% after two stages. Raffinate from the solvent extraction loading circuit generally contains 0.1–0.3 g of dissolved copper per liter, and the pH level is generally between 1.5 and 2.0.

Stripping circuits generally increase the copper content from about 35 g of dissolved copper per liter in the lean electrolyte to around 45 in the rich electrolyte. Acid contents are generally near 180 g of sulfuric acid per liter.

The most common type of solvent extractants for copper are hydroxioximes. Examples of common products include LIX64, LIX 65 N, SME529, and P50, which each have slight variations in the extractant molecule [5]. Aldoximes, especially salicylaldehyde oxime are also becoming important due to their favorable extraction kinetics and selectivity over iron [5].

> *Hydroxioximes are the primary extractants used for copper.*

Mixtures of different molecules such as ketoximes and aldoximes have been used to provide better overall performance for loading and stripping. Improvements in mixtures of copper solvent extractants have made it feasible to include fewer extraction and stripping stages than the first generation of extractants allowed.

Some recent developments have resulted in extractants for chloride based media, which have received considerable attention for copper extraction in recent years. One compound, Acorga CLX 50, is reported to be effective in high chloride environmens [5].

6.1.2.2 Gold

Solvent extraction is not the dominant method for concentrating gold in aqueous solution. Despite its limited use, it is effectively utilized in commercial practice. High loadings are possible relative to other methods. However, the high density of gold can lead to phase inversion if loadings are excessive. Stripping times are long, and solvent losses are often high. Gold can be stripped by precipitation, direct electrolysis, or by traditional stripping into a concentrated solution.

Amines, guanidine, ethers, phosphates, phosponates, and ketones can be used for gold solvent extraction. Primary, secondary, and tertiary amines, which have pKa values of 6.5, 7.5, and 6.0, respectively are available [6, 7]. The terms primary, secondary, and tertiary reference the number of hydrocarbon chains bonded to the nitrogen in the amine. Primary amines have one hydrocarbon bonded to the nitrogen, seconardy amines have two, and tertiary have three. Quaternary amines are more selective for gold than other amines. Primary amines have low selectivity and are easily stripped. Loading is best below the pKa. Stripping is most effective above the pKa. Thus, in alkaline media, loading can be difficult. Blending with solvents such as those containing phosphorous can increase the effective pKa [8].

Guanidine is a small molecule with three nitrogen atoms bonded to a central carbon atom. One of the nitrogen atoms forms a double bond with the carbon atom. Guanidine is a strong base. It complexes readily with gold cyanide in alkaline media. It has a pKa of 13.6 [9]. Consequently, it is difficult to strip.

Dibutyl carbitol (DBC) and similar ether compounds have been used to extract gold. DBC is effective in separations involving strong acids. High loading levels can be achieved [10]. The DBC solvated gold can be stripped directly into metallic gold using hot oxalic acid $(2DBCHAuCl_4 + 3(COOH)_2 \leftrightarrow 2DBC + 2Au + 8HCl + 6CO_2)$ [7].

> *A variety of solvent extractants are effective in concentrating gold. However, gold is more commonly concentrated in solution by loading and stripping activated carbon.*

Phosophorous-containing organic extractants for gold include phosphates and phoshonates such as tri-n-butyl phosphate (TBP) and di-n-butyl butyl phosphonate (DBBP). These compounds are often used with amines, but they are limited in use.

Ketones such as methyl isobutyl ketone (MIBK) and di-isobutyl ketone (DIBK) have been used for gold extraction. These compounds are selective for gold. Their use is limited by high solubility in water and stripping difficulty [7].

6.1.2.3 Other Precious Metals

Extraction of precious metals such as palladium, platinum, and silver is often very similar to gold. Many of the same extractants that are effective with gold are used for extraction of other precious metals.

In mixed platinum group metal solutions, di-n-octyl sulfide or hyxroxyoximes are often used for palladium extraction, which is commonly followed by extraction of platinum using tri-n-butyl phosphate and an amine or *N*-alkylamides [5]. Iridium extraction is suppressed by reduction of iridium to its trivalent state.

6.1.2.4 Nickel and Cobalt

Nickel is usually found with cobalt. Cobalt and nickel have very similar properties and are challenging to separate. The general approach is to extract cobalt from chloride based solutions using amine extractants or from acidic sulfate based solutions using organophosphorus acids such as CYANEX 272. In some cases cobalt is oxidized to cobalt (III) and the nickel is extracted using mixtures of hydroxyoximes and carboxylic acid.

6.1.2.5 Rare Earth Metals

Lanthanide series elements, which are generally referred to as rare earth elements, along with scandium and yttrium, are very difficult to separate. Solvent extraction is the most common method of separating these elements from solution. Extractants such as di-ethylhexyl phosphoric acid (DEHPA), phosphonic acid ester, versatic acid, amines,

> *Separation of rare earth elements is commonly performed using mult-step solvent extraction processing.*

and tri butyl phosphate (TBP) are often used with nitrate and thiocyanate solutions [5]. Effective separations generally require multiple stages under carefully controlled conditions.

6.1.2.6 Uranium

Common uranium extractants include DEHPA, TBP, and amines (trioctyl and tridecyl). DEHPA and amine use is generally associated with sulfuric acid based solutions, and TBP is more commonly associated with nitric acid based solutions [5].

6.2 Ion Exchange

6.2.1 General Ion Exchange Information

Ion exchange is a very common form of solution concentration and purification. It has been known for almost 150 years [11] with respect to certain naturally occurring soils. Some soils containing alumino-silicates are known to have ion exchange capabilities. Today, specific clays, which are layered silicate minerals, are used for ion exchange. Zeolites, which are porous silicate minerals, are also used. Porous resin beads are also used for ion exchange. Ion resin beads are usually made of a porous polymeric network. Functional groups with specific ion exchange

> *Ion exchange is commonly applied to separate and concentrate metals in addition to the well-known process of water "softening", which removes calcium and magnesium from culinary water to reduce "hard water" deposits.*

capability are placed inside pores. An example of resin beads is shown in Fig. 6.16. A cross-sectional illustration of a typical resin particle is shown in Fig. 6.17.

Fig. 6.16 Magnified view of
typical ion exchange resin
beads, which are
approximately 1–2 mm in
diameter

Fig. 6.17 Schematic diagram
of a resin particle with its
associate network of pores,
which are shown as black
lines

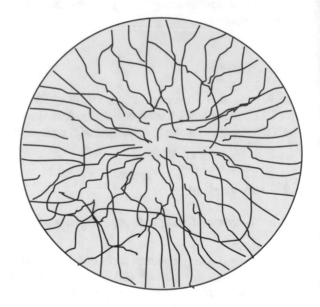

Ion exchange resin was first developed in 1935 [12]. The resulting porous beads
are capable of loading a large quantity of dissolved species. After adsorption
occurs, the beads are stripped of the loaded species. Stripping is accomplished using
a high concentration of similarly charged species. This process is essentially the
same as the solvent extraction process discussed previously. One difference between ion exchange and solvent extraction is the structure of the extraction medium.
Ion exchange does not normally require solution clarification or filtration prior to
extraction. Ion exchange resins release negligible quantities of organic matter into
the aqueous process streams. Thus, ion exchange has some advantages over liquid–
liquid extraction. However, liquid–liquid extraction is generally much faster and
easier to apply commercially.

The terminology for ion exchange is different from solvent extraction. The aqueous solution from which the ions have been extracted is called the effluent. The stripping solution is called the eluant, and the solution into which the ions have been stripped is referred to as the eluate.

The types of ion exchange resins are similar to those used for liquid–liquid extraction. However, the resin beads contain a polymer backbone such as polystyrene-divinyl benzene copolymers. The backbone also contains functional groups that perform the extraction. Because the extractant is immobilized by the polymer matrix, resins cannot participate in solvation mechanisms of extraction. Instead, the resin extracts by means of ion exchange and coordination. The most common functional groups used in resin ion exchange beads include amines, carboxylates, phosphonates, and sulfonates.

Ion exchange resins are often used in packed columns. As solution enters the column, the ion exchange begins at the entrance. Resin near the entrance has first access to the solution. Consequently, resin near the entrance fills with adsorbed ions first. During initial stages the exit region has little adsorption. During later stages the entry region is full, leaving more ions for the exit region. Eventually the resin reaches its capacity to adsorb ions. Consequently, the concentration of desired ions leaving the resin column is low initially. However, as more of the resin is filled, the exit concentration rises. The concentration of adsorbing ions leaving the column changes as a function of volume as shown in Fig. 6.18. The point at which the effluent (solution leaving the column) metal ion concentration surpasses the threshold concentration limit is known as breakthrough. The capacity of the resin to maintain the effluent concentration below the threshold limit is known as the breakthrough capacity. The breakthrough capacity is often reported on the basis of the number of bed volumes of solution.

Solution flow rate is an important parameter due to its relationship with mass transport. The effluent concentration is high with high flow rates. At high flow rates ions often do not have time to adsorb before leaving. At low flow rates ion diffusion into resin allows for greater adsorption and lower effluent concentrations.

Fig. 6.18 Typical plot of effluent metal concentration versus total bed volumes of solution flow for ion exchange resin in a column

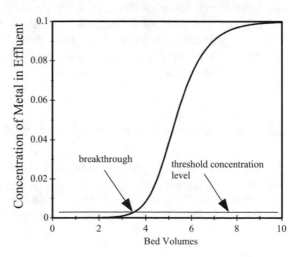

Fig. 6.19 Plot of eluate metal concentration versus total elution bed volumes for ion exchange resin column stripping

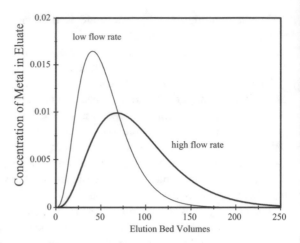

The same type of phenomenon occurs with stripping. However, as shown in Fig. 6.19, the eluate (stripping solution leaving the column) concentration increases initially as metal is removed or stripped. The eluate concentration decreases as the resin becomes depleted. The concentration peaks at higher levels with lower flow rates.

> *Ion exchange is usually applied using columns of resin through which solutions flow.*

In practice the ion exchange resin beads are used in columns or in slurries. The most common method is in columns. However, in many instances the resin is sufficiently durable to be placed in slurries. The capability of placing the resin in a slurry or pulp (resin in pulp or RIP) allows for more rapid kinetics. Resin that is used in slurries is subsequently separated from the slurry by screening or by flotation. The resin is subsequently stripped to remove the adsorbed ions.

The same principles of selectivity, extraction coefficient, and fraction extracted that apply to solvent extraction also apply to ion exchange. However, with ion exchange, the area available for extraction remains constant, and the extraction can be a very predictable function of diffusion through the pores. In contrast, the extraction with solvents is dependent upon interfacial tension, viscosities, velocities, and mixing within the solvent droplets.

6.2.2   Equilibrium Ion Exchange Adsorption Models

The ability to model adsorption equilibrium is particularly useful when designing systems that utilize ion exchange methods. Several models have been developed to mathematically describe

> *Various models can be used to describe ion exchange adsorption.*

adsorption at equilibrium, and they can be useful in understanding metal concentration processes, such as ion exchange, that involve adsorption. These models are derived on the basis of exchange between species on surfaces at or near equilibrium. Consequently, they do not provide kinetics information, which is often critical to commercial applications.

6.2.2.1 Freundlich Model

The Freundlich adsorption model is derived by assuming a site distribution function that is based on varying adsorption site free energy values. It reduces to the Langmuir equation for $n = 1$ at low concentrations and for $n = \infty$ at high concentrations. Using this model, the adsorbed concentration can be expressed as [13]:

$$C_{AS} = KC_{Tot}C_A^{1/n} \tag{6.15}$$

where C_{AS} is the concentration of adsorbate that is adsorbed on the surface, K is a constant, C_{Tot} is the total concentration of available sites, C_A is the concentration of available adsorbate, and n is an empirical parameter that is typically greater than 1. Determination of whether or not the Frueundlich model applies to a given set of data can be made by plotting ln C_{AS} versus ln C_A [ln $C_{AS} = \ln(KC_{Tot}) + (1/n)\ln (C_A)$] and evaluating the linearity of the data.

6.2.2.2 Langmuir Model

The Langmuir adsorption model assumes that all surface adsorption sites are equivalent–regardless of whether neighboring sites are occupied. (A + S $\leftrightarrow$ AS). Using the Langmuir model, the adsorbed adsorbate concentration is expressed as [14]:

$$C_{AS} = \frac{K_A C_A C_{Tot}}{1 + K_A C_A} \tag{6.16}$$

where C_{AS} is the concentration of adsorbate that is adsorbed on the surface, K_A is the adsorption equilibrium constant, C_{Tot} is the total concentration of available sites, and C_A is the concentration of available adsorbate. Determination of whether or not the Langmuir model applies to a given set of data can be made by plotting $1/C_{AS}$ versus $1/C_A$ [$1/C_{AS} = 1/(K_A C_A C_{Tot}) + (1/C_{Tot})$] and evaluating the linearity of the data.

6.2.2.3   Tempkin Model

The Tempkin adsorption model is an empirical adsorption model that considers nonuniform site distribution. According to this model, the adsorbed adsorbate concentration is expressed as [15]:

$$C_{AS} = C_{tot}K \ln(kC_A) \tag{6.17}$$

where C_{AS} is the concentration of adsorbate that is adsorbed on the surface, K is a constant, C_{Tot} is the total concentration of available sites, C_A is the concentration of available adsorbate, and k is an empirical parameter. (Note kC_A must be greater than 1).

6.2.3   Ion Exchange Adsorption Kinetics Models

Understanding adsorption kinetics is clearly an important engineering design tool. The following models describe adsorption or related kinetics in a variety of scenarios.

Diffusion based models assuming Fick's first law applies:

6.2.3.1   Pore Diffusion (Shrinking-core Model Only)

The basic rate model for pore diffusion is:

$$\frac{dn}{dt} = -\frac{AD(C_b - C_s)}{r_o - r} \tag{6.18}$$

where A is the area for diffusion, D is the diffusivity, C_b is the bulk concentration, C_s is the surface concentration, r_o is the outer radius of the absorbing solid, and r is the radius of the reaction front within the particle. The fraction reacted-based solution to this model is:

$$\frac{3r_o}{2D}\left[1 - \frac{2}{3}\alpha - (1 - \alpha)^{2/3}\right] = \frac{3C_b}{C_{tot}r_o}t \tag{6.19}$$

6.2.3.2   Film Diffusion

The basic film diffusion kinetics model based upon Fick's first law gives:

$$\frac{dn}{dt} = -\frac{AD(C_b - C_s)}{\delta} \tag{6.20}$$

where A is the area for diffusion, D is the diffusivity, C_b is the bulk concentration, C_s is the surface concentration, and δ is the boundary layer thickness. For the case of infinite volume and constant bulk concentration, the solution is expressed as:

$$\ln(1 - \alpha) = -\frac{3DC_b t}{r_o \delta C_{tot}}$$ (6.21)

For the case of finite volume the solution is:

$$\ln(1 - \alpha) = -\frac{3DC_b t}{r_o \delta} \left(\frac{C_{tot} V_{resin} + C_b V_{soln}}{C_{tot} V_{soln}} \right)$$ (6.22)

6.2.3.3 Empirical Diffusion Model

A useful empirical model for determining the rate of adsorption is:

$$\frac{dn}{dt} = -K(C_{AS} - kC_A)$$ (6.23)

where K is an empirical constant related to the more commonly used k_{1A}, C_{AS} is the concentration of adsorbed adsorbate, C_A is the available concentration of the adsorbate in solution, and k is an empirical constant related to the ability of the adsorbate to diffuse to the adsorption sites.

The solution to this kinetics model can be expressed as:

$$\frac{1}{k} \ln \left(1 - \frac{C_{AS}}{kC_A} \right) = Kt$$ (6.24)

6.2.3.4 General Empirical Adsorption Kinetics Model

A general empirical equation that describes adsorption kinetics for some systems is given as:

$$\frac{dC_{AS}}{dt} = KC_A t^n$$ (6.25)

where K is a constant, C_A is the concentration of available adsorbate in solution, t is the adsorption time, and n is an empirical constant. The solution for the general empirical adsorption kinetics model can be given as:

$$C_{AS} - C_{AS_o} = \frac{KC_A t^{n+1}}{n+1}$$ (6.26)

6.2.3.5 Empirical Reaction and Diffusion

Another useful kinetics expression that combines empirical reaction and diffusion equations is:

$$\frac{dC_{AS}}{dt} = -\frac{K''C_A V}{M} = -K'(C_{AS} - KC_A) \tag{6.27}$$

For which the solution can be expressed as:

$$\ln\frac{C_{AS}}{C_{AS_o}} = \left(\frac{KM}{K''V} + \frac{1}{K'}\right)t \tag{6.28}$$

6.2.3.6 Ion Exchange for Specific Metals

Gold ion exchange can be performed usng resins with quarternary and tertiary amines. Ion exchange resins for gold adsorption offer higher loading capability than carbon. However, ion exchange resins are more difficult to strip and handle than activated carbon. Ion exchange has not generally been successful on a commercial scale for gold extraction due to the success of activated carbon.

6.3 Activated Carbon Adsorption

Activated carbon is often made by heating a carbon source such as coconut shells to 700-1000°C with water vapor.

Carbon adsorption is a common method of species removal. Carbon is particularly adept at removing low levels of dissolved ions. The activated carbon can be made from a wide variety of organic starting materials. Items such as peach pits, coconut shells, and wood are used for activated carbon. The material is converted into activated carbon by heating it in a low-oxygen environment. Initially, the carbon source is carbonized by heating it to 500 °C with dehydrating agents to remove water and impurities [16]. The carbon is then heated to 700–1000 °C with steam, carbon dioxide, and/or air to volatilize residues, develop pore structures, and form functional groups [16]. Different types of carbon surfaces also provide active adsorption sites for dissolved ions. A sample of coconut shell carbon is shown in Figs. 6.19, 6.20, 6.21, 6.22 at three magnifications. The particles are approximately 2 mm in diameter.

Activated carbon is commonly used to extract aurocyanide from solution. The exact mechanisms are not well understood. Evidence suggests adsorption is an ion exchangeIon exchange process [17, 18]. However, for gold chloride complexes, it is believed that the adsorption process involves reduction to metallic gold at the

Fig. 6.20 Photograph of coconut shell activated carbon pieces. (The scale is centimeters)

Fig. 6.21 Magnified view (×6) of 2–3 mm diameter coconut shell activated carbon

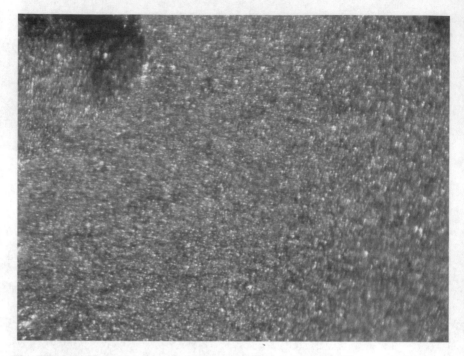

Fig. 6.22 Magnified view of coconut shell activated carbon surface (×30)

carbon surface [19]. Extraction using carbon adsorption is most often diffusionDiffusion controlled. The same theoretical treatment that applied to ion exchange kinetics applies to carbon adsorption kinetics.

Activated carbon is often utilized in a counter current manner as depicted in Fig. 6.23 to maximize adsorption. Counter current processing allows the carbon with the least adsorbed matter to contact the solution that is most depleted. Carbon that is nearly loaded contacts the solution that has the highest concentration of the desired species.

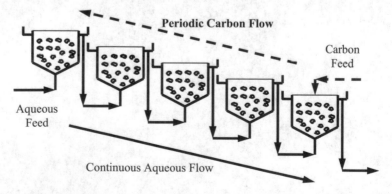

Fig. 6.23 Counter-current flow sheet for carbon adsorption

There are several processing options for activated carbon use. The process of loading carbon from leaching solution without ore or concentrate particles is known as carbon-in-column (CIC). The carbon is commonly partially fluidized by the flow of leaching solution upward through CIC columns. Locating columns at different elevations often creates the head pressure for solution flow. CIC utilization requires particle/liquid separation after leaching, prior to CIC loading.

Exposure of the carbon to a gold leaching slurry or pulp with particles present after leaching is known as the carbon-in-pulp (CIP) process. This process is often operatated similarly to CIC. However, the particles must be ground to fine particles to facilitate a screen separation of the carbon from the pulp after adsorption.

Another way of adsorb gold is to utilize carbon in a leaching slurry. This process offers an advantage of extracting gold at lower concentrations, thereby reducing gold losses from natural carbon sources in the ore. The unwanted removal of gold from the gold-bearing "pregnant" solution is referred to as "preg-robbing". Therefore, carbon-in-leach (CIL) is sometimes used for gold loading from refractory ores that contain carbonaceous matter. The CIL process results in lower gold loading and higher carbon concentrations.

Factors that influence gold adsorption include pH, temperature, gold concentration, ionic strength, and system impurities. The lower the pH is, the more selective the carbon is for gold. However, the pH must generally be greater than 10 for safety reasons. Temperature is a key parameter that is utilized for stripping because elevated temperatures favor desorption. Adsorption is more rapid when the gold concentration in solution is higher. Increasing ionic strength improves loading. Cation impurities improve extraction, whereas anionic impurities decrease adsorption. The presence of silver, mercury, and copper can reduce gold loading capacity.

Figure 6.24 shows industrial gold adsorption data plotted in an inverse adsorption concentration versus inverse solution concentration format to identify the constants needed for the Langmuir adsorption isotherm [20]. The associated constants can be used to show the model fit of the data presented in Fig. 6.25.

Gold is often stripped or eluted from carbon at elevated temperatures (95–150 °C) in a caustic cyanide solution. Stripping is often performed useing either the Zadra or Anglo American Research Lab (AARL) process. The stripping solution often con-

> *Activated carbon is often stripped of adsorbed gold by heating it in a caustic cyanide solution in a pressurized vessel.*

tains 1% NaOH and may contain some sodium cyanide [21]. The elution or stripping process is strongly influenced by temperature. Consequently, most processes tend to strip carbon above 100 °C in pressurized vessels.

Activated carbon is exposed to a variety of chemicals during loading. Organic compounds and calcium and magnesium carbonate commonly deposit on the carbon surface and restrict gold adsorption access. Consequently, activated carbon performance decreases with use unless it is regenerated. Regeneration is often

> *Activated carbon often needs to be thermally regenerated.*

Fig. 6.24 Example of plot of
gold adsorption data based on
the inverse of the gold
concentration on the carbon
versus the inverse of the gold
concentration in solution

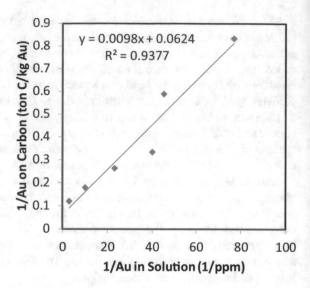

Fig. 6.25 Comparison of the
gold data in Fig. 6.24 plotted
to illustrate the fit of
adsorption to the Langmuir
adsorption model

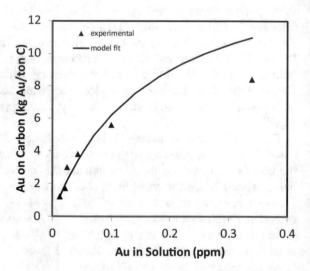

performed by thermal activation in a rotary kiln at 600–900 °C (usually around
650 °C) in steam to remove organic debris. Thermal activation is often preceded or
followed by hydrochloric acid washing to remove inorganic precipitates such as
carbonates [21]. These data show that industrial gold adsorption on carbon can fit
the Langmuir adsorption model.

Fig. 6.26 Illustration of the process of ultrafiltration or reverse osmosis. (P represents pressure on a piston that forces liquid through the membrane)

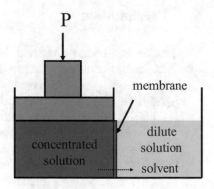

6.4 Ultrafiltration or Reverse Osmosis

Ultrafiltration is the process of filtering out solute from a solution through a membrane at high pressure. The process of ultrafiltration is illustrated in Fig. 6.26. Ultrafiltration or membrane filtration is, in fact, the same as reverse osmosis. Ionic solutions have a significant osmotic driving force for acquiring additional solute molecules. Solute molecules must have a pathway for transfer to occur. The pathway for ion transfer can be a membrane. Membranes have small pores that selectively allow ion transport. Membranes with very small pores may allow only small ions or molecules to pass. Thus, if a solution is forced through such a membrane, solute molecules are retained. Consequently, high pressures are required to reverse the osmotic process and expel solute molecules through membranes. Thus, this process is often called ultrafiltration because solute is filtered from solvent. The pressure required for ultrafiltration is the osmotic pressure. The osmotic pressure of sea water at 25 °C is 25 atm or 25 MPa (368 psi) [22]. The osmotic pressure is given as [23]:

$$P = CRT \qquad (6.29)$$

P is the osmotic pressure. C is the concentration of solute in molal. R is the gas constant. T is the absolute temperature.

Ultrafiltration and membrane filtration systems utilize thin membranes with very small pores. These pores allow water molecules to pass. However, the small pores severely restrict the passage of hydrated ions and larger molecules. Ideally, the membranes are relatively thin and are supported by more porous material.

> *Ultrafiltration or reverse osmosis is commonly used in desalination and in treating water with high dissolved salt content.*

The porous material and its support must withstand the high pressures. Membranes are made of materials such as cellulose acetate, but they exhibit tremendous resistance to flow (0.2 $cm^3 s^{-1}$ atm^{-1} m^{-2} is a typical flow rate). High resistance makes it necessary to increase surface areas dramatically to enhance flow [22].

6.5   Precipitation

Precipitation is a common method for concentrating metal content and purifying solutions. Iron is commonly removed from solution by precipitation. Ferric ions have low solubility above pH 3. Many divalent metal ions such as ferrous ions are relatively soluble at low and intermediate pH levels. This difference in solubilities as a function of pH facilitates separations. Figure 3.9. In Chap. 3 shows the relationship between metal ion solubilities and pH. The differences in precipitation of metals at specific pH levels facilitate separations. As an example, ferric ions are often separated from other divalent metal ions at pH values of 3 to 4 by selective precipitation. The ferric ions are generally precipitated as ferric hydroxide. One industrially important example of iron precipitation is found in zinc metal production.

Precipitation is a very common method for removing or recovering specific species from solution.

Precipitation can also be accomplished using gasses and/or water. Ferrous iron, for example, can be precipitated as hematite by the addition of oxygen and water:

$$2Fe^{2+} + 0.5O_2 + 2H_2O \leftrightarrow Fe_2O_3 + 4H^+ \tag{6.30}$$

In zinc metal production from zinc sulfide ore concentrate, the concentrate is often first calcined. Calcined concentrate is leached in sulfuric acid to dissolve the zinc. Iron also dissolves in the leaching process. Iron is separated from the zinc by precipitation using hydroxide compounds. The resulting iron precipitate (jarosite $[(NH_4, Na)Fe_3(SO_4)_2(OH)_6]$ or goethite $[FeO \cdot OH])$ is removed from the zinc-bearing solution. The reactions are as follows:

$$3Fe_2(SO_4)_3 + 2(NH_4, Na)OH + 10H_2O \leftrightarrow 2(NH_4, Na)Fe_3(SO_4)_2(OH)_6 + 5H_2SO_4 \tag{6.31}$$

$$Fe_2(SO_4)_3 + ZnS \leftrightarrow 2FeSO_4 + ZnSO_4 + S° \tag{6.32}$$

$$2FeSO_4 + 1/2\ O_2 + 3H_2O \leftrightarrow 2FeO \cdot OH(or\ Fe_2O_3H_2O) + 2H_2SO_4 \tag{6.33}$$

Another example of iron removal through precipitation is:

$$2Fe^{3+} + 3Ni(OH)_2 \leftrightarrow 2Fe(OH)_3 + 3Ni^{2+} \tag{6.34}$$

Precipitation is also practiced through contact reduction to produce metal. Because this is an electrochemical reaction resulting in metal recovery, it is discussed in the next chapter.

Precipitation is also used to recover by-products from process solutions. Nickel carbonate precipitation

Additional discussion of precipitation is presented in Chap. 8 in the context of environmental remediation.

6.5.1 Process and Waste Water Treatment

Microorganisms are capable of performing a variety of functions related to wastewater treatment. Reducing organisms can reduce dissolved species in solution to form compounds that are more easily removed from solution. Oxidation processes driven by microbial activity can make species more readily absorbed or complexed for removal purposes. Other microorganisms provide desirable adsorption sites that facilitate removal of toxic solution species. In other scenarios microorganisms can alter a very toxic species to a less toxic form. These capabilities of

> *Microorganisms can facilitate removal of specific species by precipitation or adsorption from solution.*

microorganisms make them attractive from a wastewater treatment perspective. Commercial applications of microorganisms in wastewater treatment are evident in a variety of environments, including mining-related areas.

Among the common approaches to utilizing bacteria in mine water treatment is that of sulfate reduction. Bacteria can reduce sulfate to sulfide and elemental sulfur. An electron donor is needed for the process. A common donor is hydrogen gas. The reactions involved include:

$$H_2SO_4 + 4H_2 \leftrightarrow H_2S^- + 4H_2O \tag{6.35}$$

The resulting hydrogen sulfide can react with metals such as lead to produce metal sulfides:

$$H_2S + Pb^{2+} \leftrightarrow PbS + 2H^+ \tag{6.36}$$

The metal sulfides precipitate readily from solutions. The resulting precipitates can be filtered and removed from the solutions.

Alternatively, sulfate can be converted to elemental sulfur:

$$H_2SO_4 + 4H_2 + 0.5O_2 \leftrightarrow S + 5H_2O \tag{6.37}$$

This type of reaction and the conversion to sulfide are advantageous for mine water treatment because they remove acid as well as sulfate. The removal of acid by this process reduces the need for other chemical compounds such as lime for neutralization. Compounds such as lime often create metal hydroxide and calcium sulfate precipitates. Sulfate reduction produces only one precipitate and water. There are no reports of large scale commercial operations that utilize this technology though it has been demonstrated at reasonable scale [24].

References

1. G.M. Ritcey, Development of industrial solvent extraction process, in *Solvent Extraction: Principles and Practice*, 2nd edn. (Marcel Dekker, New York, 2004)
2. K. Biswas, W.G. Davenport, *Extractive Metallurgy of Copper*, 2nd edn. (Pergamon Press, Elmsford, 1980), p. 322
3. G.M. Ritcey, A.W. Ashbrook, *Solvent Extraction in Process Metallurgy*. (AIME, 1978), p. 25
4. K. Biswas, W.G. Davenport, *Extractive Metallurgy of Copper*, 2nd edn. (Pergamon Press, Elmsford, 1980), p. 317
5. M. Cox, *Solvent Extraction Principles and Practice*, 2nd edn. (CRC Press, 2004), Chapter 11
6. M.J. Nicol, C.A. Fleming, R.L. Paul, The chemistry of gold extraction, in *The Extractive Metallurgy of Gold*, ed. by G.G. Stanley (South African Institute of Mining and Metallurgy, Johannesburg, 1987), pp. 831–905
7. J. Marsden, I. House, *The Chemistry of Gold Extraction*, 2nd edn. (SME, Littleton, 2006), pp. 355–358
8. P.L. Sibrell, J.D. Miller, Soluble losses in the extraction of gold from alkaline cyanide soutions by modified amines. *Proceedings of ISEC 86', Munich, International Solvent Extraction Conference*, vol. 2. (1986), pp. 187–194
9. http://en.wikipedia.org/wiki/Guanidine. Accessed May 7 2012
10. J.A. Thomas, W.A. Phillips, A. Farais, The refining of gold by a leach-solvent extraction process. Paper presented at 1st International Symposium on Precious Metals Recovery, Reno, NV, 10–14 June, 1984
11. E. Jackson, *Hydrometallurgical Extraction and Reclamation* (Ellis Horwood Limited, Chichester, 1986), p. 78
12. B.A. Jackson, E.L. Holmes, Adsorptive properties of synthetic resins. J. Soc. Chem. Ind. **54**(1 (T)), 1935
13. P.C. Hiemenz, R. Rajagopalan, *Principles of Colloid and Surface Chemistry*, 3rd edn. (Marcel Dekker, Inc., New York, 1997), p. 337
14. H.S. Fogler, *Elements of Chemical Reaction Engineering* (Prentice-Hall, Englewood Cliffs, 1986), p. 241
15. P.W. Atkins, *Physical Chemistry*, 3rd edn. (W. H. Freeman and Co., New York, 1986), p. 781
16. J. Marsden, I. House, *The Chemistry of Gold Extraction*. (Ellis Horwood, New York, 1993), p. 298
17. E. Jackson, *Hydrometallurgical Extraction and Reclamation* (Ellis Horwood Limited, Chichester, 1986), p. 101
18. P.L. Sibrell, J.D. Miller, Significance of graphitic structural features in gold adsorption by carbon. (Minerals and Metallurgical Processing, 1992), p. 189
19. J.B. Hiskey, X.H. Jiang, G. Ramodorai, Fundamental studies on the loading of gold on carbon in chloride solutions. Gold 90', ed. D. Hausen, (1990), p. 83
20. J.A. Herbst, S.W. Asihene, Modeling and simulation of hydrometallurgical processes. In *Proceedings of Hydrometallurgy*, ed. by V.G. Papangelakis, G.P. Demopoulos. (CIM, Montreal, 1993)
21. E. Jackson, *Hydrometallurgical Extraction and Reclamation* (Ellis Horwood Limited, Chichester, 1986), pp. 102–103
22. P.C. Hiemenz, R. Rajagopalan, *Principles of Colloid and Surface Chemistry*, 3rd edn. (Marcel Dekker, Inc., New York), p. 140
23. P.W. Atkins, *Physical Chemistry*, 3rd edn. (W. H. Freeman and Co., New York, 1986), p. 175
24. http://www.gardguide.com/index.php/Chapter_7. Accessed Feb 2, 2012

Problems

(6-1) Determine the extraction coefficient in a system at equilibrium that has an organic volumetric flow rate of 350 l/min, and aqueous volumetric flow rate of 450 l/min, and a fraction extracted of 0.4.

(6-2) For a four-stage solvent extraction process, calculate the concentration of metal in the organic phase in the fourth stage given that the flow rate of aqueous solution is 1000 l/min, the organic flow rate is 500 l/min, the aqueous feed concentration is 6 g/l, the aqueous raffinate solution concentration is 0.1 g/l, and the initial concentration of metal in the organic feed solution is 0.5 g/l.

(6-3) A one to one mixture of solvent and aqueous solutions results in an equilibrium metal ion (M^{2+}) concentration of 0.03 in the aqueous phase, a loaded extractant (R_2M) concentration of 0.15 m and a acid complexed extractant (RH) concentration of 0.05 m. For an equilibrium constant of 0.0005 for the metal complex formation reaction, calculate the pH_{50} value. (Assume unit activity coefficient values.)

(6-4) Calculate the selectivity index for M1 over M2 at pH 1.75 using the information in the accompanying plot as well as the assumptions that both metals have a single charge and the aqueous and organic volumes are equivalent.

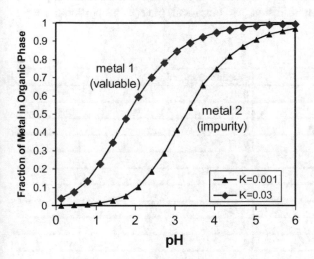

(6-5) Determine K_{conc} from the following set of solvent extraction data ($n = 1$, and pH = 2 or C_{H+} = 0.01 – keep the H^+ concentration units dimensionless).

$$\begin{array}{llllllll} C_{RM} & (g/l) & 0.029 & 0.040 & 0.081 & 0.164 & 0.50 \\ C_{M^+} & (g/l) & 0.010 & 0.013 & 0.030 & 0.071 & 0.30 \end{array}$$

(6-6) The following equilibrium data is best described by the Fruendlich, Langmuir, or Tempkin equilibrium model?

C_{AS}	C_{Abulk}
0.001	0.002
0.0025	0.005
0.0045	0.009
0.007	0.015
0.011	0.027
0.013	0.038
0.016	0.052
0.018	0.068
0.019	0.083
0.0195	0.10

(6-7) Which model describes best the following kinetics data resin loading data [pore diffusion, film diffusion, or empirical (rate is proportional to t^n) model]?

Time (hr)	Fraction absorbed
1.0	0.01
2.0	0.03
3.0	0.05
5	0.07
7	0.10
10	0.15
15	0.20
28	0.23
28	0.35
40	0.45

Chapter 7
Metal Recovery Processes

> *Metals are recovered from hydrometallurgical solutions in metallic form by electrochemical reduction.*

Key Chapter Learning Objectives and Outcomes:
Understand the principles and practices of electrowinning in aqueous media
Be able to calculate basic electrowinning parameters
Understand the principles and practices of electrorefining
Know how cementation is performed
Know how metals can be recovered in solution as metals

7.1   Electrowinning

Electrowinning is the electrolytic process of "winning" or recovering dissolved metal using an applied potential. An example of an electrowon copper plate is shown in Fig. 7.1. This process is practiced extensively in the metals industry. Copper, zinc, and gold, as well as other metals are produced by this process.

Electrowinning utilizes an applied potential to drive electrochemical reactions in the desired direction. An external power source supplies the potential and current. An inert anode is used to complete the circuit and the necessary counter reaction to metal recovery. Metal is recovered at the cathode. Ions or molecules are reduced at the cathode. Molecules or ions are oxidized at the anode. An electrolyte or ion conducting medium must exist between the anode and cathode. Water containing dissolved ions is a common electrolyte. Commercial electrowinning in aqueous solutions often involves acid. The hydrogen ions in the acid as well as the counter ions (usually sulfate) provide most of the solution conductivity. The rate of electrowinning is determined by electrochemical kinetics as discussed in Chap. 4. The process of electrowinning is depicted in Fig. 7.2.

> *Electrowinning is the most common method of recovering metal from solution in its metallic form.*

The primary electrowinning parameters are potential and current. Current is often tracked through current density. Current density is a more practical term in an

© The Minerals, Metals & Materials Society 2022
M. L. Free, *Hydrometallurgy*, The Minerals, Metals & Materials Series,
https://doi.org/10.1007/978-3-030-88087-3_7

Fig. 7.1 Photograph of a
copper cathode sheet
produced by electrowinning

Fig. 7.2 Schematic diagram
of a typical
hydrometallurgical
electrowinning process

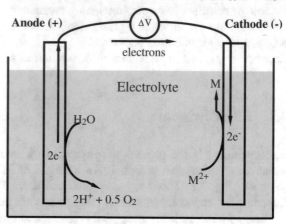

industrial setting. Potential and current density are related to thermodynamics and
application parameters. Potential and current density are affected by solution and
other resistances as well as the deposition area. The current density is a direct
measure of the reaction kinetics.

In the case of electrochemical reactions, the rate of reaction may be increased by
several orders of magnitude [1] by applying the proper voltage. Thus, in electrowinning, the voltage plays an important role in determining the overall reaction
rate. There are, however, limitations imposed by the
solution media. Solution limitations, such as mass
transport discussed in Chap. 4, often prevent the
practical application of high voltages.

The effect of potential is shown in Fig. 7.3 for
copper electrowinning. Note that the anodic reaction
occurs for the half-cell reaction with the highest

> *Electrowinning consists
> of applying a potential or
> voltage and an associated
> current between a positive
> (anode) and negative
> (cathode) electrode pair.*

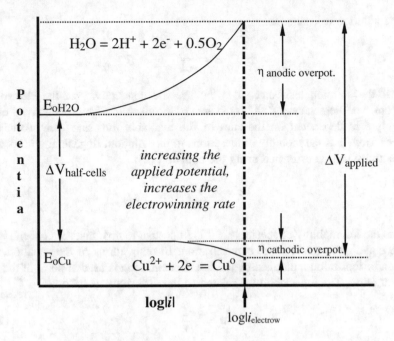

Fig. 7.3 Schematic diagram for electrowinning (E versus Log$|i|$)

equilibrium potential. The anodic reaction is the decomposition of water to hydrogen ions and oxygen. The cathodic reaction is the reduction of metal to its metallic state. If the applied potential is greater than the difference between the two half-cell reactions, electrodeposition of the metal will occur. The rate of electrodeposition depends on the applied potential and the associated electrochemical reaction kinetics. A significant overvoltage, η, is needed to allow the water decomposition to occur at a reasonable rate. The overpotential for the metal deposition is generally not as large.

The applied voltage is related to the rate of the electrowinning as indicated in the diagram. If the voltage is not greater than the difference between the half-cell voltages plus the reaction overpotentials and solution and contact resistance voltage drops, no electrowinning occurs. The applied potential can be expressed mathematically as:

$$\Delta V_{applied} = \Delta V_{half-cells} + \Delta V_{cathodicoverpot.} + \Delta V_{anodicoverpot.} + \Delta V_{soln+cont+misc.}$$

$$(7.1)$$

which is more commonly written using different terms as:

$$V_{applied} = E_{anodic} - E_{cathodic} + \eta_{anodic} + \eta_{cathodic} + IR_{solution} + IR_{other} \qquad (7.2)$$

in which E is the specified half-cell potential, I is the overall current, η is the overpotential, and R is the resistance of the medium specified ($V = IR$—Ohm's Law).

The solution resistance is:

$$R = \frac{d}{\sigma A} \tag{7.3}$$

R is the solution resistance. σ is the specific conductivity of the electrolyte (Ω^{-1} cm^{-1} or Siemen cm^{-1}). "A" is the area of the electrode. The conductivity of an electrolyte is dependent on the sum of the available ions and their associated charges as well as the mobility of these ions in the solution. In general, the specific conductivity can be estimated using the equation [2]:

$$\sigma = F \sum_i z_i c_i u_i \tag{7.4}$$

u_i is the ion mobility of species "i". It is important to note that the concentration in the equation is the free ion concentration. In other forms of the equation, the fraction of ionization multiplied by the salt concentration is used in place of the ion concentration. The ion mobility is related to the diffusivity as follows:

$$u = \frac{zFD}{RT} \tag{7.5}$$

Example 7.1 Calculate the mobility of hydrogen ions in a solution in which it has a diffusivity of 9.31×10^{-5}cm^2/sec.

$$u_j = \frac{zFD}{RT} = \frac{(1)\left(96485\frac{C}{\text{mole}}\right) 9.31 \times 10^{-5}\frac{\text{cm}^2}{\text{sec}}}{8.314\frac{J}{\text{mole K}} 298K \frac{VC}{J}} = 3.63 \times 10^{-3}\frac{\text{cm}^2}{V \text{ sec}}$$

Utilization of this equation requires ion diffusivity or mobility data. The accompanying table (Table 7.1) provides important ion mobility, molar conductivity (λ), diffusivity, and charge information for common ions.

The diffusivity and mobility are functions of viscosity and temperature. A common equation relating diffusivity to viscosity and temperature is the Stokes-Einstein relation [2]:

$$D = \frac{kT}{6\pi r\mu} \tag{7.6}$$

Viscosity is a strong function of temperature. Correspondingly, the diffusivity or mobility of ions changes significantly with temperature. [5]

$$\mu(Pa \cdot \text{sec}) = 2.414 \times 10^{-5}\left(10^{\left(\frac{247.8 K}{T-140 K}\right)}\right) \tag{7.7}$$

Table 7.1 Dilute solution ion data table for aqueous media at 298 K (based on data in references [3] and [4]). (Note that a Siemen, S, is equivalent to the inverse of an Ohm.)

Ion	$|z|$	$|u|(cm^2\ sec^{-1}\ V^{-1})$	λ (S·cm²/equiv.)	D (cm²/s)
H^+	1	3.625×10^{-3}	349.82	9.31×10^{-5}
Na^+	1	0.519×10^{-3}	50.11	1.33×10^{-5}
K^+	1	0.762×10^{-3}	73.52	1.96×10^{-5}
Li^+	1	0.401×10^{-3}	38.69	1.03×10^{-5}
NH_4^+	1	0.761×10^{-3}	73.4	1.95×10^{-5}
Cu^{2+}	2	0.560×10^{-3}	54	0.72×10^{-5}
OH^-	1	2.050×10^{-3}	197.6	5.26×10^{-5}
Cl^-	1	0.791×10^{-3}	76.34	2.03×10^{-5}
$CH_3O_2^-$	1	0.424×10^{-3}	40.9	1.09×10^{-5}
NO_3-	1	0.740×10^{-3}	71.44	1.90×10^{-5}
SO_4^{2-}	2	0.827×10^{-3}	79.8	1.07×10^{-5}
HSO_4^-	1	0.520×10^{-3}	50	1.33×10^{-5}

The resistance in an electrowinning cell is inversely related to conductivity. Thus, in order to have low resistance, the conductivity of the solution must be high. Consequently, salts and acids with high degrees of ionization and high mobility are desired. Sulfuric acid is a common acid in electrolyte solutions. Although sulfuric acid ionizes readily only to HSO_4^- and H^+, these ions have good molar conductivity. Other acids such as HCl and HNO_3 ionize readily and have very good ion molar conductivities. Ions that form strong neutral complexes such as acetic acid have low molar conductivities. Strong acids are also desirable from a mobility perspective because H^+ ions are much more mobile than most ions. As discussed in connection with ion mass transport, the fraction of current carried by individual ions is related to the transference number. The transference number is directly related to ion mobility.

> *Electrical resistance in an electrowinning cell is based on the electrolyte conductivity.*

Another factor that influences the resistance in the cell is the distance between the anode and cathode. This distance is usually only a few centimeters. The combination of increasing deposit thickness, edge strips, shorting, and mechanical harvesting requirements generally results in distances greater than 2 cm. However, shorter distances reduce solution resistance.

Other resistances can form in association with electrode contacts. These resistances can be caused by corrosion of contact surfaces, dirt, salt, etc.

In copper electrowinning, the difference between the cathodic ($E^\circ = 0.34$ V) and anodic ($E^\circ = 1.23$ V) potentials is approximately 0.9 V. The anodic overpotential is usually 0.2–1.0 V. The cathodic overpotential is usually around 0.1 V. The solution voltage drop is on the order of 0.1 V. The other voltage drops across connectors and wires are around 0.05 V. Thus, the applied potential is usually between 1.5 and 2.5 V for copper electrowinning [6]. The applied potential for zinc electrowinning is often between 3.0 and 3.7 V due to the much lower zinc reaction potential ($E° = -0.76$ V) and higher current densities, which result in higher overpotentials.

The purity of the cathode product is dependent on the solution species concentrations. Purity is also dependent on the applied potential. Impurity metals with potentials higher than the desired metal (more noble) deposit preferentially. The rate of impurity metal deposition depends on the applied potential and the impurity's electrochemical kinetics parameters. Figure 7.4 illustrates the relationship between applied potential and impurity deposition. Figure 7.4 shows three applied potential regimes. The first applied potential option is a low voltage. A low voltage may be sufficient to deposit only the more noble metal. An optimum applied voltage will result in the desired metal as well as some more noble metal. A high applied potential will result in rapid deposition of the desired metal along with some of the less noble metal and some more noble metal. Avoidance of more noble metal impurities requires removal prior to electrowinning.

More noble metal impurities will deposit according to Butler-Volmer and mass transport kinetics. Most often, impurity deposition occurs at low concentrations at the limiting current density. The deposition of lead in zinc during zinc

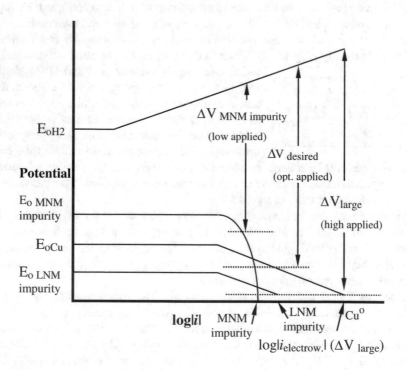

Fig. 7.4 Possible effect of less noble metal (LNM) and more noble metal (MNM) impurities on copper electrowinning current

electrowinning is a good example of this scenario. In such cases the impurity fraction can be calculated using the limiting current density. The following equation is for lead in zinc, which can also be applied to other metals:

$$C_{Pb,in,Zn} = \frac{i_{l,Pb} n_{Zn} A_{w,Pb}}{i_{total} \frac{\beta}{100} n_{Pb} A_{w,Zn}} \tag{7.8}$$

In the case of zinc electrowinning the presence of more noble impurities such as lead allow for the determination of the boundary layer thickness. Correspondingly, the boundary layer thickness can be calculated as:

$$\delta = \frac{n_{PbPb}^2 F D_{Pb} C_{b,Pb} n_{Zn} A_{w,Pb}}{C_{Pb,in,Zn} i_{total} \frac{\beta}{100} A_{w,Zn}} \tag{7.9}$$

Example 7.2 If the concentration of lead in a zinc cathode is 10 ppm in a cell operating at 395 A/m^2 and 89% efficiency and the associated diffusivity of Pb^{2+} ions is 1×10^{-5} cm^2/sec, the concentration of Pb^{2+} ions in solution is 0.133 mg/l or 6.4×10^{-7} mol/l, calculate the boundary layer thickness.

$$\delta = \frac{(2)\frac{296485\,\text{Coul}}{\text{mole}}\left(\frac{1\times10^{-5}\,\text{cm}^2}{s}\right)\frac{6.4\times10^{-7}\,\text{mol}}{1000\,\text{cm}^3}\left(\frac{207.2\,\text{g}}{\text{mol}}\right)}{(10\times10^{-6})395\frac{A}{\text{m}^2}\frac{\text{Coul}}{A}\frac{1\,\text{m}^2}{10,000\,\text{cm}^2}\frac{89}{100}2\left(\frac{65.4\,\text{g}}{\text{mol}}\right)} = 0.0111\,\text{cm}$$

As impurity concentrations are decreased, the associated potentials decrease. Decreased concentrations also result in lower limiting current densities. The rate of deposition is often at the mass transport limiting current density. More noble impurities are particularly problematic for zinc electrowinning. The potential for zinc deposition is much lower than other common impurity metals such as iron, lead, and cadmium. Even sub-ppm levels of contaminants such as lead can result in significant contamination levels in the cathode as demonstrated in the boundary layer example. Consequently, in the case of zinc electrowinning electrolyte, more noble metal impurities are commonly removed using zinc powder. More noble metal impurities cement on the zinc and cause the zinc to dissolve. Cementation will be discussed in a subsequent section.

> *Impurities electrodeposit if the potential and electrolyte composition are not properly controlled.*

Impurity metals that have lower potentials (less noble) than the desired metal are less troublesome. However, if the applied voltage is high, less noble metal may codeposit with the desired metal, making it less pure. In copper electrowinning, less noble metals such as iron generally do not codeposit because of their much lower potentials. However, as the less noble metal concentration increases, the associated potential increases, increasing the possibility of codeposition.

In copper eletrowinning, ions such as bismuth, arsenic, and antimony are common impurity ions with potentials near that of copper (standard potentials [activities = 1]: $BiO^+ + 2H^+ + 3e^- \leftrightarrow Bi^o + H_2O$, $E_o = 0.32$ V; $HAsO_2 + 3H^+ + 3e^- \leftrightarrow As^o + 2H_2O$, $E_o = 0.25$ V; $SbO^+ + 2H^+ + 3e^- \leftrightarrow Sb^o + H_2O$, $E_o = 0.21$ V [7]). Consequently, they present impurity problems if their concentrations or the applied potential become too high. Other less noble metals such as nickel and iron are usually found in abundance. However, iron and nickel do not deposit at the cathode because of their low potentials (-0.25 and -0.44 V, respectively). Normally, the concentrations of the common species in copper electrowinning are: $H_2SO_4 = 140$–220 g/l; Cu = 35–50 g/l; As = 0.3–5 g/l; Sb = 0.1–0.4 g/l; Bi = 0.1–0.4 g/l; Ni = 2–20 g/l; Fe = 0.5–10 g/l [6, 8].

Iron can be a troublesome acceptor/donor of electrons. Ferric iron can accept electrons provided at the cathode. The electrons used to reduce ferric ions are wasted. The current used for reactions other than the desired reaction reduce current efficiency. Iron can also be a corrosive nuisance. Its presence is often controlled. However, high levels of iron can often be tolerated without significant adverse effects. The presence of some iron may be beneficial to producing smooth copper cathode sheets.

Other common impurities are often above acceptable levels. Consequently, removal of such impurities is needed. More noble metals such as gold, platinum, and silver are rarely found at significant levels, except in precious metal operations.

Most copper electowinning is performed using insoluble lead-antimony or lead-calcium anodes [8]. Anodes made with precious metal coatings are promising alternatives [9]. Dimensionally stable anodes (DSA) that contain precious metal coatings, reduce overvoltages and eliminate lead.

Cathode starting sheets are usually made of stainless steel. Metal is deposited on the starting sheets and mechanically stripped after 4–10 days of plating.

Cathode quality is more difficult to control in zinc electrowinning than it is in copper electrowinning. The standard zinc potential (-0.763 V) is lower than many of the impurity metals that can be found in zinc electrolyte solutions. Despite the difficulty, 80% of the primary zinc production is derived by electrowinning [10]. Typical conditions are: $H_2SO_4 = 125$–175 g/l; Zn = 50–90 g/l; anodes are lead (99.25%) silver (0.75%); Temperature = 30–38 °C; current density = 320 A/m^2;

Undesirable reactions such as hydrogen ion and ferric ion reduction consume current and contribute to electrowinning inefficiency.

applied voltage = 3.2 V; energy consumption = 3400 kWhr/tonne [10]. One important challenge in zinc electrowinning is hydrogen evolution. The low potential for zinc electrowinning necessitates some hydrogen evolution from water decomposition. Thus, hydrogen gas removal is important for safety reasons. Minimizing hydrogen evolution is important to current efficiency. Most of the loss in current efficiency in zinc electrowinning is associated with water decomposition at the cathode. Trace levels of impurities such as Ni, Co, Cu, Ge, and Sn are particularly problematic for zinc because they codeposit with zinc and enhance hydrogen evolution.

The keys to having high purity metal at the cathode are proper control of the applied potential and impurity levels. The applied potential needs to be high enough to achieve optimal deposition rates. However, less noble metal deposition must be avoided. Also, reactions such as hydrogen evolution need to be avoided. Consequently, as illustrated in Fig. 7.10, the potential must be adjusted accordingly.

Impurity levels are usually controlled by "bleeding". Bleeding is the process of removing a small fraction of solution for impurity removal. Generally, in copper electrowinning the "bleed" solution is sent to a liberator cell. Much of the remaining copper is electrolytically removed in the liberator cell. After the passing through the first liberator cell, the bleed solution is often sent to a second liberator cell. Excess copper is removed along with some impurities in this step. In many applications metals are precipitated from bleed solutions. Other impurity removal steps such as ion exchange or solvent extraction are also used. The solution can then be recycled. Two reasonable presentations of impurity control are found in

> *Electrolyte impurities are removed by "bleeding" or removing some electrolyte when impurity levels reach unacceptable levels. Bled electrolyte is replaced with more pure electrolyte.*

References [11] and [12]. It should be noted that many operations employ different methods of impurity removal. Well-run copper electrowinning operations produce cathode copper of 99.99% purity.

A variety of additives are used in electrowinning. Leveling agents such as guar or modified starch are used to control cathode surface quality. Antimony is used in zinc electrowinning to counteract the polarizing effect of additives such as glue. Dissolved cobalt (100–200 ppm) is added to extend the life of lead anodes and reduce the overvoltage. Other additives are often used to suppress acid misting associated with oxygen gas bubbles bursting at the surface. Plastic balls are also used to suppress acid misting. Electrical shorting between electrodes is detected using gauss meters and infrared cameras. Shorting is manually corrected by breaking short-causing nodules or realigning electrodes.

The energy requirements for electrowinning operations are enormous. Energy is expressed as:

$$\text{Energy} = tP = EIt \tag{7.10}$$

P is the power. E is the applied potential, t is the time, and I is the current. Energy divided by the mass deposited based on Faraday's law results in:

$$\frac{\text{Energy}}{\text{Mass}} = \frac{EIt}{\frac{ItA_w}{nF}} = \frac{nFE}{A_w} \tag{7.11}$$

However, this equation assumes 100% efficiency. In other words it assumes all electrons are used to deposit metal. In practical electrodeposition some of the electrons are used for other reactions. Thus, the mass of metal deposited is related to the current efficiency. The current efficiency, β, is the percentage of current used for metal deposition. Consequently, the energy (kWhr/tonne) needed to electrowin the metal is expressed as:

$$\text{Energy}(\text{kWhr/tonne}) = \frac{E(n)\,26{,}800}{Aw\left(\frac{\beta(\%)}{100}\right)} \tag{7.12}$$

> *Energy consumption in electrowinning is based primarily on cell voltage, metal properties and current efficiency.*

Note that tonne represents kg the metric tonne (1000). Copper electrowinning generally consumes between 1300 and 2600 kWhr/tonne for the typical values of applied potentials (1.5–2.4 V) and efficiencies (85–98%).

Other important parameters for electrowinning operations include the rate of deposition and current efficiency. The mass rate of deposition is calculated by:

$$R_{\text{dep.}} = \frac{I\beta A_{\text{w}}}{100nF} = \frac{i\beta A_{\text{w}}A}{100nF} \tag{7.13}$$

In terms of mass deposited, the equation is:

$$m_{\text{dep}} = \frac{iAt\beta A_{\text{w}}}{100nF} \tag{7.14}$$

The current efficiency, β is given as:

$$\beta(\%) = \frac{m_{\text{dep}}nF}{ItA_{\text{w}}}(100\%) \tag{7.15}$$

m_{dep} is the mass deposited on the cathode.

Current efficiency is affected significantly by parasitic reactions in some systems. Parasitic reactions are unwanted reactions that consume current. In zinc electrowinning, hydrogen evolution is the main parasitic reaction. Hydrogen evolution on pure zinc is very slow due to a very low equilibrium exchange current density. However, codeposited impurities such as Ni, Co, Cu, Ge, and Sn can significantly increase hydrogen evolution on zinc. In copper electrowinning, ferric ion reduction and oxidation are problematic. The effect of parasitic reactions on the current efficiency can be calculated. An alternative formula for current efficiency based on parasitic reaction current density is:

$$\beta(\%) = 100\frac{i_{\text{total}} - i_{\text{parasitic}}}{i_{\text{total}}} \tag{7.16}$$

Example 7.3 Consider the example of copper electrowinning at 300 A/m². If under the electrowinning conditions, the parasitic reaction of ferric reduction occurs at its limiting current density. Thus, assuming an approximate diffusivity of ferric ion, which is actually found as ferric sulfate, of $6 \times 10^{-6} cm^2/s$, a boundary layer thickness of 0.013 cm, and a concentration of 2 g of ferric ions per liter (0.0358 mol/l) leads to an estimated limiting current density (neglecting electromigration) of:

$$i_l = \frac{nFDC_b}{\delta} = \frac{(1)\left(96,485 \frac{Coul}{mol}\right)\left(6 \times 10^{-6} \frac{cm^2}{s}\right)\left(\frac{0.0358 \, mol}{1000 \, cm^3}\right)}{0.013cm}$$

$$i_l = 0.001594 \frac{Coul}{cm^2s} \frac{A}{\frac{Coul}{s}} \frac{10,000 \, cm^2}{m^2} = 15.94 \frac{A}{m^2}$$

Substitution of the limiting current density for the parasitic current density leads to:

$$\beta(\%) = 100 \frac{i_{total} - i_{parasitic}}{i_{total}} = 100 \frac{300 - 15.94}{300} = 94.7\%$$

Conventional metal electrowinning is performed using flat metal sheets. Electrowinning is generally performed well below the limiting current density. Current densities above about half of the limiting current density often result in rough deposits. Rough deposits often contain anomalous growth. Often the anomalous growth is in the form of nodules. Nodules are generally hemispherical growths that can be large enough to contact an adjacent anode, causing an electrical short. Current densities usually vary between 200 and 600 A/m^2 for commercial tank houses. Tank house is the term used to describe large electrowinning facilities. Tank houses usually contain hundreds of individual tanks or cells.

Each cell typically contains around 45–50 cathodes and 46–51 anodes. Cathodes often consist of 316L stainless steel sheets attached to a conductive support bar on which the sheet hangs during operation. Stainless steel cathodes also contain polymeric edge strips (Fig. 7.5).

> *Commercial electrowinning cells are commonly made of polymer/concrete composite and often contain 45-50 cathodes and 46-51 anodes that are at least 1 meter tall and 1 meter wide.*

These strips prevent growth around the sides, thereby facilitating stripping. Figure 7.6 shows commercial electrodes with edge strips after copper plating. Some stainless steel cathodes contain polymeric or wax bottom strips to facilitate dual sheet harvesting. Other stainless steel electrodes have a tapered bottom to facilitate a fold of the two faces. Cathodes and anodes are often 1–1.5 m in length and 1 m wide. Each electrode has a total surface area of 2–3 m^2. Electrodes are generally placed with a 2–4 cm gap between electrodes. The electrodes are placed in an alternating pattern. An illustration of a cathode sheet and a set of electrodes in a cell is shown in Fig. 7.5. Figure 7.7 shows a view of sets of electrodes in adjacent commercial cells. Figure 7.8 shows a view of an electrowinning tank house full of cells. Each of the cells has a hood to collect evolving acid mist. The mist is the

top view of cell with alternating anodes and cathodes

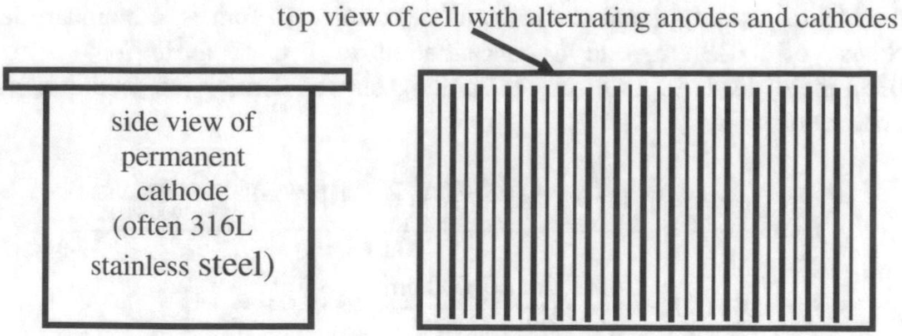

side view of
permanent
cathode
(often 316L
stainless steel)

Fig. 7.5 Side view of permanent cathode and top view of electrochemical cell with alternating anode and cathode sheets. Bold lines represent cathodes

Fig. 7.6 Picture of copper cathode electrodes with yellow edge strips

result of oxygen bubbles that burst at the surface. The oxygen bubbles are created at the anode. Plastic balls are often used to suppress the acid misting as seen in Fig. 7.7.

Other methods of electrowinning are also utilized for metal production. In some cases metal is electrowon at high current densities to produce metal particles. In other situations spouted bed electrochemical cells are used. Metal is also produced as a continuous foil for some applications. Other metal recovery circuits use high surface area electrodes. High surface area electrodes are ideal for recovery from

Fig. 7.7   View of sets of electrodes in adjacent electrowinning cells along with small plastic balls

Fig. 7.8   View inside an electrowinning tank house with cells covered by hoods used to collect off gasses

dilute solutions. Metal is also recovered using spiral sheet electrodes known as Swiss rolls. Other types of cells that utilize a variety of mass transport and high surface areas are also used.

Gold electrowinning is commonly performed using steel wool cathodes and punched stainless steel plate anodes. General conditions include 2–4 V, 100–400 g Au/l input, 1–10 g Au/l output, 0.5–2% NaCN, 0.5–2% NaOH, and 50–90 °C with very low effciencies. [13] Cells can be operated to produce a precious metals sludge that is easier to refine than precious metal plated steel wool. Cyanide is destroyed at the anode (see Chap. 9 for more details and reactions), and hydroxide and hydrogen gas are produced at the cathode in addition to metal reduction.

7.2   Electrorefining

The electrorefining of metals is similar to electrowinning. However, electrorefining utilizes the– desired metal of intermediate purity (95–99.5%) as the anode.

> *Electrorefining is used to purify metal such as smelted copper to achieve desired high purity levels.*

Electrorefining consists of electrolytically dissolving the desired metal from the anode. The dissolved metal is electrodeposited on the cathode as shown in Fig. 7.9. Electrorefining results in deposited metal of higher purity than the anode. Much of the high purity copper produced in the world is produced by electrorefining. The anode for electrorefining is made from copper smelting operations.

Electrorefining is less energy intensive than electrowinning. The potential difference between the two half-cell reactions is zero in electrorefining. Consequently, no energy is consumed to overcome differences in half-cell potentials (see Fig. 7.10) In addition, the normally large water hydrolysis overpotential is eliminated. The overpotentials for dissolution and plating of the metal are generally small. The net outcome of these factors in copper electrorefining is a five–tenfold decrease in the energy requirement. The applied voltage is generally between 0.2 and 0.35 V. The current efficiency is often above 98%.

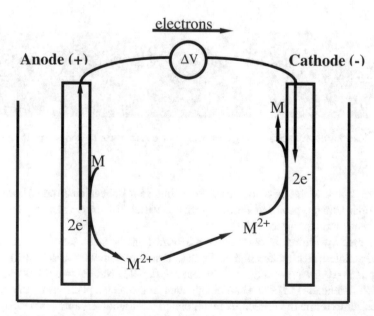

Fig. 7.9 Schematic diagram of the electrorefining process

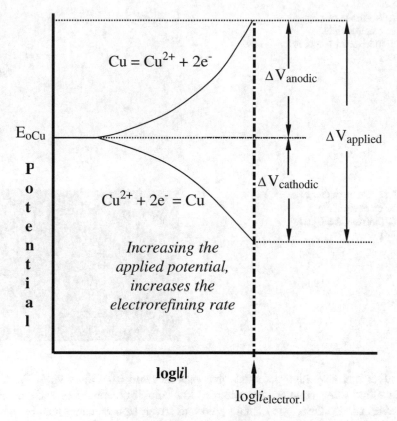

Fig. 7.10 Schematic diagram of the relationship between electrorefining potentials and the absolute value of the logarithm of the current density or rate of electrorefining per unit area

The anodes used in electrorefining consist of blister copper with its inherent nickel, lead, bismuth, antimony, and arsenic impurities. These impurities are often recirculated through anode scrap. In the case of antimony and arsenic they may be intentionally added. The arsenic and antimony play a beneficial role in producing better cathodes [14]. The presence of impurities leads to the formation of inclusions within the copper matrix as shown in Fig. 7.11. As the anode dissolves, many of these inclusions, which are dominated by copper oxides, also dissolve. However, some of the inclusions do not dissolve. Inclusions that do not dissolve form anode slime as shown in Fig. 7.12. Slime is a term used to describe very fine particles. Slime tends to settle to the bottom of cells. Unsettled slime can become incorporated into cathodes. Some of the slime consists of gold

> *Electrorefining consumes a fraction of the energy needed for electrowinning because the cell voltage is relatively small.*

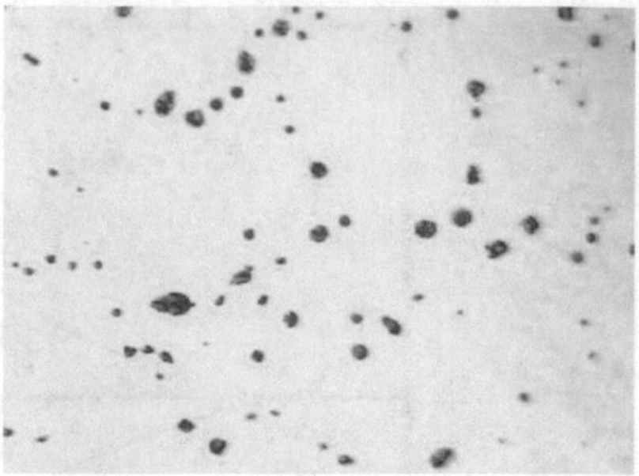

Fig. 7.11 Anode micrograph
(copper electrorefining
×400), (horizontal length is
200 μm)

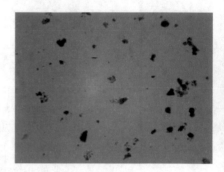

Fig. 7.12 Slime micrograph
(copper electrorefining
×400), (horizontal length is
200 μm)

and silver that have high potentials that prevent them from dissolving. Thus, the anode slime often contains undesirable, low-solubility impurities such as lead. However, anode slimes also contain gold and silver. Consequently, anode slime is recovered and processed to recover the desirable contents and treat the undesirable material.

Gold electrorefining is often performed at 60 °C in hydrochloric acid solution (80–100 g/l) with 80–100 g/l dissolved gold at 800 A/m^2. [13] Gold anodes of 99.6% purity are often used. Deposition from the $AuCl_4^-$ species is targeted because $AuCl_2^-$ species disproportionates to form $AuCl_4^-$ and metallic gold, which forms undesirable sludge particles [13].

Reagents such as thiourea and glue are used in most refineries to create better cathode surfaces. Based upon information in the literature [15, 16] it appears that the glue preferentially adsorbs onto dendrites that extent out from the main surface, thereby retarding their growth. Glue also increases nucleation, which reduces roughness [17]. Reduced growth of protruding copper dendrites results in a smoother, more dense cathode surface. Thiourea is believed to enhance the formation of nuclei at the cathode surface, allowing higher current densities and lower overpotentials [16].

7.3 Cementation or Contact Reduction

Cementation of metals has been practiced for centuries to recover dissolved metals from the aqueous phase. The basic principle of cementation is contact reduction. In other words the electrons from a less noble metal are given up to a more noble dissolved metal. The electron exchange is made as the more noble metal ion contacts the less noble metal surface. This interaction results in reduction of the dissolved metal to the metallic state. The other consequence of the interaction is dissolution of the metal that is less noble. This process is economically practiced on a small scale for copper recovery using iron scrap metal. It is more commonly practiced for precious metal recovery using zinc powder. Zinc is used commercially to recover gold in the Merrill-Crowe process. The Merrill-Crowe process is performed in cyanide solution. The associated potential verus pH diagram for zinc and gold in a cyanide solution is shown in Fig. 7.13. Zinc powder is used commercially to purify zinc electrowinning electrolytes.

> *Cementation is used on a smaller scale than electrowinning to recover metals such as gold or to purify solutions such as zinc electrolyte.*

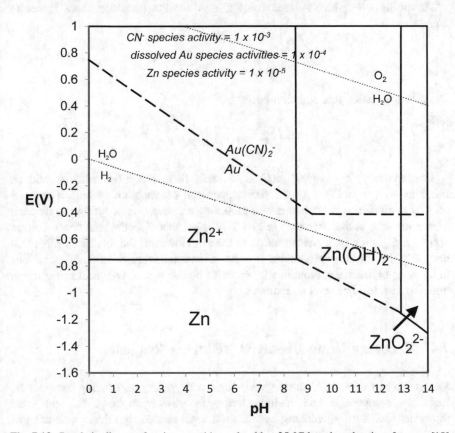

Fig. 7.13 Pourbaix diagram for zinc, cyanide, and gold at 25 °C based on data in references [18]

The Merrill-Crowe process for gold recovery is generally performed near ambient conditions. Gold content in Merrill-Crowe process solutions is often between 0.5 and 10 ppm [13]. Treatment by this process generally requires a clarified solution that is deaerated. Deaeration reduces zinc losses and unwanted reactions. Zinc powder is added at 5–30 times the stoichiometric requirement for Au and Ag [13]. The precipitation of gold onto zinc is usually completed in a few minutes with removal efficiencies greater than 99% [13].

Contact reduction requires electron transfer and mass transport. Mass transport is usually rate limiting for contact reduction. As discussed in Chap. 4, mass transport limited flux can be expressed mathematically as:

$$J = -k_l C_b \tag{7.17}$$

This equation can be rearranged to:

$$\frac{dn}{dt}\frac{1}{A} = -k_l C_b \tag{7.18}$$

Assuming a one-liter volume basis allows for the transition from moles to concentration:

$$\frac{dC_b}{C_b} = -A_{(perliter)} k_l dt \tag{7.19}$$

After integration this equation becomes:

$$\ln\left(\frac{C_b}{C_o}\right) = -A_{(perliter)} k_l t \tag{7.20}$$

Consequently, by plotting $\ln(C_b/C_o)$ versus time, the relationship should be linear as shown in Fig. 7.14 for the copper-iron cementation system. The mass transfer coefficient can be determined assuming a constant area during the reaction from plots such as that presented in Fig. 7.14. Note that C_b is the bulk concentration at time t. C_o is the bulk concentration at time 0. However, the relationship is often not linear. Nonlinearity can be due to either a mass-transfer limitation of metal ions to the reaction interface. Nonlinearity can also be due to area enhancement through tree-like, dendritic product structures.

7.4 Recovery Using Dissolved Reducing Reagents

Metals can be recovered from solution using a wide variety of reducing compounds. Reducing compounds can include hydrogen gas, hypophosphite, and even formaldehyde. Ideally, reducing agents should not reduce significant quantities of

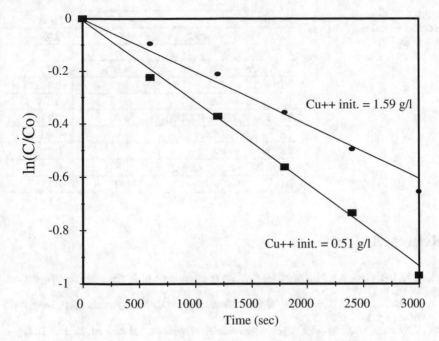

Fig. 7.14 Plot of $\ln(C/C_o)$ versus time for the copper/iron cementation system obtained using data from Reference [19]

water to hydrogen gas. The production of hydrogen gas poses safety challenges and consumes energy. Avoiding hydrogen evolution restricts effective use of reducing agents to more noble metals. Metals such as gold, silver, copper, and nickel are commonly plated using reducing agents.

One effective reducing agent used in the metals industry is hydrogen gas. Hydrogen gas is used to recover some metals in autoclaves where it is possible to achieve high partial pressures of hydrogen gas at elevated temperatures to expedite the recovery process. The resulting metal consists of relatively pure particles.

> *Metals can be recovered as powders or coatings from solution using reducing reagents.*

Other applications for reduction include electrodeposition in the microelectronics industry as well as the manufacturing of coatings, which are discussed in more detail later in the text. A general set of electrochemical reactions is shown in Appendix G. Some examples of metals that can be reduced by solution species and related reactions are shown in Table 7.2. Reactions with high potentials can be driven in the reduction direction by reactions with lower potentials. Values in Table 7.2 are based on standard potentials, which are often different than potentials in applications. The data in the table show that most metals can be reduced from ions to metals using BO_3^{3-}. Gold ions can be reduced from solution using compounds as simple as oxalic acid, $(COOH)_2$.

Table 7.2 Selected
electrochemical reactions and
potentials

Reaction	E_{rxn} (V)
$AuCl_4^- + 2e^- \leftrightarrow AuCl_2^- + 2Cl^-$	0.93
$Fe^{3+} + e^- \leftrightarrow Fe^{2+}$	0.77
$Cu^{2+} + 2e^- \leftrightarrow Cu$	0.34
$Ni^{+2} + 2e^- \leftrightarrow Ni$	-0.250
$Co^{+2} + 2e^- \leftrightarrow Co$	-0.277
$2H_2CO_3 + 2H^+ \leftrightarrow (COOH)_2$	-0.39
$Cd^{+2} + 2\,e^- \leftrightarrow Cd$	-0.403
$Fe^{+2} + 2e^- \leftrightarrow Fe$	-0.410
$Au(CN)_2^- + e^- \leftrightarrow Au + 2CN^-$	-0.57
$BO_3^{3-} + 7H_2O + 7e^- \leftrightarrow BH_4 + 10OH^-$	-0.75

References

1. J.O.M. Bockris, A.K.N. Reddy, *Modern Electrochemistry*, vol. 2, (Plenum Press, New York, 1970), p. 1143
2. J.O.M. Bockris, A.K.N. Reddy, *Modern Electrochemistry*, vol. 1 (Plenum Press, New York, 1970), pp. 373–382
3. J. Newman, K.E. Thomas-Alyea, *Electrochemical Systems*, 3rd edn. ed. by N.J. Hoboken (John Wiley, 2004), p. 284
4. A.J. Bard, L.R. Faulkner, *Electrochemical Methods: Fundamentals and Applications* (John Wiley and Sons, 2001) p. 67
5. G. Elert. *Viscosity. The Physics Hypertextbook.* https://Hypertextbook.com. http://hypertextbook.com/physics/matter/viscosity/. Retrieved 14 Sept 2010
6. A.K. Biswas, W.G. Davenport, *Extractive Metallurgy of Copper*, 2nd edn. (Pergamon Press, Elmsford, 1980), pp. 326–328
7. A.K. Biswas, W.G. Davenport, Extractive Metallurgy of Copper, 2nd edn. (Pergamon Press, Elmsford, 1980). p. 299
8. P.M. Tyroler, T.S. Sanmiya, D.W. Krueger, S. Stupavsky, Copper electrowinning at INCO's copper refinery, in *The Electrorefining and Winning of Copper*, ed. by J.E. Hoffmann, R.G. Bautista, V.A. Ettel, V. Kudryk, R.J. Wesely, (TMS, Warrendale, 1987), pp. 421–435
9. M. Moats, K. Hardee, C. Brown, Mesh-on-lead anodes for copper electrowinning. JOM **55** (7), 46–48 (2003)
10. E. Jackson, *Hydrometallurgical Extraction and Reclamation* (Ellis Horwood Limited, Chichester, 1986), pp. 217–221
11. T. Shibata, M. Hashiuchi, T. Kato, Tamano refinery's new processes for removing impurities from electrolyte, in *The Electrorefining and Winning of Copper,* ed. by J.E. Hoffmann, R.G. Bautista, V.A. Ettel, V. Kudryk, R.J. Wesely, (TMS, Warrendale, 1987), p. 99–116
12. K. Toyabe, C. Segawa, H. Sato, Impurity control of electrolyte at sumitomo niihama copper refinery, in *The Electrorefining and Winning of Copper*, ed. by J.E. Hoffmann, R.G. Bautista, V.A. Ettel, V. Kudryk, R.J. Wesely, (TMS, Warrendale, 1987), pp. 117–128
13. J. Marsden, I. House, *The Chemistry of Gold Extraction*, 2nd edn. (SME, Littleton, 2006), pp. 367–461
14. V. Baltazar, P.L. Claessens, J. Thiriar, Effect of arsenic and antimony in copper electrorefining, in *The Electrorefining and Winning of Copper*, ed. by J.E. Hoffmann, R.G. Bautista, V.A. Ettel, V. Kudryk, and R.J. Wesely, (TMS, Warrendale, 1987), pp. 211–222
15. A.K. Biswas, W.G. Davenport, *Extractive Metallurgy of Copper*, 2nd edn. (Pergamon Press, Elmsford, 1980), pp. 312–313

16. K. Knuutila, O. Forsen, A. Pehkonen, The effect of organic additives on the electrocrystallization of copper, in *The Electrorefining and Winning of Copper*, ed. by J.E. Hoffmann, R. G. Bautista, V.A. Ettel, V. Kudryk, R.J. Wesely, (TMS, Warrendale, 1987), pp. 129–143

17. M.L. Free, R. Bhide, A. Rodchanarowan, N. Phadke, Evaluation of the effects of additives, pulsing, and temperature on morphology of copper electrodeposited from halide media. ECS Trans. **2**(3), 335–343 (2006)

18. N.P. Finkelstein, The chemistry of the extraction of gold from its ores, in *Gold Metallurgy on the Witwatersrand*, ed. by R.J. Adamson, (Cape and Transvaal Printers, Ltd., Cape Town, 1972), pp. 284–351

19. J.D. Miller, Cementation, in *Rate Processes of Extractive Metallurgy*. ed. by H.Y. Sohn, M.E. Wadsworth (Plenum Press, New York, 1979), pp. 197–244

Problems

(7-1) What is the hourly *rate* of copper production in an electrowinning operation with 50,000 cathodes (each with 1 m^2 on *each* side) operating with a current density of 300 Amps/m^2 assuming 96% efficiency? ($n = 2$).

(7-2) What is the energy consumption per kilogram of zinc in an electrowinning operation with a current density of 250 Amps/m^2 assuming 95% efficiency, $n = 2$, and a voltage drop of 3.0 V across the cell? (2.589 kWhr/kgZn).

(7-3) If industrial electrical energy costs 5 cents per kWhr, how much does the energy cost to produce one kilogram of copper in a copper refinery operating with a voltage drop of 250 mV and 98% efficiency? (answer: $0.0108/ Kg Cu).

(7-4) The following data were obtained in a copper cementation test:

Conc (g Cu/l)	Time (min.)
0.72	20
0.52	40
0.38	60

Determine the mass transfer coefficient assuming a 1 cm^2 area for cementation, and an initial copper concentration of 1.00 g/l.

(7-5) Determine the current efficiency in an operation that deposits 200 metric tons of nickel per day using a current of 8,500,000 amps. ($n = 2$) (answer = 89.5%).

(7-6) Calculate the solution resistance (in ohms) and voltage drop associated with 1 m^2 electrodes spaced 3 cm apart that operate at 300 Amps/m^2 in a solution that has a specific conductivity of 0.8 Ω^{-1} cm^{-1}. (Recall that $E = IR$).

Chapter 8
Metal Utilization

> *Metals are used in aqueous media in many ways that include pipes and vessels as well as batteries and fuel cells. Their behavior in applied settings is very important.*

Key Chapter Learning Objectives and Outcomes:
Understand the ways in which metals are used in aqueous media
Understand how batteries and fuel cells operate on hydrometallurgical principles
Understand the application of hydrometallurgy in electroforming and electromachining
Know basic forms of corrosion

8.1 Introduction

Metals can be manufactured and are often utilized in aqueous media. Metals are commonly used in non rechargeable alkaline batteries. They are also used in most types of rechargeable batteries. Metals are the backbone of several types of aqueous-based fuel cells, which are used to generate electricity more efficiently than traditional gas turbine units. Metal parts can be manufactured in aqueous media by many techniques. Utilization of metals in aqueous media results in corrosion. The importance of aqueous processes involving the production and use of metals is well known.

8.2 Batteries

Batteries utilize the coupling of two or more electrochemical reactions that occur spontaneously. Batteries operate in the reverse direction to electrowinning. In electrowinning, electrochemical reactions are forced to proceed against the natural thermodynamic tendency. Electrowinning requires energy input. In batteries, electrical energy is extracted using separate conducting electrodes. The associated reactions proceed due to free energy in the system. Batteries convert chemical free energy to electrical energy.

© The Minerals, Metals & Materials Society 2022
M. L. Free, *Hydrometallurgy*, The Minerals, Metals & Materials Series,
https://doi.org/10.1007/978-3-030-88087-3_8

8.2.1   Primary Batteries (Non-rechargeable)

Batteries that are not rechargeable are primary batteries. Common types include zinc-manganese dioxide and zinc-silver oxide batteries. Most commercial primary batteries contain an alkaline electrolyte. However, in the case of batteries involving lithium, the electrolytes are nonaqueous. Nonaqueous media are needed to avoid water hydrolysis at the low potentials associated with lithium. Nonaqueous batteries will not be covered in this text.

Primary batteries cannot be recharged because of irreversible reactions or water hydrolysis. Some primary batteries have reactions that cannot be reversed repeatedly. Consequently, they are not used as rechargeable batteries. Other primary batteries have reactions with high or low potentials that are close to water hydrolysis potentials. Recharging aqueous batteries near water hydrolysis potentials leads to oxygen and/or hydrogen production. The production of hydrogen and oxygen can present safety concerns. In addition, the production of oxygen and hydrogen reduces the energy efficiency of recharging.

8.2.1.1   Zinc-Manganese Dioxide Alkaline Cells

The most prevalent primary batteries are the traditional alkaline, dry cell batteries that are based upon reactions involving manganese dioxide at the cathode and zinc metal at the anode. The half-cell reactions are:

Cathodic reaction:

$$2MnO_2 + 2H_2O + 2e^- \leftrightarrow 2MnO(OH) + 2OH^- \quad E_o = 0.132 \qquad (8.1)$$

Anodic reaction:

$$Zn(NH_3)_2^{2+} + 2H_2O + 2e^- \leftrightarrow Zn + 2OH^- + 2NH_4^+ \quad E_o = -1.17 \qquad (8.2)$$

And the overall reaction is written as:

$$Zn + 2MnO_2 + 2NH_4^+ \leftrightarrow 2MnOOH + Zn(NH_3)_2^{2+} \quad E_{ocell} = 1.30 \text{ V} \qquad (8.3)$$

The term dry cell is misleading. Dry cell electrodes are placed in a wet paste. The paste consists of particles such as flour and starch as well as aqueous electrolyte. The electrolyte consists of around 28% ammonium chloride, 16% zinc chloride, and the balance water [1]. The particles immobilize the electrolyte. Thus, the particles retain water and prevent leakage. Consequently, the battery acts as a "dry" cell from a handling point of view. The cathode consists of fine particles of carbon (acetylene black) and manganese dioxide. The anode consists of zinc in the form of foil or fine particles [1]. The normal overall cell voltage is usually around 1.6 V. However, this voltage varies depending on the electrolyte composition.

8.2.1.2 Zinc-Silver Oxide Batteries

Zinc-silver oxide batteries utilize the following half-cell reactions:

Cathodic reactions:

$$Ag_2O + H_2O + 2e^- \leftrightarrow 2Ag + 2OH^- \quad E_o = 0.346 \text{ V} \tag{8.4}$$

$$2AgO + H_2O + 2e^- \leftrightarrow Ag_2O + 2OH^- \quad E_o = 0.571 \text{ V} \tag{8.5}$$

Anodic reaction:

$$Zn(OH)_2 + 2e^- \leftrightarrow Zn + 2OH^- \quad E_o = -1.239 \text{ V} \tag{8.6}$$

Note that two possible cathodic reactions are given, which lead to two overall reactions:

$$Zn + Ag_2O + H_2O \leftrightarrow Zn(OH)_2 + 2Ag \quad E_{ocell} = 1.585 \text{ V} \tag{8.7}$$

$$Zn + 2AgO + H_2O \leftrightarrow Zn(OH)_2 + Ag_2O \quad E_{ocell} = 1.810 \text{ V} \tag{8.8}$$

The zinc-silver oxide batteries utilize an alkaline electrolyte. They can be recharged and used as secondary batteries under proper conditions.

8.2.2 Secondary Batteries (Rechargeable Batteries)

Batteries that can be readily recharged are known as secondary batteries. The most common type of secondary battery is the lead acid battery. The lead acid battery is used primarily in automobiles. Other types include nickel cadmium, metal hydride, and lithium ion batteries.

8.2.2.1 Lead-Acid Batteries

Lead acid batteries have been widely used in automobiles for many years. Lead acid batteries have high charge storage capacities, high rechargeability, low cost, and low maintenance. Lead-acid batteries consist of high surface area lead plates. These plates are at least partially coated with porous layers of either lead dioxide (anode) or lead sulfate (cathode). The associated chemical reactions are:

Cathodic reaction:

$$PbO_2 + H_2SO_4 + 2H^+ + 2e^- \leftrightarrow PbSO_4 + 2H_2O \quad E_o = 1.636 \text{ V} \qquad (8.9)$$

Anodic reaction:

$$PbSO_4 + 2H^+ + 2e^- \leftrightarrow Pb + H_2SO_4 \quad E_o = -0.295 \text{ V} \qquad (8.10)$$

Overall cell reaction:

$$PbO_2 + 2H_2SO_4 + Pb \leftrightarrow 2PbSO_4 + 2H_2O \quad E_{ocell} = 1.931 \text{ V} \qquad (8.11)$$

These reactions take place in a sulfuric acid electrolyte. As seen from the reactions, the lower potential reaction is reversed during discharge. Both half-cell reactions consume acid. Consequently, the state of the cell charge can be analyzed based upon

> *Lead acid batteries are a very significant part of the battery industry because of reliability and low cost.*

the acidity. The acidity is related to the specific gravity of the solution. Thus, many lead acid battery testing devices are based on specific gravity.

8.2.2.2 Nickel-Cadmium Batteries

Nickel-cadmium rechargeable batteries were commonly found in cordless power tools. These batteries require low maintenance and provide reasonable electrical power output. The cadmium in these batteries creates an environmental issue. Thus, battery disposal is an important concern with these batteries. However, the low maintenance, reasonable power density (10–35 Wh/kg for sealed cells [2]), consistent voltage, high overcharge tolerance, low natural discharge rate, and long life attributes of this system made it one of the most common types of rechargeable batteries [2, 3]. One disadvantage to nickel-cadmium batteries is charge memory. These batteries need to be fully discharged prior to recharging. Otherwise, they tend to loose their charging capacity. The associated electrochemical reactions are [2]:

Cathodic reaction:

$$NiO(OH) + H_2O + e^- \leftrightarrow Ni(OH)_2 + OH^- \quad E_o = 0.490 \text{ V} \quad [\text{Ref. 3}] \qquad (8.12)$$

Anodic reaction:

$$Cd(OH)_2 + 2e^- \leftrightarrow Cd + 2OH^- \quad E_o = -0.809 \text{ V} \quad [\text{Ref. 3}] \qquad (8.13)$$

Overall reaction:

$$Cd + 2NiO(OH) + 2H_2O \leftrightarrow Cd(OH)_2 + 2Ni(OH)_2 \quad E_{ocell} = 1.30 \text{ V} \quad [\text{Ref. 3}]$$
$$(8.14)$$

Nickel-cadmium batteries are made in sealed and unsealed varieties. The sealed versions require no maintenance. However, sealed versions do not perform as well as their vented counterparts. The electrolyte often consists of an aqueous solution of potassium hydroxide (20–28 wt %) and some lithium hydroxide (1–2 wt %) [2]. Most sealed nickel-cadmium batteries can be recharged several hundred times [2].

8.2.2.3 Metal Hydride Batteries

Metal hydride batteries are widely used in rechargeable batteries for electronic devices and electric vehicles.

Metal hydride batteries are very similar to nickel-cadmium batteries except that the cathodic reaction involves the reaction of metal and water to form a metal hydride as indicated by the reactions shown:

Cathodic reaction:

$$NiO(OH) + H_2O + e^- \leftrightarrow Ni(OH)_2 + OH^- \quad E_o = 0.490 \text{ V} \quad [\text{Ref. 4}] \quad (8.15)$$

Anodic reaction:

$$M + H_2O + e^- \leftrightarrow MH + OH^- \quad E_o = -0.828 \text{ V} \quad [\text{Ref. 4}] \quad (8.16)$$

Overall reaction:

$$MH + NiO(OH) \leftrightarrow M + Ni(OH)_2 \quad E_o = 1.318 \text{ V} \quad [\text{Ref. 4}] \quad (8.17)$$

The discovery in the 1960s that some metal alloys, of classes AB_2 and AB_5 (A is a rare earth element such as lanthanum and B is a transition metal such as nickel) can absorb over a thousand times their own volume of hydrogen [5] allowed for the development of metal hydride batteries. The anodic reaction at the positive electrode is also used in the nickel-cadmium batteries. Nickel hydroxide is designed to be the limiting reagent. Therefore, overcharging results in oxygen production rather than hydrogen. Excess oxygen can react with the metal hydride to form water and pure metal. This reaction prevents pressure build-up. Metal hydride batteries are capable of slightly higher power output than nickel-cadmium batteries. Metal hydride batteries can be recharged more than 1,000 times with proper care. Power densities approach 70 Wh/kg [5]. Metal hydride batteries, however, suffer from significant natural discharging. Metal hydride batteries utilize an electrolyte similar to that in nickel-cadmium batteries.

8.2.3 General Battery Information

Many of the modern electrodes used in batteries are made by a variety of manufacturing techniques. In some cases electrode plates are cast and rolled. Some electrodes are made by mixing a desired powder with a removeable diluent powder such as NH_4CO_3. The mix is pressed into shape and sintered. The sintering process vaporizes the diluent powder. The final product is very porous. Other electrodes are made by depositing the desired metal or alloy on a fibrous network. Deposition takes place by electroless, chemical vapor, or electroplating deposition techniques.

Battery application voltages often necessitate cell stacking. Often application potentials are 3, 6, or 12 V. These voltages are too large for normal electrochemical cells. Consequently, single cells are stacked in alterating layers. Each layer is separated by porous polymer sheets or meshes. Electrodes are connected to the opposing electrode of the adjacent stack. The cell stacks create a greater potential output. A 9-V battery requires 6 alternating stacks of typical 1.5 V cells.

Battery performance is based on charge storage density, discharge current density, the duration of discharge, cycle life, and natural discharge rate. As shown in Fig. 8.1 the rate of discharge and charging are potential dependent. As the rate of discharge increases, the discharge potential decreases. It approaches zero at the natural corrosion or reaction rate. This point represents the short-circuited condition. Similarly, if the potential to charge a battery increases, the rate of charging increases. If the battery is charged at excessive potentials, unwanted reactions occur. In aqueous systems excessive potentials result in oxygen and hydrogen evolution. Oxygen and hydrogen can be removed using catalysts. Thus, catalysts can prevent explosions during charging. It is more efficient, however, to avoid gas production by charging at optimal potentials. As shown in Fig. 8.1, charging is performed at similar or slower rates than discharging to minimize such deleterious reactions.

8.3 Fuel Cells

Fuel cells are electrochemical cells that utilize fuel to produce electricity. Fuel cells are effectively batteries supplied continuously with fuel. The general principle is simple.

The energy available from simple fuels such as methane, methanol, and hydrogen gas is utilized to provide energy for the production of electricity. Hydrogen in the fuel is combined with oxygen in the air to produce water. Other products can also form. Some general reactions for aqueous-based fuel cells are as follows:

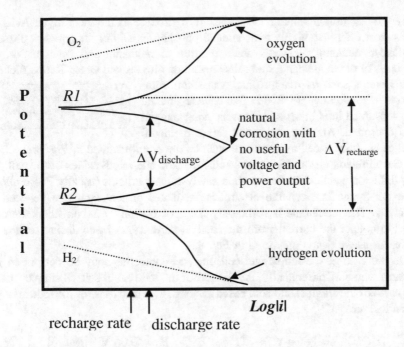

Fig. 8.1 Schematic diagram illustrating the relationships between potentials, current densities and the interrelationship between various reactions. *R1* and *R2* represent the anodic and cathodic battery reactions, respectively, and O_2 and H_2 represent the respective oxygen and hydrogen reactions involving water hydrolysis. Note that the solid, curved lines represent the net electrochemical output, whereas the dashed lines that are associated with the oxygen and hydrogen reactions represent the theoretical output of these reactions that is normally masked by the battery half-cell reactions except at the extreme potential portions of the diagram as indicated

Cathodic Reactions:

$$O_2 + 4H^+ + 4e^- \leftrightarrow 2H_2O \quad E_o = 1.23 \text{ V} \quad (\text{pH } 0) \tag{8.18}$$

$$2H_2O + O_2 + 4e^- \leftrightarrow 4OH^- \quad E_o = 1.23 \text{ V} \quad (\text{pH } 0) \tag{8.19}$$

Anodic Reactions:

$$4H^+ + 4e^- \leftrightarrow 2H_2 \quad E_o = 0.000 \text{ V} \quad (\text{pH } 0) \tag{8.20}$$

$$2H_2O + 2e^- \leftrightarrow 2OH^- + H_2 \quad E_o = 0.000 \text{ V} \quad (\text{pH } 0) \tag{8.20}$$

$$CO_2 + 6H^+ + 6e^- \leftrightarrow CH_3OH + H_2O \quad E_o = 0.031 \text{ V} \quad (\text{pH } 0) \tag{8.21}$$

There are five basic types of aqueous-based fuel cells. They are acidic, alkaline, direct methanol, polymer electrolyte, and redox fuel cells. Alkaline fuel cells utilize a very concentrated solution of potassium hydroxide as the electrolyte. Acid fuel cells most often use highly concentrated phosphoric acid in the aqueous phase.

Direct methanol fuel cells use a mixture of methanol and water vapor. The products of direct methanol fuel cells are carbon dioxide and water. Redox fuel cells use systems such as titanium and vanadium complexes in a redox couple. Redox fuel cells regenerate the consumed ions in separate vessels using oxygen and hydrogen. Alkaline, acid, and direct methanol fuel cells utilize a cell arrangement such as the one illustrated in Fig. 8.2.

> *Fuel cells operate like continuous-feed batteries.*

Gas diffusion electrodes are used for cathodic and anodic sides of the cell. The gas diffusion electrodes often contain catalyst nanoparticles (usually Pt-Ru alloys). The membrane between the electrodes facilitates proton exchange between the catalysts. The membrane is commonly made of Nafion®. Carbon fibers are often used to collect the current from the catalyst particles. A basic design for the gas diffusion electrodes is presented in Fig. 8.3.

In the case of the redox fuel cell, the gas contacting may be performed in a separate vessel. Consequently, the design for the redox fuel cells is simpler than in the cases of the alkaline and acid based systems. The reactions for the redox couples often include:

$$Ti(OH)^{3+} + H^+ + e^- \leftrightarrow Ti^{3+} + H_2O \quad E_0 = 0.06 \text{ V} \quad [\text{Ref. 6}] \qquad (8.22)$$

Fig. 8.2 Schematic diagram of aqueous based fuel cell

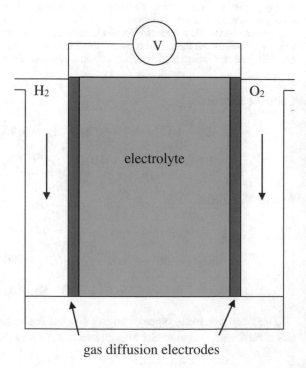

gas diffusion electrodes

Fig. 8.3 Schematic illustration of the gas diffusion electrode portion of many aqueous based fuel cells. Note that the catalyst particles, current collection system, and structural backing of the electrode, which often consists of graphite-based material, are not shown

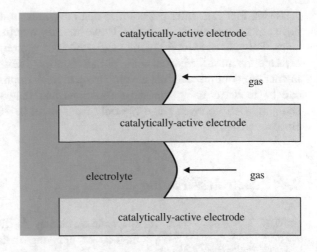

$$VO_2^+ + 2H^+ + e^- \leftrightarrow VO^{2+} + H_2O \quad E_o = 1.00 \text{ V} \quad [\text{Ref. 6}] \qquad (8.23)$$

The corresponding regeneration reactions in separate vessels are:

$$2Ti(OH)^{3+} + H_2 \leftrightarrow 2Ti^{3+} + 2H_2O \quad (\text{using } H_2 \text{ over Pt-Al}_2O_3 \text{ at } 60\,^\circ C) \quad [\text{Ref. 6}] \qquad (8.24)$$

$$2VO^{2+} + 0.5O_2 + H_2O \leftrightarrow 2VO_2^+ + 2H^+ \quad (\text{in nitric acid at } 75\,^\circ C) \quad [\text{Ref. 6}] \qquad (8.25)$$

There are many advantages and disadvantages to fuel cells. The alkaline fuel cells have high power density (0.3 A/cm²). They can operate at reasonable temperatures (25–200 °C). They achieve reasonable efficiency (36% @ 0.15 A/cm²). However, they suffer from extreme intolerance to carbon monoxide and carbon dioxide. Thus, additional gas purification is a prerequisite [6].

Acidic fuel cells (usually a concentrated 95% phosphoric acid solution) achieve high power density (0.3 A/cm²). Acidic fuel cells have better efficiency (40% @ 0.15 A/cm² if off-gas heat is utilized) than alkaline cells. Acidic fuel cells also have much greater tolerance to carbon dioxide and carbon monoxide. However, acidic fuel cells operate at moderate temperatures (200 °C). They also require pressurized vessel equipment [6].

Direct methanol fuel cells have reasonable power density (0.2 A/cm²). They operate at modest temperatures (80 °C). They utilize methanol rather than hydrogen gas as a fuel source. However, hydrogen gas can be produced from methane or methanol by steam reforming process. Direct methanol fuel cells are slow and have low efficiency (30%). They are also prone to methanol leakage to the membrane-separated anode compartment [7].

Redox fuel cells offer possibilities that are similar to the other cells. These cells have a simplified electrode design. However, they require two separate regeneration steps. Polymer electrolyte membrane (PEM) or proton exchange membrane fuel cells are the most likely fuel cells for automobiles. Their popularity will increase as the durability of the polymer electrolyte medium is improved. The cell performance can be modeled using the same equations and diagrams used for battery discharging. Cell voltages range typically from about 0.4–0.9 V depending upon the current density.

8.4 Electroless Plating

> *Electroless plating is the electroplating of metal without an applied voltage and a counter electrode. It consists of metal deposition onto substrates using a dissolved reducing agent.*

Electroless plating is a common and effective method of applying metallic coatings to surfaces without the use of an externally applied potential. This type of plating has inherent advantages and disadvantages. Advantages include the ability to coat intricate parts, obtain excellent surface quality, and utilize substrate versatility. Disadvantages are high reagent consumption, limited metal and alloy selection, and low efficiency.

Electroless plating occurs simultaneously on all exposed surfaces. Deposition occurs at approximately the same rate everywhere—including interior surfaces. Consequently, the coating distribution is often more uniform in irregular sections than is typically achieved by traditional electroplating.

Proper solution control, combined with appropriate additives, results in excellent coating quality. In fact, electroless coating controll is sufficient to allow for its use in the microelectronics industry, where deposition standards are stringent and coated features are complex.

Electroless deposition can be used to coat materials such as plastic and ceramic objects as well as metal surfaces. In order for deposition to occur, an electrically conductive surface must be present. Naturally insulating materials such as ceramics and plastics must be pretreated with a conductive additive.

In order for electroless deposition to occur, a reduction and oxidation process must occur. The reduction process involves metal reduction from the dissolved state to a metallic state. In contrast, the oxidation process involves the oxidation of a species in solution. Consequently, both the metal ions and reductant species are consumed during the reaction, making replenishment a costly necessity.

Many metals cannot be plated effectively or economically by electroless means. Metals with low potentials are not effectively reduced by most common reducing agents. Also, some low potential metals have potentials below the potential of water hydrolysis reactions. Consequently, water decomposition occurs simultaneously with metal deposition. This makes the process inefficient for some metals.

Electroless plating requires the presence of metal ions, which become reduced. It also requires a suitable reducing agent. Moreover, the reducing agent must have a lower electrochemical potential than the metal ions in solution. Common metals include silver, copper, nickel, and iron. Most reduction takes place using formaldehyde, sodium hypophosphite, hydrazine, formic acid, or dimethylamine borane (DMAB) [8]. Some of the important reactions and their associated potentials are [8]:

$$Cu^{2+} + 2e^- \leftrightarrow Cu \quad E_o = 0.337 \text{ V} \tag{8.26}$$

$$Ni^{2+} + 2e^- \leftrightarrow Ni \quad E_o = -0.250 \text{ V} \tag{8.27}$$

$$HCOOH + 2H^+ + 2e^- \leftrightarrow HCHO + H_2O \quad E_o = -0.0283 \text{ V} \quad \text{(at pH 0)} \tag{8.28}$$

$$HCOO^- + 2H^+ + 2e^- \leftrightarrow HCHO + 3OH^- \quad E_o = -1.070 \text{ V} \quad \text{(at pH 14)} \tag{8.29}$$

$$HPO_3^{2-} + 2H_2O + 2e^- \leftrightarrow H_2PO_3^- + 3OH^- E_o = -1.650 \text{ V (at pH 14)} \tag{8.30}$$

Effective electroless plating generally requires one or more additives to overcome inherent difficulties. Complexing agents such as sodium or potassium tartrate or ethylenediamine tetraacetic acid (EDTA) are often added. Such additives maintain constant, low metal ion availability at intermediate and alkaline pH conditions. There presence prevents precipitation at intermediate pH levels. Solution pH buffers containing phosphates or carboxylates (acetate or citrate) are also used. These buffers enhance pH stability during plating. Stabilizers such as thiourea, oxygen, or 2-mercaptobenzothiazole are added. These additives modify deposits to meet specific deposition needs. Accelerants such as cyanide are often added to enhance the rate of the anodic reaction. The anodic reaction often limits the overall rate of deposition. Brighteners such as glue, gelatin, or ammonium thiosulfate are often added to reduce the grain size of the deposit and minimize anomalous growth.

In some electroless deposition systems such as electroless deposition of copper in microelectronic manufacturing, it is necessary to add activators as well as additional processing steps to ensure acceptable deposits [8–10].

8.5   Electrodeposited Coatings

A wide variety of metal alloy and metal matrix composite coatings can be applied to conductive substrates by electrodeposition. The process is essentially the same as electrowinning or electrorefining. An applied voltage is used to force metal ions to become reduced and deposited at a conductive cathode surface. Coatings can be electrodeposited on small, large, intricate, and simple parts. An example of copper-coated surface used in the microelectronics industry is shown in Fig. 8.4.

The process of electrodeposition can be performed by connecting the desired part as the cathode and an inert counter electrode as the anode. In many small part coating operations a conductive, perforated barrel or basket is rotated in an electrolyte solution. The rotation is performed while in electrical contact with both the small parts inside and the power supply. The parts inside become cathodes. The parts as well as the container are coated with metal. The removed metal ions are often replenished using an anode of similar composition. Alternatively, the metal in solution must be replenished through chemical additions. The reagents produced at the anode surface must be periodically removed.

> *Electrodeposited coatings are commonly applied in metal finishing and electronics industries.*

Electrodeposited coatings often have excellent qualities. If electrodeposition is performed properly, the resulting coatings have excellent adhesion, good hardness, low porosity, and aesthetic appeal. Common coatings include chromium, zinc, nickel, copper, silver, and gold. When appropriate solutions

Fig. 8.4 MRAM chips with thin electrodeposited copper on a silicon wafer

Fig. 8.5 Scanning electron microscope image of the cross-section of an electrodeposited zinc-nickel alloy matrix composite coating containing alumina particles

are used, alloy coatings such as zinc-nickel and iron-nickel can be easily produced at appropriate potentials. Metal matrix composite coatings with conductive and/or nonconductive particles can also be produced by electrodeposition methods as shown in Fig. 8.5.

8.6 Electroforming

Electroforming is the process of producing or forming metal parts by electrodeposition. This process is very similar to producing electrodeposited coatings. The primary difference is that parts, rather than coatings are made. Electroformed parts are made by electrodepositing the desired metal on a conductive mandrel or mold. Often electroformed parts are more than one millimeter thick. Electrodeposited coatings are usually only a few micrometers thick. In order to control the electrolyte composition, electroforming is usually accomplished using anodes of the same composition as the deposited metal. Because electroformed parts must be removed from the original mold or mandrel after production, the mandrels are often made of low melting point alloys, which can be melted out of the electroformed part. Alternatively, they are made of metals such as aluminum than can be chemically removed. Chemical removal is often made using strong basic or strong acid. The dissolution chemicals are formulated to avoid part damage.

Electroforming can result in very intricate parts that have extremely fine surface finishes. The ability to produce high quality finishes on parts is a primary advantage of electroforming. Consequently, it is used for critical optical components such as mirrors and reflectors. Other noteworthy applications for electroformed parts include precision sieves, fuel cell electrodes, printed circuit board foil, apertures for inkjet printers, holograms, digital recording masks, injection-molding equipment, tools, holographic stampers, heat sinks for missile cones, searchlights, jewelry, erosion shields for helicopter blades, components for nuclear energy research, dental implants, and micromechanical devices [11, 12].

The process of electroforming has some inherent advantages and disadvantages. Electroforming allows considerable flexibility in controlling the hardness of the deposit. However, the choice of metals is generally limited to Cu, Ni, Fe, Cr, Pb, Ag, and Au [11]. Parts produced by electroforming are easy to produce on a large scale. The process is slow and the associated production costs are high compared to other methods (30 times the cost of the contained metal) [11].

8.7   Electrochemical Machining

Metal parts are often intricate in design. Metal casting can be used to obtain intricate shapes for many parts. However, it is difficult to achieve uniform properties in all portions of intricate cast parts. In addition, part casting rarely results in surface finishes. Cast part tolerances are often not acceptable for finished products. Thus, machining is an important step in achieving the desired final part tolerances, shapes, and qualities. Mechanical machining is the most common form of machining. However, mechanical machining results in some

> *Electroforming is used to make high precision products such as mirrors for small, niche markets.*

undesirable surface defects and low quality surface finishes. Mechanical machining is extremely difficult to perform on interior sections or intricate shapes. In contrast, electrochemical machining can be used to produce intricate shapes at fast production rates. Electrochemical machining results in outstanding surface finishes without significant surface defects. Consequently, electrochemical machining is a very important metal part manufacturing technique.

The basic concept behind electrochemical machining is the controlled, rapid dissolution of metal. This process uses an applied potential to dissolve the metal. The applied potential is usually between 2 and 40 V depending upon the metal and solution environment. Current densities can approach $1,000,000$ A/m^2. The electrolyte consists of high levels of dissolved salt (around 10%) to facilitate charge transfer. Sodium chloride is the most common salt additive. The electrolyte is either acidic or basic depending upon the metal and its dissolution behavior. Alkaline pH values allow the dissolving metal to precipitate as colloidal particles. Precipitation of dissolved metal minimizes metal deposition on the cathode. Complexing compounds may also be added to minimize metal deposition.

Fig. 8.6 Schematic diagram
of an electrochemical
machining set up

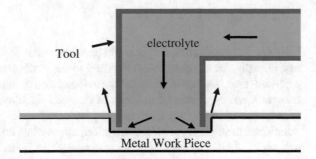

Resistive heating can cause the electrolyte to boil during electrochemical machining. The resistance is maintained at low levels by minimizing the distance between the tool and the work piece. This distance is often only a few percent of one millimeter. However, even with the short separation distance and high salt content, the resistive heating must be controlled.

Electromachining involves rapid metal dissolution using an applied voltage to remove metal in desired areas is done in a machine shop. However, electromachining leaves no mechanical surface damage.

Thus, a high flow rate of electrolyte must be provided. High fluid flow rates provide cooling and metal transport. Fresh electrolyte is, therefore, injected through a hollow, cathodically-biased machining tool at pressures between 30 and 500 PSI [13]. A schematic diagram of the tool and work piece is illustrated in Fig. 8.6.

The combination of rapid fluid flow, high potentials, tight tolerances, and appropriate solution chemistries in electrochemical machining lead to high precision parts (± 15 μm) and smooth surfaces (as low as 0.1 μm) [13].

8.8 Corrosion

Metal corrosion is nearly always an electrochemical process. Consequently, corrosion reactions follow the same principles that were discussed previously. For example, the corrosion potential and rate are determined using a mixed potential analysis. Corrosion rates are often measured using polarization resistance data. Consequently, most of this section will focus on a brief introduction to types of corrosion that are unique to specific scenarios. These types include general corrosion, pitting corrosion, crevice corrosion, galvanic corrosion, intergranular corrosion, environmentally induced cracking, corrosion-assisted wear, microbiologically influenced corrosion and dealloying.

Corrosion is the reverse of metal recovery. Corrosion damage is expensive and challenging to mitigate. Corrosion is a hydrometallurgical process that affects everyone.

8.8.1   Uniform Corrosion

Uniform or general corrosion is common when water is available. Examples include rust on a pipe or a sheet of steel exposed to the atmosphere. A sample of a piece of cast iron that has experienced uniform corrosion is shown in Fig. 8.7. Uniform corrosion can be evaluated in terms of rate and likelihood of occurrence using the principles and techniques discussed in previous chapters. Uniform corrosion can be minimized through proper alloy selection, appropriate coating application, inhibitor use, and cathodic protection. Cathodic protection is based on the electrochemical fundamentals discussed previously.

8.8.2   Pitting Corrosion

Pitting is the process of localized corrosion that results in the formation of pits. An example of pitting corrosion is shown in Fig. 8.8. Metals such as aluminum and stainless steel form stable reaction product films. These films passivate the metal. However, these films are susceptible to pitting if the environment destabilizes the film. Passive films are destabilized at high oxidizing potentials. They are also destabilized by chloride and other ions. Such ions react with the passive layer to form alternative complexes that are less stable and often slightly soluble. Weaknesses in the passive film can form when exposed to high potentials or

Fig. 8.7 Uniform corrosion of cast iron (Ni-Hard 4)

Fig. 8.8 Magnified (10x) surface of a 316L stainless steel plate that experienced pitting in chloride based aqueous media

destabilizing ions. Weak areas can expand and deepen to penetrate the underlying metal. These events occur in localized areas. The localized areas of corrosion experience accelerated corrosion due to the accumulation of dissolved ions. Hydrolysis reactions increase localized acidity and reduce solution resistance. Some of the dominant reactions in a general stainless steel alloy pitting scenario are:

$$FeOOH + Cl^- \leftrightarrow FeOCl + OH^- \qquad (8.31)$$

$$FeCl_2 + 2H_2O \leftrightarrow Fe(OH)_2 + 2HCl \qquad (8.32)$$

$$O_2 + 4H^+ + 4e^- \leftrightarrow 2H_2O \qquad (8.33)$$

$$Fe \leftrightarrow Fe^{2+} + 2e^- \qquad (8.34)$$

Pitting can be reduced by minimizing exposure to chloride ions, reducing the temperature, reducing the oxidation potential in solution, and by the selection of alloys that do not tend to exhibit pitting tendencies.

8.8.3   Crevice Corrosion

Crevice corrosion consists of corrosion in crevices. Crevices between metal pieces or metals and coatings accelerate corrosion. The environment inside the crevice causes the corrosion acceleration. The exposed entry to the crevice has more oxygen than the interior. The crevice interior traps water and concentrates ions. The environment inside the crevice is similar to the environment inside a pit. Consequently, pitting and crevice corrosion share similar mechanisms. Both pitting and crevice corrosion are accelerated by the accumulation of ions. Metal hydrolysis reactions within the localized environment are also important. An example of an aluminum alloy that has undergone a combination of crevice corrosion and pitting is shown in Fig. 8.9. Crevice corrosion can be reduced by designing to eliminate unnecessary crevices. It is also minimized by reducing water accumulation. Better bonding between coatings and metal substrates also reduces crevice corrosion.

8.8.4   Galvanic Corrosion

Galvanic corrosion can occur when two dissimilar metals are electrically connected and exposed to the same corrosive solution. Galvanic corrosion results in preferential corrosion of the less noble metal. Figure 8.10 illustrates the effect of galvanic corrosion. The rate of corrosion of the less noble metal depends upon the environment. Other factors that affect the rate are the cathodic counter reaction, and the

Fig. 8.9 Scanning electron microscope image of the cross section of an aluminum alloy (2024—T6) that experienced a mixture of pitting and crevice corrosion in choride-based aqueous media. Note that the black in the image is the epoxy mount, the dark grey is the corrosion reaction product, and the light grey is the aluminum substrate

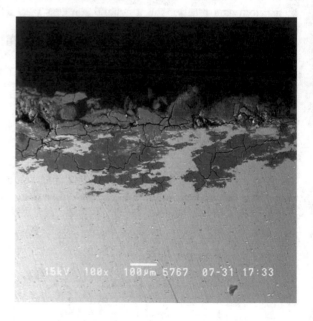

Fig. 8.10 Schematic illustration of galvanic corrosion occurring when two dissimilar metals are in electrical contact

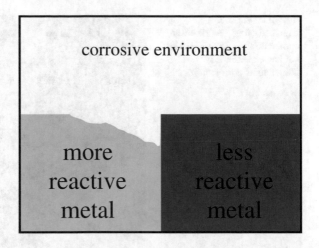

ratio of the noble and less noble metal areas. Increasing the ratio of the noble to less noble metal accelerates the corrosion of the less noble metal.

Galvanic corrosion is also used to protect metals. As an example, zinc metal coatings are used to protect steel substrates. The coating of steel with zinc is generally referred to as galvanization. The coupling of the dissimilar metals reduces the rate of corrosion of the more noble metal. However, the same coupling accelerates the less noble metal corrosion. Galvanic corrosion can be reduced by avoiding electrical coupling of dissimilar metals. Galvanic corrosion is only a problem when coupled dissimilar metals are exposed to the same corrosive environment.

8.8.5 Intergranular Corrosion

Intergranular corrosion results when metals with susceptible grain boundary regions are exposed to corrosive environments. Grain boundaries can become susceptible to corrosion by specific heat treatments. Heat treatments that lead to grain boundary precipitates often sensitize metal to intergranular corrosion. The formation of grain boundary precipitates depletes adjacent areas of important elements. Many metals are susceptible to intergranular corrosion. Stainless steel will be used to illustrate the issues involved for most metals.

In some stainless steels chromium reacts to form iron-chromium carbides along grain boundaries at intermediate temperatures (400–800 °C) and short times (5–100 s). Grain boundary precipitation depletes chromium near the grain boundaries. The process of grain boundary depletion is often referred to as sensitization. Chromium provides resistance to corrosion. Consequently, chromium depleted regions are more susceptible to corrosion. Additives such as titanium or niobium are often added to stabilize some stainless steels (alloys 321, 347, and 348). These

Fig. 8.11 View of a
magnified cross-section of an
aluminum alloy (5083-H131)
that has experienced
intergranular corrosion

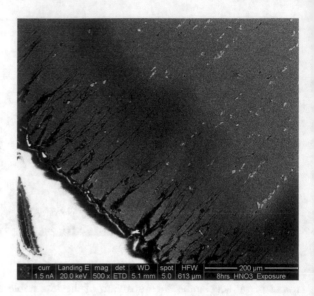

stabilizers remove carbon and minimizing sensitization that can easily occur during welding. In addition, the base metal, grain boundary precipitates, and depleted zones near the grain boundaries have different potentials, making localized galvanic corrosion a potential problem that can accelerate the corrosion. Consequently, intergranular corrosion results in deep penetration along grain boundaries as illustrated in Fig. 8.11. Intergranular corrosion of stainless steels can be reduced by stabilizers and low carbon. Appropriate heat treatments can prevent sensitization. Reducing the corrosivity of the environment can also reduce intergranular corrosion.

8.8.6 *Environmentally-Induced Cracking*

Corrosion often enhances other forms of metal degradation such as cracking. There are three main types of cracking that are associated with or accelerated by corrosion: stress corrosion cracking, corrosion fatigue cracking, and hydrogen embrittlement.

8.8.6.1 Stress Corrosion Cracking

Stress causes metal atoms become more susceptible to corrosion. Stress also tends to rupture passive films. Stress tends to be more pronounced at the tips of cracks. Combining the influence of stress, cracks, and a corrosive environment leads to accelerated crack propagation and part failure. Stress corrosion cracking (SCC) tends to occur predominantly along grain boundaries as shown in Fig. 8.12.

Fig. 8.12 Magnfied image (100x) of an etched piece of low carbon steel from a stress corrosion cracking failure. (The image horizontal distance is approximately 1 mm)

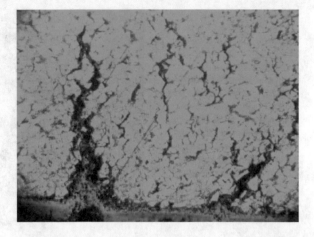

Stress corrosion cracking is reduced by proper alloy selection, minimization of stresses, reducing the corrosivity of the environment, and by utilizing metals with larger grains.

8.8.6.2 Corrosion Fatigue Cracking

Corrosion fatigue cracking occurs when corrosion is combined with metal fatigue to accelerate part failure. An example of corrosion fatigue cracking (CFC) is shown in Fig. 8.13. Only parts with cyclical loading experience fatigue. Cyclical loading of parts with high loads tends to cause existing cracks to grow at significant rates. Crack growth rates are a function of the load level and frequency. Crack growth rates can be increased greatly by corrosion. Corrosion fatigue is reduced by reducing the stress load, increasing the load frequency above a minimum threshold level, and by minimizing the corrosivity of the environment.

Fig. 8.13 Photograph of the cross-section of a 4 cm diameter steel shaft that failed by corrosion fatigue

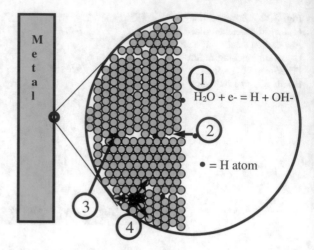

Fig. 8.14 Schematic illustration of the process of hydrogen induced cracking. The numbers represent the steps associated with hydrogen induced cracking or hydrogen embrittlement: (1) hydrogen reduction, (2) hydrogen atom diffusion, (3) hydrogen gas molecule formation, (4) hydrogen gas accumulation and rupture causing bubble formation

8.8.6.3 Hydrogen-Induced Cracking

Metals are susceptible to hydrogen induced cracking. One common way in which hydrogen induces cracking is depicted in Fig. 8.14, and it begins with (1) Reduction of hydrogen at the metal surface; (2) diffusion of hydrogen atoms into the metal, which occurs more readily along grain boundaries; (3) combination of hydrogen atoms to form hydrogen gas molecules, which are too large to diffuse out of the metal as easily as they entered; and (4) accumulation of hydrogen gas molecules to form high pressure gas bubbles that cause metal cracks and failure. Hydrogen induced cracking can be reduced by heating parts to allow hydrogen to diffuse out, reducing exposure to hydrogen ions, and by selecting alloys with larger grains and reduced concentrations of alloying elements.

8.8.7 Corrosion-Assisted Wear

8.8.7.1 Erosion Corrosion

Erosion corrosion is corrosion accelerated by rapid flow of corrosive fluid. The fluid may be only mildly corrosive such as water. An example of a pump impeller that has experienced erosion corrosion is shown in Fig. 8.15. Softer metals are generally more susceptible to erosion corrosion than hard metals. Less noble metals are usually more likely to experience erosion corrosion than more noble metals. It is believed that the rapid flow of fluid over the surface contributes to reduced passive film stability and accelerated corrosion. The process is aggravated when the flow velocity leads to a pressure that is low enough to cause water vapor bubbles to form locally and later collapse. The collapse of these bubbles creates high localized flow

Fig. 8.15 Centrifugal pump
impeller with extensive
erosion corrosion

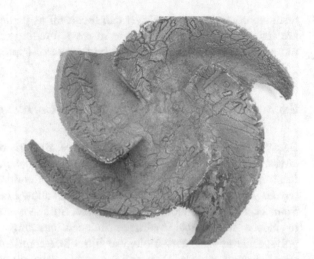

rates. The formation and collapse of these bubbles under these conditions is called cavitation. Erosion corrosion can be reduced by reducing flows and by selecting stronger, more corrosion resistant alloys.

8.8.7.2 Fretting

Fretting is the result of combining corrosion and physical abrasion of a corrosion product layer. An example of fretting of a bolt is shown in the lower part of Fig. 8.16. The upper bolt in the Fig. 8.16 shows signs of corrosion without wear assistance. The lower bolt has been abraded as well as corroded. The abrasion was accelerated by the corrosion. In fretting the part that undergoes corrosion forms a corrosion product layer that is physically removed by contacting surfaces. The

Fig. 8.16 Fretting corrosion
(lower bolt) compared to
uniform corrosion observed
with the upper bolt. Note the
reduction in diameter in the
middle section of the lower
bolt that was caused by
fretting

resulting corrosion product layer can break off as particles. The resulting particles are abrasive and further contribute to wear. Fretting can be reduced by selecting more corrosion resistant alloys and by the use of appropriate lubricants.

8.8.8   Microbiologically Influenced Corrosion (MIC)

Bacteria can affect corrosion in several ways. Bacteria can create a more oxidizing environment by oxidizing species to a higher oxidation state (e.g. ferrous ions to ferric ions). They can lower the concentration of corrosion inhibiting ions such as ferrous ions (lower ferrous ion concentrations allow iron to corrode more rapidly). Some bacteria can reduce sulfate to sulfide. Sulfide is a corrosive ion that also tends to induce more rapid hydrogen induced cracking. Other bacteria can form polysaccharide and other biological films that contribute to the formation of differential aeration cells. Differential aeration cells are areas with different oxygen availabilities that lead to potential differences. Bacteria can also contribute to the formation of ferric oxides and hydroxides. These compounds often form scale structures known as tubercules. Tubercules create a microenvironment similar to crevice corrosion and pitting. MIC can be reduced through control of the environment and through the use of biocides. MIC can also be reduced by the selection of more corrosion resistant alloys.

8.8.9   Dealloying

Dealloying is the process of selective removal of a metal constituent from an alloy. At room temperatures, dealloying beyond a few surface layers occurs by dissolution of the alloy. Dissolution is followed by redeposition of the more noble metal. Another form of dealloying involves removal of a less noble phase that is continuous from the surface. One common form of the first type of dealloying is the dezincification of high-zinc brass alloys. The second form is also known as graphitization. Graphitic cast irons are susceptible to graphitization. This form of dealloying leaves only graphite near the surface. A representation of dealloying is presented in Fig. 8.17. Dealloying is minimized by selecting resistant alloys such as low zinc brass or non graphitic cast iron. It can also be minimized by modifying the solution environment to lower oxidation potentials.

Fig. 8.17 Schematic
representation of the
cross-section of a metal that
has experienced dealloying

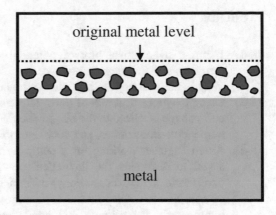

References

1. C.A. Vincent, B. Scrosati, *Modern Batteries: An Introduction to Electrochemical Power Sources*, 2nd edn. (Wiley, New York, 2001)
2. C.A. Vincent, B. Scrosati, *Modern Batteries: An Introduction to Electrochemical Power Sources*, 2nd edn. (Wiley, New York, 2001), pp. 164–166
3. Technical Marketing Staff of Gates Energy Products, Inc., *Rechargeable Batteries: Application Handbook*, (Butterworth-Heinemann, Oxford, 1992), p. 15
4. Technical Marketing Staff of Gates Energy Products, Inc., *Rechargeable Batteries: Application Handbook*, (Butterworth-Heinemann, Oxford, 1992), p. 21
5. A. Vincent, B. Scrosati, *Modern Batteries: An Introduction to Electrochemical Power Sources*, 2nd edn. (Wiley, New York, 2001), pp. 175–180
6. E. Barendrecht, Electrochemistry of fuel cells, Chapter 3, in *Fuel Cell Systems*, ed. L.J.M. J. Blomen, M.N. Mugerwa (Plenum Press, New York, 1993), pp. 73–119
7. K. Scott, W. Taama, J. Cruickshank, Performance of a direct methanol fuel cell. J. Appl. Electrochem. **28**, 289–297 (1998)
8. P. Bindra, J.R. White, Fundamental aspects of electroless copper plating, Chapter 12, in *Electroless Plating: Fundamentals and Applications*, ed. by G.O. Mallory, J.B. Hajdu (American Electroplaters and Surface Finishers Society, Orlando, 1990), pp. 289–395
9. C.J. Weber, H.W. Pickering, K.G. Weil, *Surface Development during Electroless Copper Deposition*, ed. by M. Paunovic, D.A. Scherson. Proceedings of the Third Symposium on Electrochemically Deposited Thin Films, vol. 96–19 (1997), pp. 91–102
10. S. Lopatin, Y. Shacham-Diamond, V.M. Dubin, P.K. Vasudev, B. Zhao, J. Pellerin, *Conformal Electroless Copper Deposition for Sub-0.5 μm Interconnect Wiring of Very High Aspect Ratio*, ed. by M. Paunovic, D.A. Scherson. Proceedings of the Third Symposium on Electrochemically Deposited Thin Films, vol. 96–19 (1997), pp. 91–102
11. M.J. Sole, Electroforming: methods, materials, and merchandise. JOM, 29–35 (June 1994)
12. S.G. Bart, *Historical Reflections on Electroforming, Symposium on Electroforming—Applications, Uses, and Properties of Electroformed Metals*, (ASTM Special Technical Publication No. 318, ASTM, 1962), pp. 172–183
13. A.E. De Barr, D.A. Oliver, *Electrochemical Machining* (MacDonald, London, 1968), pp. 51–67

Problems

(8-1) Using a hypothetical plot of potential versus $\log|i|$, show how battery charging and discharging voltages vary as a function of current density. Why is the charging voltage always higher than the discharging voltage?

(8-2) Using a hypothetical plot of potential versus $\log|i|$, show how a typical fuel cell voltage is related to the output rate. What happens to cell voltage as the output current increases, and what does this change do to the power output?

(8-3) As an engineer working for a company that produces fuel cells, you are asked to determine the theoretical power output at 100 mA/cm^2 for a phosphoric acid hydrogen/oxygen fuel cell system that is designed to operate at room temperature with atmospheric pressures of oxygen and hydrogen in the cathode and anode compartments, respectively in an electrolyte containing 0.5 m H_3PO_4. Assume that the exchange current densities for the cathodic and anodic reactions are both 0.0001 mA/cm^2 and that both reactions have symmetry factors of 0.5. Assume the current efficiency is 100% and that solution and contact resistances are negligible.

(8-4) Describe each of the main types of corrosion discussed in Chap. 8.

Chapter 9
Environmental Issues

> *Environmental issues are often as important as technical and economic feasibility in determining project viability.*

Key Chapter Learning Objectives and Outcomes:
Understand the importance of environmental issues in hydrometallurgy
Understand the historical context of environmental regulations
Understand how hydrometallurgy can be used in an environmentally responsible way
Understand basic technologies for dealing with environmental issues

9.1 Introduction

Over the past few decades, the public awareness of environmental issues has played a dramatic role in shaping environmental law. In the United States, environmental laws have been enacted for more than 100 years as shown in Table 9.1. Since 1970, the number of environmental regulations in the United States has increased. The intent of the regulations is excellent, and the regulations have resulted in cleaner living for all. In most instances the regulations have been implemented over reasonable time periods, allowing industries to meet the regulations without excessive economic hardship.

9.2 United States Environmental Policy Issues

9.2.1 National Environmental Policy Act (NEPA)

The National Environmental Policy Act was established to impose environmental regulations and guidelines to help improve our environment. This policy is considered to be our national charter with respect to environmental policy. The essence of the policy is to ensure environmental issues are considered in federal policy decisions and that the public is informed of such considerations.

© The Minerals, Metals & Materials Society 2022
M. L. Free, *Hydrometallurgy*, The Minerals, Metals & Materials Series,
https://doi.org/10.1007/978-3-030-88087-3_9

Table 9.1 Major environmental legislation in the United States

Year	Law	References
1899	Rivers and Harbors Act	[1]
1924	Oil Pollution Act	[2]
1948	Clean Water Act (Water Pollution Control Act)	[3]
1949	Federal Insecticide Fungicide & Rodenticide Act	[3]
1954	Atomic Energy Act	[1]
1955	Air Pollution Control Act	[1]
1963	Clean Air Act	[1]
1967	Clean Air Act (Amend.)	[1]
1970	Clean Air Act (Amend.)	[1]
1970	National Environmental Policy Act (NEPA)	[1]
1970	Creation of Environmental Protection Agency (EPA)	[1]
1970	Occupational Safety and Health Act (OSHA)	[2]
1972	Clean Air Act (Amend.)	[4]
1972	Clean Water Act	[4]
1972	Federal Insect. Fungi. And Rodenticide Act (Amend.)	[4]
1974	Safe Drinking Water Act	[2]
1975	Hazardous Materials Transportation Act	[4]
1976	Toxic Substances Control Act (TSCA)	[4]
1976	Resource Conservation and Recovery Act (RCRA)	[4]
1977	Clean Air Act (Amend.)	[4]
1977	Clean Water Act (Amend.)	[4]
1978	Federal Insect. Fungicide & Rodenticide Act (Amend.)	[4]
1980	Low-level Radiation Waste Policy Act	[2]
1980	Resource Conservation and Recovery Act (Amend.)	[4]
1980	Comprehensive Env. Response, Compens., & Liab. Act	[4]
1984	Resource Conservation and Recovery Act (Amend.)	[4]
1986	Superfund Amendments & Reauthorization Act (SARA)	[4]
1986	Safe Drinking Water Act (Amend.)	[2]
1986	National Environmental Policy Act (Amend.)	[2]
1986	Compreh. Env. Resp., Compens. & Liab. Act (Amend.)	[2]
1987	Clean Water Act (Amend.)	[4]
1987	Occupational Safety and Health Act (Amend.)	[2]
1990	Clean Air Act (Amend.)	[4]
1990	Hazardous Waste Operations and Emergency Response	[4]
1990	Oil Pollution Act (Amend.)	[2]
1990	Hazardous Materials Transportation Act (Amend.)	[2]
1990	Pollution Prevention Act	[4]

9.2.2 Clean Air Act (CAA)

The most important thrust of the CAA was to authorize the EPA to establish National Ambient Air Quality Standards (NAAQS) for specific pollutants (suspended particles less than 10 μm or PM10, sulfur oxides, nitrogen dioxide, lead, carbon monoxide, hydrocarbons, and ozone). Also, the CAA allowed EPA to set up standards for all new sources of pollution (New Source Performance Standards—NSPS) in addition to the National Emission Standards for Hazardous Air Pollutants (NESHAPS). The air pollutants included

> *Environmental regulations have played a major role in reducing pollution and contamination associated with industrial processes.*

asbestos, benzene, beryllium, inorganic arsenic, mercury, radionuclides, and vinyl chloride. Since 1970, several major amendments have been added. The ammendments make the regulations stricter for a much larger number of pollutants. The amendments also include greater penalties for noncompliance.

9.2.3 Clean Water Act (CWA)

This act established criteria for discharges of water containing pollutants into streams and rivers. Potential new sources of pollutants are required to obtain permits with strict provisions prior to discharge. Violation of the CWA can result in large fines and imprisonment.

9.2.4 Resource Conservation and Recovery Act (RCRA)

This act began as an amendment to the Solid Waste Disposal Act. It is designed to prevent mismanagement of hazardous wastes from source to final disposal. This cradle-to-grave concept of hazardous waste management includes three main elements [5]:

> *RCRA is one of the major environmental regulation acts that influences modern industrial processing involving chemicals.*

- Description of waste and identification of responsible individual
- System for positively tracking hazardous wastes from cradle-to-grave
- Promotion of proper waste management to protect human health and the environment

RCRA regulates solid hazardous materials from generation to disposal. It includes intermediate settings such as storage, transportation, and treatment of hazardous wastes. (Note that a solid waste is defined as anything that has reached the end of its useful life and is to be discarded [6]). As part of this regulatory process, RCRA established a set of guidelines for determining whether or not a material is hazardous. A material is considered to be hazardous if it is ignitable, corrosive, reactive, or toxic. RCRA guidelines are used in the determination. Note that radioactive materials are regulated through the Nuclear Regulatory Commission and the Department of Energy.

There are lists of specific toxic materials. The P-list identifies acutely toxic materials that are more closely regulated. The U-list contains chemicals that are toxic, but not considered to be acutely toxic. RCRA also contains two lists of specific and nonspecific hazardous waste sources. The first of these lists is the K-list which describes specific industrial processes whose wastes are hazardous. The other list, the F-list, identifies several types of hazardous waste streams that are found in a wide variety of industrial processes rather than a specific indus-

> *Materials are commonly classified based on the type of hazard they pose to humans and the environment.*

trial process. If a material is not listed in the P-, U-, K-, or F-list, it may still be considered toxic if after undergoing the TCLP test (see reference [7]), the level of toxic species in solution is greater than the limits established in list D (see Table 9.2). Generally the metal concentrations in the D list vary from 100 ppm for barium to 0.2 ppm for mercury [8]. Finally, there are some exceptions that are excluded from RCRA regulations. These exceptions include sources such as sewer, empty containers, waste pesticides from individual farmers that are disposed of on his/her property, some mineral wastes, and others [9]. It should also be noted that these regulations are subject to change, and interested individuals should always check current regulations from both the state and federal environmental agencies prior to evaluation.

9.2.5 Toxic Substances Control Act (TSCA)

TSCA regulates the distribution of new chemicals (not on the original list of 66,000 substances). Those wishing to distribute new chemicals must submit a Premanufacture Notice (PMN) to the EPA. As part of TSCA, allegations of harmful effects must be recorded for 30 years if they involve humans or 5 years if they involve the environment.

Table 9.2 Inorganic species and concentrations from list D (see Ref. [8])

Arsenic	5.0 ppm
Barium	100 ppm
Cadmium	1.0 ppm
Chromium	5.0 ppm
Lead	5.0 ppm
Mercury	0.2 ppm
Selenium	1.0 ppm
Silver	5.0 ppm

9.2.6   Comprehensive Env. Response, Compens., and Liability Act (CERCLA)

CERCLA, also known as "Superfund" was established following many highly-publicized environmental disasters. CERCLA regulates the release of hazardous substances from hazardous waste sites. It also establishes legal liability and provides for monetary means for clean up. This act targets contaminant removal and site remediation. Clean up is funded by responsible parties as well as potentially responsible parties that can be identified. Because clean up costs are usually extremely high, the liabilities are high for hazardous material disposal, so it is wise to know and follow the rules.

9.2.7   General Water Discharge Regulations

Industries that utilize hydrometallurgical processing often need to discharge process water. Process water enters local streams, lakes, and oceans only after appropriate treatment. Prior to discharge, proper permits must be secured. Discharge permits are negotiated typically between the states and the permitee based upon EPA and state regulations. Most often, discharges into local water resources must meet national drinking water standards as given in Table 9.3. Some exceptions to such regulations are granted periodically depending upon individual company circumstances and local water resource conditions. Often the exceptions deal with surges and overall quantities diecharged. Discharges are generally subject to wet tests involving minnows and/or water fleas that must survive in the solution for a predetermined period of time that is typically 48 h.

Table 9.3 National drinking water standards. *Source* http://www.epa.gov, 2011) primary standards (Maximum Contaminant Level (MCL)

Contaminant	MCL (mg/L)
Antimony	0.006
Arsenic	0.01
Barium	2
Beryllium	0.004
Cadmium	0.005
Chromium (total)	0.1
Copper	1.3
Cyanide (as CN-)	0.2
Fluoride	4.0
Lead	0.015
Inorganic Mercury	0.002
Nitrate (as N)	10
Selenium	0.05
Thallium	0.0005

9.3   Metal Removal and Remediation Issues

Changes in environmental regulations have resulted in the development of many new technologies to *reduce*, *reuse*, and *recycle* materials as well as technologies to remove, stabilize, and alter hazardous materials. The basic approach to removing toxic metals from solution is the same as that applied to metal removal and solution purification discussed in Chaps. 6 and 7. This approach includes ion exchange, precipitation, carbon adsorption, ultrafiltration, and, biosorption.

> *Plant solution discharges are commonly regulated to meet drinking water standards.*

Stabilization of metals is accomplished by forming stable precipitates. Alternatively, more soluble precipitates can be encased in a glass-like slag. This process of glass-like encapsulation is called vitrification. Reducing toxicity of materials is done by converting them to a less toxic form. Electrochemical oxidation in solution, precipitation, or thermal oxidation at high temperatures can be used for toxicity reduction of some compounds.

9.3.1   Metal Removal Technologies

Ion exchange is one of the most widely used methods of removing dissolved toxic substances from solution. Often, ion exchange takes place using a resin, impregnated with a functional group that attracts specific ions. After adsorption occurs, the resin is stripped into a separate solution at a much higher concentration. The stripped resin is then reused for further adsorption as discussed in Chap. 6. The stripping solution, which contains a much higher concentration of the toxic ion(s) in a reduced volume, must then be toxicity reduced or stabilized before disposal.

Biosorption is essentially another form of ion exchange, although in some cases there may be other reactions that distinguish it from ion exchange. Many dead biological species are capable of absorbing significant amounts of toxic ions as shown in Table 9.4. Although Table 9.4 contains the most effective absorbing species, it is evident that biomass is an important material for toxic metal removal. One application of biomass that has been immobilized in a polymeric bead is

> *Metal removal is often performed by precipitation. Ion exchange and adsorption methods are also used.*

Biofix® beads developed by the U. S. Bureau of Mines. These beads contain immobilized biomass that is used as an ion exchange medium [10]. However, once the toxic material is adsorbed, the biomass or stripped solution must be treated.

Materials such as clay, zeolites, activated carbon, and activated alumina can be used to remove dissolved toxic species from solution. Again, the mechanism of toxic species removal is arguably ion exchange, except under unusual conditions. As with the other methods of removal, the material must be detoxified or stabilized before disposal.

Table 9.4 Biosorption Capabilities of Killed Biomass [after Ref. 11]

Metal	Uptake (mg/g)	Biomass type
Ag	86–94	freshwater alga
Cd	215	*Ascophyllum nodosu*
Co	100	*Ascophyllum nodosum*
Cr	118	Bacillus biomass
Cu	152	*Bacillus subtilis*
Hg	54	*Rhizopus arrhizus*
Ni	40	*Fucus vesiculosus*
Pb	601	*Bacillus subtilis*
Pd	436	freshwater alga
U	440	*Streptomyces longwoodensis*

Precipitation is another effective method of removing toxic metal ions from solution. In contrast to the other methods, however, precipitation can be a method of both removing and stabilizing the toxic species. The use of precipitation to stabilize toxic species will be discussed in the next section.

9.3.2   Stabilization and Toxicity Reduction Technologies

Both vitrification and precipitation are methods of stabilizing toxic species. Only precipitation will be discussed here because vitrification involves high temperatures in non aqueous media. Precipitation generally involves cations and anions attracting each other to form neutral compounds that are often nearly insoluble. Insoluble precipitate particles can be separated from solution by gravitational forces or filtration. Coulomb's law states that the attractive force between two point charges is proportional to the multiple of the respective charges. Thus, a cation and anion with double charges, such as Ca^{2+} and SO_4^{2-} will experience an attractive force 4 times greater than an anion/cation pair with single charges such as Na^+ and Cl^- assuming equal separation distances. This general principle suggests that the resulting precipitates of the pair with the double charges ($CaSO_4$) will have a reduced tendency to be soluble relative to the pair with the single charges (NaCl). As might be expected based upon Coulomb's law, the solubility of NaCl is high, whereas $CaSO_4$ has low solubility. In general, the higher the charge of the species and counter ion, the more stable the precipitate will be. There are, however, many very significant exceptions to this rule due to hydration effects, steric factors relating to ion size, specific ion properties, and solution conditions. The data in Table 9.5 illustrate the general trend of greater solubility with smaller charges. The data also show the reduced solubility with higher charges, although there are clearly some exceptions such as the sulfates.

Some metal ions such as As, Se, Cr, W, and Mo have natural valencies of 3, 5, and 7. These valences could make them very capable of forming stable precipitates.

Table 9.5 Solubility Products of Selected Metal Complexes (Log K—complete dissociation from neutral precipitates) (Data calculated from References [12–16])

	CO_3^{2-}	Cl^-	OH^-	NO_3^-	PO_4^{3-}	SO_4^{2-}	S^{2-}
Ag^+	−11.09	−9.76	−2.065	0.227		−4.91	−50.1
K^+	4.15	0.898	11.41	−0.05			12.01
Na^+	0.732	1.58	7.40	1.15		−0.19	12.19
Cd^{2+}	−11.29	−0.440	−14.35				−25.8
Co^{2+}	−9.98		−14.9	7.75		7.36	−21.3
Cr^{3+}			−29.8				
Cu^{2+}	−9.63	3.79	−19.32			2.65	−36.1
Fe^{2+}	−10.68		−15.1			−0.48	−18.1
Fe^{3+}			−38.8		−17.9		
Hg^+	−11.68	−17.60				−5.96	
Hg^{2+}		−15.46	−21.92			−8.19	−52.7
Ni^{2+}	−6.78		−15.83			2.91	−19.4
Pb^{2+}	−13.13	−4.79	−14.38		−54.2	−7.88	−27.5

Their tendency to combine with oxygen results in the formation of oxyanions. These anions are generally MO_4^{3-}, MO^+, MO_2^-, or HMO_4^{2-}, $H_2MO_4^-$, etc. The charge on the complex is usually highest when the oxidation state is highest. In the case of arsenic, the highest charge on the ion complex (−3) occurs when arsenic is in its highest oxidation state (+5). However, in anionic form, it requires a cation, such as a metal cation, for precipitation. As presented in Table 9.6, one of the most stable precipitates for arsenic is ferric arsenate. In ferric arsenate both ions have a charge of 3. The stability of this complex is due, at least in part, to the large electrostatic attraction between the ions with these high charges. The anion and cation have the same charge, which leads to greater stability and more rapid precipitate formation.

Stable precipitates tend to form when the metals are in their highest oxidation states. Thus, it is often useful to oxidize the both the metal anion complex and the metal cation to their highest oxidation states prior to precipitation. Oxidation is often accomplished using oxidants such as hydrogen peroxide, sodium or calcium hypochlorite, bacterial oxidation, ozone, etc. This same approach generally applies to all metals

> *Metal ions are sometimes oxidized prior to precipitation to facilitate more stable precipitates.*

Table 9.6 Solubility products of arsenic complexes with metal ions (log K for complete dissociation) [12, 17]

	Ag^+	Ba^{2+}	Ca^{2+}	Cu^{2+}	Mg^{2+}	Zn^{2+}	Fe^{3+}
AsO_4^{3-}	−22.0	−17.2	−18.3	−34.8	−19.3	−29.1	−19.2
$HAsO_4^{2-}$		−5.1	−3.5	−7.0	−2.9	−6.5	
AsO_2^-		−4.3	−6.8	−11.6		−12.5	

that have stable valencies greater than 3. It also applies to most metals with more than one oxidation state.

Most often, solubility information is presented using solubility products. Solubility produces require additional interpretation in order to arrive at more meaningful numbers. The general form of precipitation is:

$$yM^{n+} + nX^{y-} \leftrightarrow M_yX_n \tag{9.1}$$

where M^{n+} is the metal ion and X^{y-} is the anion. The resulting equilibrium constant can be written as:

$$K = \frac{a_{M_yX_n}}{a_{Mn+}^y \, a_{Xy-}^n} \tag{9.2}$$

The solubility product is the product of the species that dissociate or dissolve from the precipitate, thus:

$$K_{sp} = a_{Mn+}^y \, a_{Xy-}^n \tag{9.3}$$

Thee solubility product assumes that the precipitate M_yX_n is present in solution with an activity of one. If the product of the species activities (or concentrations in dilute solutions) is less than the solubility product, no precipitate will be present, and Eq. 9.1 no longer applies.

If the precipitate is placed in pure water, the presence of X^{y-} and M^{n+} will be dependent upon the stoichiometry of the precipitate. Moreover, a substitution of X^{y-} in terms of M^{n+} can be made using concentration units together with the stoichiometry. When the activity coefficients are equal to one, this substitution $(C_{Xy-} = nC_{Mn+}/y)$ into Eq. 9.3 leads to:

> Metals are commonly precipitated using hydroxides and sulfides.

$$K_{sp} - C_{Mn+}^y \left[\left(\frac{n}{y} \right) C_{Mn+} \right]^n \tag{9.4}$$

Solving Eq. 9.4 for the concentration of residual metal in solution leads to:

$$C_{Mn+} = \left[K_{sp} \left(\frac{n}{y} \right)^{-n} \right]^{1/(n+y)} \tag{9.5}$$

However, it should be noted that this approximation assumes the water into which the precipitate is placed has no other ions present besides OH^- and H^+ and that neither M^{n+} nor X^{y-} has pH dependence. The latter assumption is usually not a good assumption, thus Eq. 9.5 serves only as an approximation for determining saturation concentrations. Equation 9.5 shows the strong dependence upon the species charge as well as the thermodynamic solubility product.

Metal removal and stabilization can also be performed at high temperatures in the presence of air to form relatively stable metal oxides. Generally, the metal is adsorbed on a carrier medium prior to being placed in an incinerator at high temperatures in the presence of air. In addition, metal compounds can be removed and stabilized at high pressures and moderate temperatures in aqueous media using autoclaves.

Metal detoxification is much more difficult to accomplish than stabilization. Most toxic metals are toxic regardless of their oxidation state. Thus, toxic metals are always potential toxins when exposed to harsh chemical environments. Many toxic metals with multiple valencies have varying toxicity levels depending upon the valence.

In contrast to toxic metals, toxic organic compounds such as cyanide, can often be readily detoxified through various oxidation pathways. Cyanide is most often detoxified by first oxidizing it to cyanate, which is much less toxic than cyanide, then to ammonia and carbon dioxide by exposure to low concentrations of H^+ ions in aqueous media as shown in Eqs. 9.6–9.8. The oxidation to cyanate can be accomplished using ultraviolet light, bacteria, ozone, hydrogen peroxide (H_2O_2), sodium hypochlorite (NaOCl), or other oxidizing chemicals. Relevant reactions include:

$$CN^- + OCl^- \leftrightarrow CNO^- + Cl^- \tag{9.6}$$

$$H^+ + OCN^- \leftrightarrow HCNO \tag{9.7}$$

$$HCNO + H_2O \leftrightarrow NH_3(g) + CO_2(g) \tag{9.8}$$

$$CN^- + H_2O_2 \leftrightarrow CNO^- + H_2O$$

Alternatively, cyanide can be destroyed using electrolysis. Cyanide destruction can occur at the anode during electrowinning of precious metals according to the reactions: Fig. 9.1

$$CN^- + 2OH^- \leftrightarrow CNO^- + H_2O + 2e^- \; (E^\circ = 0.79 \text{ V})$$
$$\underline{CNO^- + 2H_2O \leftrightarrow NH_3 + CO_2 + OH^-}$$
$$CN^- + H_2O + OH^- \leftrightarrow NH_3 + CO_2 + 2e^- (E^\circ = 0.75 \text{ V})$$

9.3.3   Recycling of Toxic Entities

Another environmentally responsible way to utilize entities that are toxic is to minimize consumption by maximizing utilization and recovery. Metal plating and finishing facilities often recycle valuable entities in waste streams at the associated

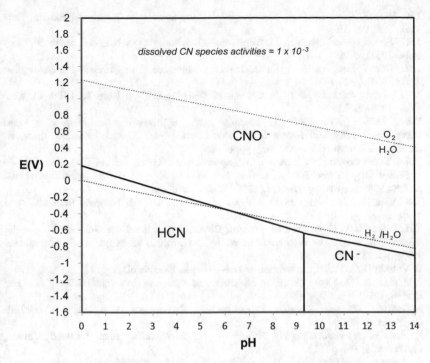

Fig. 9.1 Eh–pH or Pourbaix diagram for cyanide in water based on data in the appendices

facility or they contract with outside companies to recover valuable entities from waste streams using a variety of solution concentration and recovery technologies discussed previously. Acid can be recovered

> *Reducing, reusing, and recycling are common ways of reducing waste.*

from waste streams using ion exchange resin. Toxic ions such as cyanide can also be recovered from waste streams by converting it to hydrogen cyanide, removing it in a gaseous form, then recovering it as free cyanide in alkaline media. Thus, it is important to remember three commonly applied terms in connection with environmental issues surrounding aqueous metal processing, production, and utilization: reduce, reuse, and recycle. Furthermore, the adage that dilution is no solution to pollution should also be remembered and applied.

References

1. J.W. Vincoli, *Basic Guide to Environmental Compliance* (Van Nostrand Reinhold, New York, 1993), p. 11
2. J.G. Speight, *Environmental Technology Handbook* (Taylor and Francis, Bristol, 1996), p. 262

3. J.W. Vincoli, *Basic Guide to Environmental Compliance* (Van Nostrand Reinhold, New York, 1993), p. 19
4. J.W. Vincoli, *Basic Guide to Environmental Compliance* (Van Nostrand Reinhold, New York, 1993), p. 20
5. J.W. Vincoli, Basic Guide to Environmental Compliance (Van Nostrand Reinhold, New York, 1993), p. 110
6. J.W. Vincoli, *Basic Guide to Environmental Compliance* (Van Nostrand Reinhold, New York, 1993), p. 115
7. D.C. Seidel, *Laboratory Procedures of Hydrometallurgical-Processing and Waste-Management Experiments, Information Circular 9431* (United States Department of the Interior, Bureau of Mines, 1995), pp. 63–65
8. United States Government, Resource Conservation and Recovery Act (RCRA) of 1976, Code of Federal Regulations, Title 40, section 261, as amended—55 Federal Regulation 11862, March 1990. (see main government EPA website for updates)
9. J.W. Vincoli, *Basic Guide to Environmental Compliance* (Van Nostrand Reinhold, New York, 1993), pp. 119–121
10. T.H. Jeffers, C.R. Ferguson, P.G. Bennett, Biosorption of metal contaminants from acidic mine waters, in *Mineral Bioprocessing*, ed. by R.W. Smith, M. Misra, (TMS, Warrendale, 1991), pp. 289–298
11. B. Volesky, Z.R. Holan, Biosorption of heavy metals. Biotechnol. Prog. **11**, 235–250 (1995)
12. R.M. Garrels, C.L. Christ, *Solutions, Minerals, and Equilibria* (Jones and Bartlett Publishers, Boston, 1990)
13. E. Jackson, *Hydrometallurgical Extraction and Reclamation* (Ellis Horwood Limited, Chichester, 1986), p. 151
14. E. Jackson, *Hydrometallurgical Extraction and Reclamation* (Ellis Horwood Limited, Chichester, 1986), p. 157
15. E. Jackson, *Hydrometallurgical Extraction and Reclamation* (Ellis Horwood Limited, Chichester, 1986), p. 162
16. D.D. Wagman, W.H. Evans, V.B. Parker, R.H. Schumm, I. Halow, S.M. Bailey, K.L., Churney, R.L. Nuttall, The NBS tables of chemical thermodynamic properties, J. Phys. Chem. Ref. Data, **11**, supplement # 2, National Bureau of Standards, Washington, D.C.
17. R.G. Robins, Arsenic hydrometallurgy, in *Arsenic Metallurgy Fundamentals and Applications*, ed. by R.G. Reddy, J.L. Hendrix, P.B. Queneau, (TMS, Warrendale, 1988), pp. 215–247

Problems

(9-1) You have a solution containing 500 ppm of dissolved lead as Pb^{2+}. You have three choices for precipitating the lead from solution: (a) add sodium carbonate, (b) add lime (CaO), which hydrates to form $Ca(OH)_2$ or (c) sodium sulfate. Determine the equilibrium concentration of dissolved lead (Pb^{2+}) using each of these three compounds at a level of 10 g/l at 298 K. (neglect pH effects due to the carbonate; assume unit activity coefficients; and, assume complete dissociation of the salts).

(9-2) For each of the resulting lead precipitates from problem 8.1, determine whether they would thermodynamically remain stable enough to be considered non toxic by EPA standards using the TCLP test (assume the TCLP test can be modeled by simply calculating pH dependent equilibria where the resulting pH would be either 2.5 or 10.5 depending upon the material acidity or basicity—do not consider the acetate ion concentration or the sodium concentrations) at 298 K. (answer for part a is toxic at pH 2.5, not toxic at pH 10.5, so it is material property dependent).

(9-3) (a) Determine the dissolved mercury concentration when 5 g/l of sodium sulfate is present along with 300 ppm of mecurous ions at 298 K. (b) How does this change if the mercury is oxidized to the mercuric state prior to sulfate addition? (assume unit activity coefficients) (answer for part a is 300 ppm).

(9-4) Determine whether the equilibrium concentration of Cd^{2+} exceeds the EPA limit (5 ppm) for toxicity when a large amount of $Cd(OH)_2$ is placed in a small container of water with an equilibrium pH of (a) 2.5 and (b) 11.5 (assume the activity coefficients are one).

Chapter 10
Process Design Principles

> *Commercial-scale processes must be designed based on correct fundamental principles as well as practical constraints.*

Key Chapter Learning Objectives and Outcomes:
Understand the principles used to design processes
Know the basic flow sheet segments used in hydrometallurgical processing
Know the basic types of hydrometallurgical flow sheets for a variety of metals

A variety of factors need to be considered carefully before designing industrial processes. Sometimes small process details have significant consequences if they are not properly included in the design of a process. Thus, it is important to design processes in an organized, stepwise fashion to avoid overlooking important details. Useful design steps include establishing the overall objectives clearly, determining the basic flow sheet elements or segments, determining specific options for each segment, determining what information will be needed to design each segment, obtaining the necessary information, designing the overall process, and evaluating and validating the overall design using computer models and experimental data.

> *Good designs begin with a clear understanding of the overall objectives.*

10.1 Determination of Overall Objectives

The overall process objectives must be established clearly if the design is to be effective. Some of the important details that need to be determined are the production goals in terms of both quantity and quality. Other details include the expected life of the project and the necessary return for the investors that are likely to finance the project. The expected life of the project will be determined largely by the projected market and the product price, as well as by the size of the available resource from which the product will be made. Many of the overall objectives will be set by managers and company executives. Other objectives will result from regulatory constraints. However, regardless of who determines the objectives, they

© The Minerals, Metals & Materials Society 2022
M. L. Free, *Hydrometallurgy*, The Minerals, Metals & Materials Series,
https://doi.org/10.1007/978-3-030-88087-3_10

must be understood clearly by the designers before initiating the design of the process. This chapter is focused on designing processes for metal extraction and recovery rather than metal use or part manufacturing.

10.2 Determination of Basic Flow Sheet Segments

In the context of aqueous processing of metals there are three direct and two indirect flow sheet segments that need to be considered for most processes. The three direct segments are extraction, concentration, and recovery. The indirect segments are comminution and mineral concentration. The focus of this chapter will be on the hydrometallurgical processes, but some discussion will be

> *Most overall processes can be subdivided into smaller processing segments.*

provided with regard to the indirect segments, because they are often critical to hydrometallurgical processes.

As examples of basic flow sheet segment determination, consider the five examples of main process objectives, A, B, C, D, E, and the corresponding flow sheet segments that are necessary to achieve the objectives in Table 10.1.

The establishment of the basic flow sheet segments that need to be utilized is simple because few options are available. However, it should be noted that assembling the individual processing steps within each basic segment is far more complicated. Also, each segment may involve repeated steps. For example the concentration segment may involve two separate streams—one for the metal product stream and another for the tailings stream that will likely involve the removal of toxic species.

10.3 Survey of Specific Segment Options

Overall process goals can be achieved using more than one approach for each extraction, concentration, and recovery segment. Frequently, the most appropriate specific segment option will be determined by the nature of the minerals, metals, and solutions that are involved. The following sections describe some of the possibilities.

Table 10.1 Main process objectives (assuming ores have been properly preconcentrated and presized)

A	B	C	D	E
ZnO	As removal	Al_2O_3 ore	Copper anode	Gold ore
50% Zn	3 g/l As	40% Al	99.4% Cu	2 ppm Au
To	To	To	To	To
99.99%	50 ppb	99.9% Al_2O_3	99.995% Cu	99.99%

10.3.1 Basic Corresponding Hydrometallurgical Flow Sheet Segments

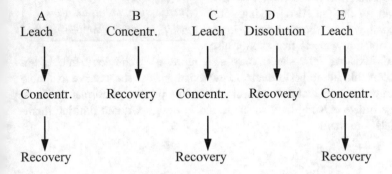

10.3.2 Extraction

Hydrometallurgical extraction of metals from minerals is accomplished by leaching. The goal is to accomplish the leaching in a timely, technically sound, and economically viable manner. Bacteria are often the least expensive approach when their capabilities are adequate. Acid leaching tends to be the most viable means of extraction for common metal-bearing minerals. Ammonia is also a possible extraction medium for nickel and copper ores. Cyanide is a highly effective and inexpensive lixiviant for gold and silver extraction.

Important parameters that need consideration are mineralogy, temperature, pressure, particle size, and reagent additives. Mineralogy or the mineral chemistry will ultimately determine the method and conditions needed for extraction. Other parameters are crucial to optimal performance. Increasing the temperature accelerates the reaction, creates additional expense, and may be necessary to breakdown the metal-bearing mineral. Increasing pressure allows for higher operating temperatures, and it allows for higher concentrations of gases such as oxygen that often provide the driving force for oxidizing reactions. Decreasing the particle size increases the rate of extraction and allows for smaller reactors, but additional size reduction also drives up the processing costs. Reagent additives are often helpful or even necessary in metal extraction processes. Additional details for design parameters are found in Chaps. 4 and 5.

One important element in process design is the examination of current flow sheets. Existing flow sheets show what processing steps are performed on an industrial scale. A summary of existing extraction processing steps for common metals is presented in Table 10.2. Additional, more specific information is provided for selected metals at the end of the chapter. However, in examining existing processing techniques it is important to realize that many published flow sheets do not provide adequate detail for design. Also, relying completely on existing flow sheets reduces innovation and adaptation aspects that should also

> *Designs are often based on processes that have a proven track record of success.*

be considered. Existing flow sheets may also represent technology that is not capable of acceptable future performance. Few existing flow sheets have available state-of-the-art technology that may be critical to future process performance. Thus, wise engineers utilize existing flow sheets to aid in design without limiting themselves completely to them.

Table 10.2 Hydrometallurgical processing steps that have been used for selected metals (some metals are not processed by hydrometallurgical steps)

Metal	Traditional Hydrom. Extr. Steps
Aluminum	Aluminum is extracted from most aluminum ores by pressure leaching using a sodium hydroxide solution
Beryllium	Beryllium can be leached from bertrandite ores using sulfuric acid
Cadmium	Recovered typically as a flue-dust by-product of zinc, nickel, and copper production. Cadmium is often leached from flue-dust using sulfuric acid
Cobalt	Extracted by the same methods as nickel, except that alkaline pressure leaching has also been successful in cobalt extraction
Copper	Dump and heap leaching using sulfuric acid and often bacteria. In-situ leaching with sulfuric acid. Pressure leaching in ammoniacal solutions
Gold	Heap and dump leaching using cyanide. Refractory sulfide ores are often pretreated by biooxidation or roasting. Concentrate leaching using cyanide
Iron	Steel production does not rely on hydrometallurgical extraction
Lead	Nearly all lead is extracted by pyrometallurgical methods
Magnesium	Hydrometallurgical processing is used to remove impurities prior to molten salt electrowinning at high temperatures
Molybdenum	Molybdenum is not typically extracted by hydrometallurgical means
Nickel	Nickel is extracted by chloride, sulfuric acid, and ammoniacal leaching. In some cases nickel matte from initial smelting is leached
Platinum	Platinum can be extracted using a wide variety of techniques that can include leaching in chlorides, sulfuric acid, aqua regia, and cyanide
Silver	Silver is extracted from silver ores by cyanide leaching. By-product silver from electrolytic slimes is often recovered by leaching in oxidizing acid solution. By-product silver from lead production is often extracted by pyrometallurgical methods
Tin	Nearly all tin is extracted by pyrometallurgical processing. Hydrometallurgical processing has been used as an intermediate step for impurity removal

(continued)

Table 10.2   (continued)

Metal	Traditional Hydrom. Extr. Steps
Tungsten	Wolframite and scheelite can be extracted using carbonate and sodium hydroxide solutions at elevated temperatures by pressure leaching
Uranium	Uranium is most commonly leached from the ore using carbonate solutions when carbonates are present, or sulfuric acid when acid consuming minerals such as carbonates are not present
Zinc	Nearly all zinc is extracted from zinc sulfide that is first roasted, then leached in sulfuric acid

The list in Table 10.2 shows that the extraction processing steps are typically quite simple. In most cases extraction involves only one leaching step, although it is important to remember that mineral separation and mineral size reduction steps are often necessary precursors as shown in Fig. 10.1. However, more complicated extraction processing occurs for some industries such as the gold industry, which often utilizes an acid leaching pretreatment process for sulfide ores that is followed by neutralization and subsequent leaching in cyanide solution as illustrated in Fig. 10.2. The neutralization step is necessary to make the transition from an acidified medium to the alkaline medium that is necessary for cyanide leaching (cyanide converts to the highly toxic hydrogen cyanide gas in acidic solutions).

Another important part of most hydrometallurgical processes is the separation of the material that is leached from the resulting leach solution as seen in Figs. 10.1 and 10.2. Often this can be accomplished within a process by simple, one-step filtration. However at the completion of leaching processes the resulting solids are often disposed of after being rinsed thoroughly to remove residual leach solution, which is often unsafe from an environmental perspective. One common approach to providing clean solids for disposal is counter-current decantation as shown in Fig. 10.3.

> *It is common practice to effectively wash leaching residue and recover leaching solution using a counter-current decantation circuit.*

10.3.3   Concentration

Hydrometallurgical concentration of metals is often accomplished by carbon adsorption, ion exchange, and solvent extraction, each of which is followed by or includes a stripping step. Another important method of concentration is precipitation. The most important parameters that need to be evaluated are chemical

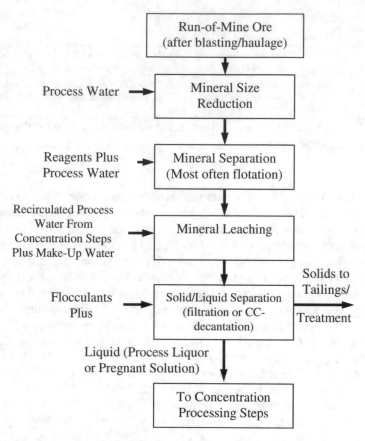

Fig. 10.1 Common steps in a metal extraction flow sheet segment

environment, temperature, flow rates, loading/stripping kinetics, product decay, product losses, compatibility with extraction and recovery steps, selectivity, and sensitivity to changing process conditions. Details for determining and evaluating most of these parameters are presented in Chaps. 6 and 9. However, some comments regarding these parameters are appropriate here.

Losses for carbon adsorption and ion exchange processes are commonly associated with material degradation. Such losses, which may or may not be significant, should be considered. For carbon adsorption systems, attrition of carbon results in fine material that often reports to the tailings along with some of the adsorbed product. In ion exchange systems, attrition can be a problem, but osmotic shock caused by sudden changes in environment is often a more significant cause of degradation that can lead to the generation of loaded fine particles of resin that are lost in effluent and elution solutions.

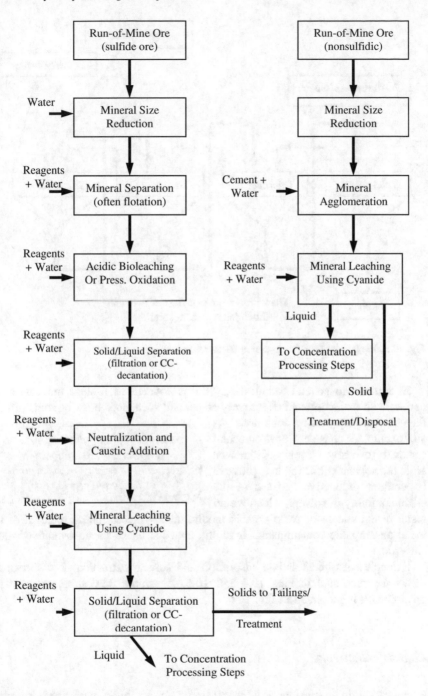

Fig. 10.2 Typical gold extraction segment for sulfidic and nonsulfidic ore flow sheets for nonrefractory ore

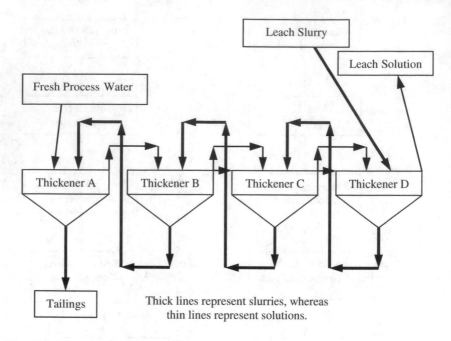

Fig. 10.3 Flow sheet for counter-current decantation segment

In addition to general technical feasibility, concentration steps must be compatible with extraction and or recovery steps, and such steps must be robust in the face of changing process conditions. As an example, consider environmental removal of low levels of metals. Utilizing solvent extraction may prove to be counter productive because of the slight solubility of solvent, which would

> *Metals are commonly concentrated using solvent extraction, ion exchange, and carbon adsorption as discussed previously.*

result in the release of some organic media, thereby necessitating removal of the metal and organic contamination from the aqueous phase for an overall win-loss situation.

Examples of typical carbon adsorption and solvent extraction flow sheet segments are presented in Figs. 10.4 and 10.5. A sample flow sheet segment for precipitation is presented in Fig. 10.6.

10.3.4 Recovery

Recovery in hydrometallurgical processing is most often performed by electrowinning, although in many cases precipitation results in a final product that is nearly always non metallic except in the case of gold and copper cementation

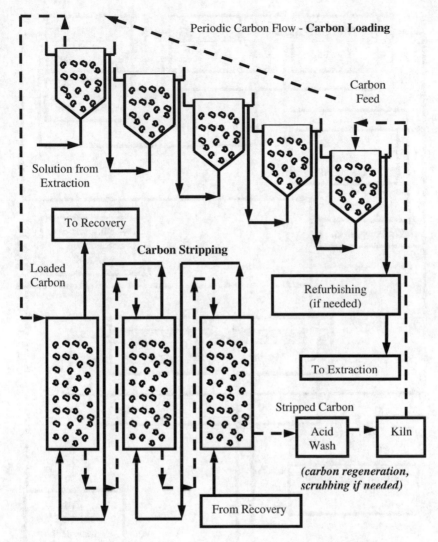

Fig. 10.4 Flow sheet for typical carbon adsorption concentration process segment

processes. The principal issues associated with electrowinning and cementation are discussed in Chap. 7. A sample flow sheet segment for recovery which involves electrowinning is presented in Fig. 10.7.

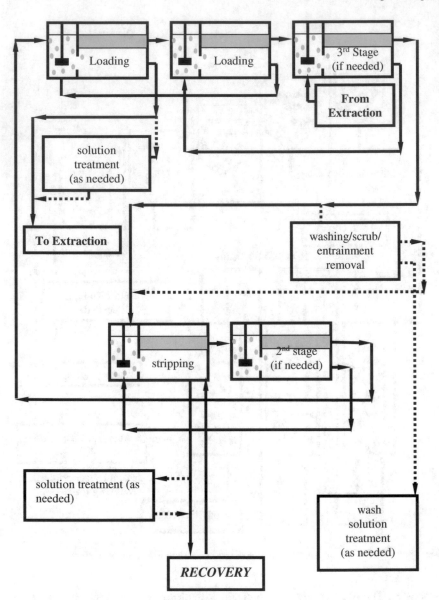

Fig. 10.5 Typical solvent extraction concentration flow sheet segment

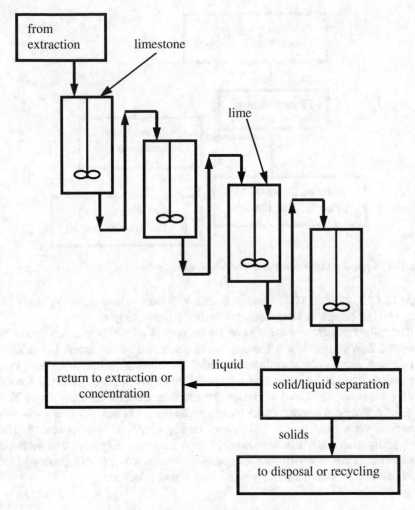

Fig. 10.6 Typical precipitation concentration flow sheet segment

10.4   Overall Flow Sheet Synthesis

After establishing the detailed individual flow sheet segments for the various extraction, concentration, and recovery process streams, the overall flow sheet must be developed. The overall flow sheet is easy to develop once the individual segments are completed, since the overall sheet consists of connecting the individual segments. Because typical overall flow sheets consist of several segments, no attempt will be made here to show completed flow sheets. Also, there is little need for showing overall flow sheets because the processing details are present in the individual segments. Note that the individual flow sheet segments shown in

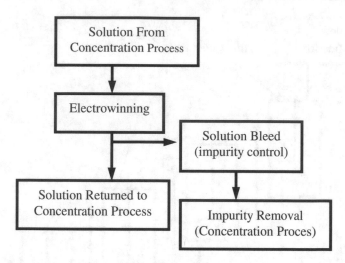

Fig. 10.7 Typical electrowinning recovery flow sheet segment

Figs. 10.1, 10.2, 10.3, 10.4, 10.5, 10.6 and 10.7 show where other segments connect, making complete flow sheet synthesis relatively simple.

The completed flow sheet should be compared carefully with other existing industrial flow sheets. Any differences should be noted and explored. Considerable information can be obtained through a critical comparison. Industrial flow sheets often contain small segments that are specific to a given set of processing conditions. Perhaps an additional thickener is present or an extra leaching stage. These types of differences suggest that a more simple design is not adequate for a given operation, yet a more simple design may be appropriate in another setting. Thus, after performing such a comparison, it will likely be apparent that additional information is necessary to address questions associated with differences in flow sheets as well as information regarding equipment selection.

10.5   Procurement of Additional Information

After synthesis of the overall flow sheet it will be necessary to obtain more information. Preparation of an outline is an important part of planning that will facilitate a more organized and successful design. Specific information will be needed for each processing step. Such information might include flow rates, temperatures, solids content, chemistry, retention time, volume, pipe size, pump capacity, reagent and water addition rates, utilization projections, etc.

Important sources for the needed details are vendors and design firms. Vendors are particularly helpful because of their interest in selling equipment. However, vendor information should be verified with industrial experience. Obtaining the most current and accurate information possible is paramount to successful design.

10.5.1   *Material Balances*

$$\text{Mass Flow In} - \text{Mass Flow Out} = 0 \text{ (If Steady-State)}$$

The countercurrent decantation circuit shown previously will be used an example. A material balance for each unit is:

Unit A: $C_B Q_{Bunder} + 0 - C_A(Q_{Aover} + Q_{Aunder}) = 0$

Unit B: $C_C Q_{Cunder} + C_A Q_{Aover} - C_B(Q_{Bover} + Q_{Bunder}) = 0$

Unit C: $C_D Q_{Dunder} + C_B Q_{Bover} - C_C(Q_{Cover} + Q_{Cunder}) = 0$

Unit D: $C_L Q_L + C_C Q_{Cover} - C_D(Q_{Dover} + Q_{Dunder}) = 0$

Rearranged, these equations lead to:

Unit A: $C_A[-(Q_{Aover} + Q_{Aunder})] + C_B[Q_{Bunder}] + C_C[0] + C_D[0] = 0$ (10.1)

Unit B: $C_A[Q_{Aover}] + C_B[-(Q_{Bover} + Q_{Bunder})] + C_C[Q_{Cunder}] + C_D[0] = 0$ (10.2)

Unit C: $C_A[0] + C_B[Q_{Bover}] + C_C[-(Q_{Cover} + Q_{Cunder})] + C_D[Q_{Dunder}] = 0$ (10.3)

Unit D: $C_A[0] + C_B[0] + C_C[Q_{Cover}] + C_D[-(Q_{Dover} + Q_{Dunder})] = -C_L Q_L$ (10.4)

Expressed in matrix table form these equations become:

$-(Q_{Aover} + Q_{Aunder})$	Q_{Bunder}	0	0	C_A	0
Q_{Aover}	$-(Q_{Bover} + Q_{Bunder})$	Q_{Cunder}	0	C_B	0
0	QBo	$-(Q_{Cover} + Q_{Cunder})$	Q_{Dunder}	C_C	0
0	0	QCo	$-(Q_{Dover} + Q_{Dunder})$	C_D	$-C_L Q_L$

In matrix format these can be expressed as:

$$[Q][C] = [B] \tag{10.5}$$

which is solved using the inverse matrix:

$$[Q]^{-1}[Q][C] = [Q]^{-1}[B] \tag{10.6}$$

which rearranges to:

$$[C] = [Q]^{-1}[B] \tag{10.7}$$

which can be easily solved using a matrix solver.

10.5.2 Economic Assessment

As engineers are keenly aware, economics drive industry. Companies are not in business for public service. Companies are in business to make money. Economics are critical to process design, and the next chapter was written to address its relevance to engineering.

10.6 Selected Industrial Flow Sheet Examples

10.6.1 Aluminum

Aluminum is produced primarily from bauxite ore, which consists of diaspore [AlO (OH)], gibbsite [Al(OH)$_3$], and boehmite [AlO(OH)]. An overall flow sheet is presented in Fig. 10.8. After the ore is ground to an appropriate size in a ball mill, the ore is dissolved in a hot caustic solution inside of an autoclave at between 150 and 220 °C. During aluminum mineral dissolution, iron oxides, titanium dioxide,

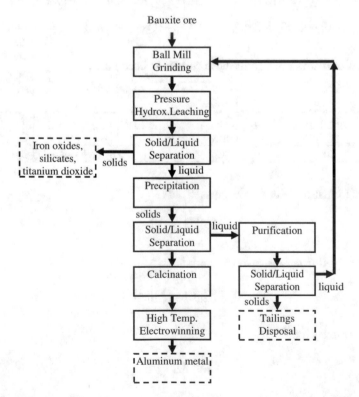

Fig. 10.8 Basic flow sheet elements for aluminum production from bauxite ore. Adapted from information in Ref. [2]

and quartz do not dissolve. Thus, these gangue minerals are separated from dissolved aluminum by filtration. Other entities such as some silicate-based clays may dissolve to form slightly soluble sodium silicate (Na_2SiO_3). Organic matter such as humus, lignin, cellulose, and proteins tend to decompose to form oxalic acid. Dissolved silica often forms a precipitate with aluminum (sodium alumina silicate) that removes dissolved silica but decreases the aluminum yield [1]. Oxalic acid can react with dissolved metals to form metal oxalates such as calcium oxalate.

The solution from the digestion is sent to a precipitation system, where hydrated alumina is precipitated in a relatively pure form. The hydrated alumina is then calcined to produce anhydrous alumina (Al_2O_3), which is converted to aluminum metal in an electrowining cell. Aluminum electrowinning is performed at high temperatures, and it commonly utilizes carbon anodes to reduce the energy requirement.

10.6.2 Copper

Copper ores are most commonly dominated by copper sulfides. The most common copper sulfide is chalcopyrite. Chalcopyrite is most commonly processed by comminution, flotation, smelting, converting, and electrorefining to produce copper metal. Thus, the dominant overall general flow sheet for chalcopyrite based ores is shown on the left-hand side of Fig. 10.9.

However, there are many copper oxide deposits that can be leached in heaps followed by solvent extraction and electrowinning to produce copper metal. In many cases copper oxides, mixed with secondary sulfides such as chalcocite, can be leached in heaps as well. A common general flow sheet for copper oxide/secondary sulfide leaching and metal recovery is shown on the right-hand-side of Fig. 10.9.

Because general copper ore processing has been discussed in Chap. 5, it will not be discussed in detail here. However, it should be noted that there are various processing options. Most options are similar to those shown in Fig. 10.9. However, in cases where copper is recovered with other metals, more interesting flow sheets are used. An example of a more unusual copper processing flow sheet is shown in Fig. 10.10.

10.6.3 Gold

Gold ores are processed in a variety of ways depending on the ore mineralogy and gold grade as discussed in Chap. 5. A general overview of gold ore processing is presented in Fig. 10.11. High grade ores are generally processed into fine particles

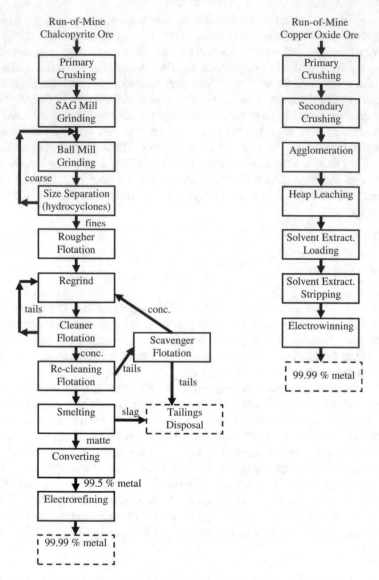

Fig. 10.9 Simplified examples of common flow sheets for copper sulfide (chalcopyrite) and copper oxide ore processing

and leached using cyanide. Refractory high-grade sulfide ores are usually pretreated by roasting, pressure oxidation, or biooxidation to make the gold amenable to extraction. Low grade ores are often heap leached, although they are sometimes leaching in tanks. Variations of processing flow sheets are significant as illustrated by the examples shown in Figs. 10.12, 10.13, 10.14, 10.15 and 10.16.

Fig. 10.10 Overview of general Sepon processing flow sheet elements. Based on data in Ref. [3]

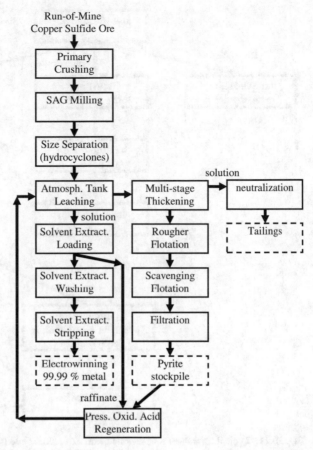

10.6.4 Nickel and Cobalt

Nickel and cobalt are commonly found together and processed by the same methods. Nickel is much more prevalent than cobalt and tends to dominate the processing method, so this discussion will focus mostly on nickel, yet the methods and outcomes are nearly the same for cobalt.

Most nickel (60%) is used to produce high grade stainless steels, although a significant amount of nickel is used in nickel based alloys, miscellaneous alloys, and electrodeposited coatings [9].

Nickel is most often obtained from sulfide ores through flotation and subsequent smelting, which is commonly followed by hydrometallurgical refining. Only a small fraction of total nickel production (about 7%) is produced directly through hydrometallurgical extraction and processing [10]. However most nickel resources are available in nickel laterite deposits. Nickel laterite deposits generally have an upper layer of limonite near the surface, a middle layer of saprolite, and a lower layer of pendodite. Each layer is generally only a few meters thick. Low iron

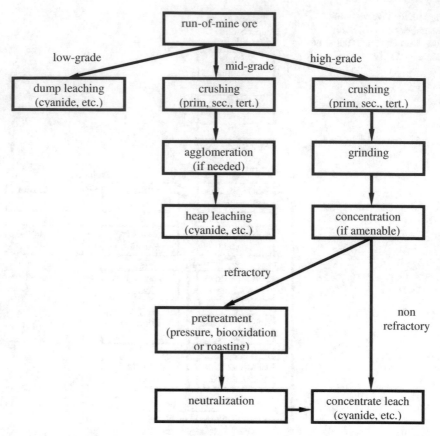

Fig. 10.11 Typical flow sheet for precious metal extraction. Note that in some cases, pretreatment of low and mid-grade ores can be made by bioleaching in a heap or dump leaching setting, followed by neutralization and traditional cyanide leaching for gold extraction

containing laterite ores that are rich in saprolite are usually processed by pyrometallurgy based methods to produce ferronickel (30% Ni, 70% Fe) that can be used directly in iron alloy production such as stainless steel [10]. Nickel ore associated with limonite or smectite minerals with high iron content is usually processed by hydrometallurgy based methods [10].

Most hydrometallurgical processing of limonite and smectite based nickel ore is performed using high pressure acid leaching (HPAL) at 50 bars and 250 °C using sulfuric acid. A general flow sheet is shown in Fig. 10.17. Impurities are commonly separated by pH adjustment and the associated precipitation. Nickel and cobalt can be precipitated using 5–7 bar of H_2S to produce sulfide products that are sent to smelters [11].

A small amount of nickel is recovered from platinum group metal processing facilities. These facilities often use flotation and smelting to produce a matte. The PGM rich residue that results from matte leaching is generally ground, then it is

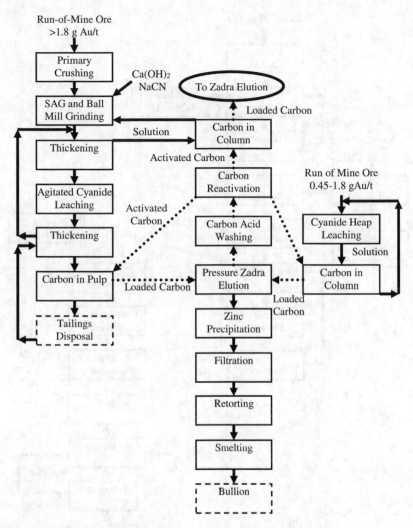

Fig. 10.12 Example of common processing scenario for an oxide ore: chimney creek process flow sheet (1990). After Ref. [4]

dissolved in spent copper electrowinning solution with oxygen gas during initial stages of leaching at 140 °C and 4–6 bar [12]. Nickel is dissolved first. The nickel rich solution is then purified. Nickel powder is often used to remove copper. Leaching then continues to dissolve the copper. Selenium and tellurium are often removed from the copper rich solution using sulfur dioxide. Copper can then be recovered by electrowinning. The nickel rich solution is then processed to remove iron at 150 °C and 6 bar. Nickel is recovered from this solution by hydrogen gas addition. (190 °C, 28 bar). Cobalt is eventually recovered from the residual solution through hydrogen reduction.

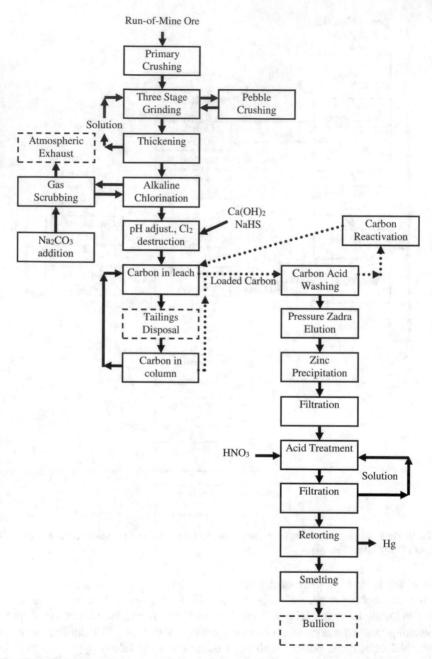

Fig. 10.13 Refractory gold ore chlorination process example: Jerritt Canyon process flow sheet (pre-1990). After Ref. [5]

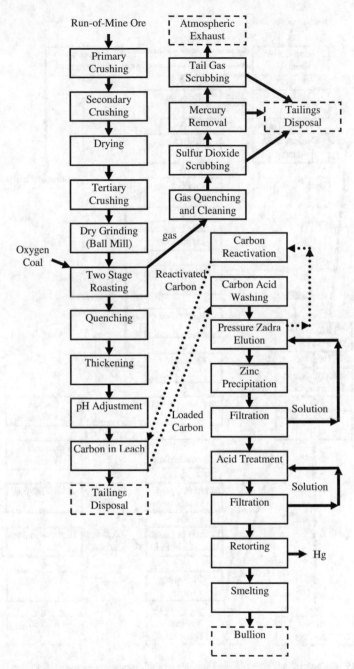

Fig. 10.14 Refractory sulfide gold ore roasting process example: Jerritt Canyon flow sheet (2004). Adapted from Ref. [6]

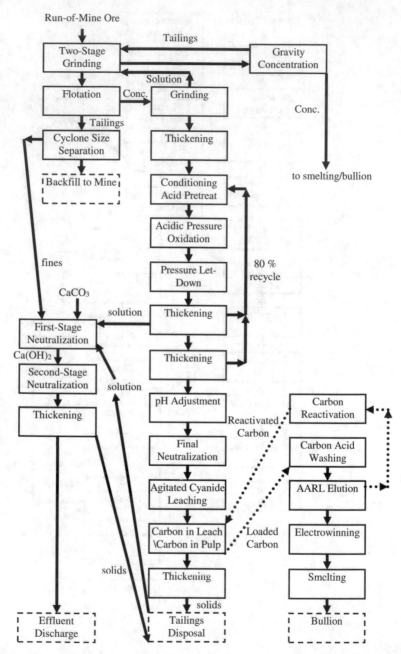

Fig. 10.15 Refractory sulfide gold ore pressure oxidation process example: Sao Bento flow sheet (1990). Adapted from Ref. [7]

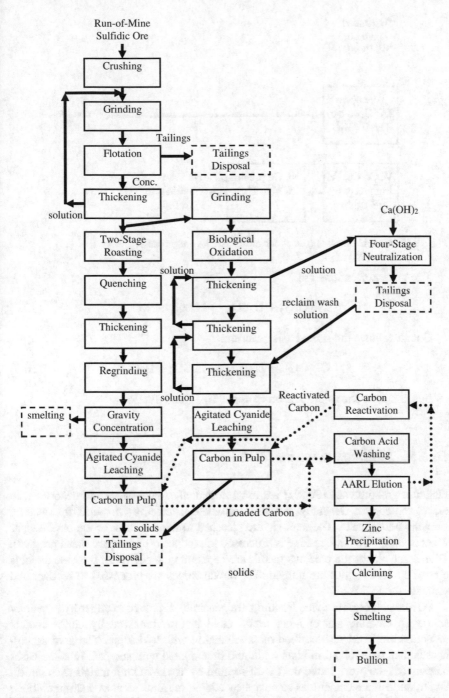

Fig. 10.16 Refractory sulfide gold-ore biooxidation process: fairview plant flow sheet overview (1990). After Ref. [8]

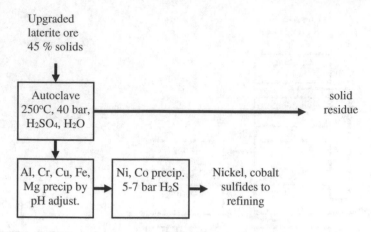

Fig. 10.17 General processing steps for nickel laterite ore processing. After Ref. [11]

The general reactions are:

$$Ni(OH)_2 + H_2SO_4 \leftrightarrow NiSO_4 + H_2O \qquad (10.8)$$

Cobalt follows the same basic reaction:

$$Co(OH)_2 + H_2SO_4 \leftrightarrow CoSO_4 + H_2O \qquad (10.9)$$

Specific flow sheets are shown in Figs. 10.18 and 10.19.

10.6.5   Platinum Group Metals

Platinum group metals (PGMs) are most commonly found in sulfide deposits with nickel and copper. Generally, the platinum is concentrated by flotation, followed by smelting into a matte. The matte is then leached to remove most of the base metals. The resulting PGM rich residue is processed to produce separate pure metal products.

In some plants, the precious metals are captured in anodes after the base metal is converted. The anodes are refined. The anode slimes are processed to recover and separate the PGMs.

PGM residues from matte leaching are generally dissolved in either hydrochloric acid with chlorine gas or aqua regia. Gold is then removed by either contact reduction the PGM residue feed or precipitation with hydrazine. The gold precipitate is then redissolved in aqua regia and precipitated with sucrose. In some cases base metals are precipitated next with sodium hydroxide. Other metals such as Ru, Os, Ir, and Rh are sometimes removed by adding reagents such as sodium chlorate or sodium bromate with pH adjustments. Platinum is precipitated from the PGM solution as ammonium hexachloroplatinate using ammonium chloride. The

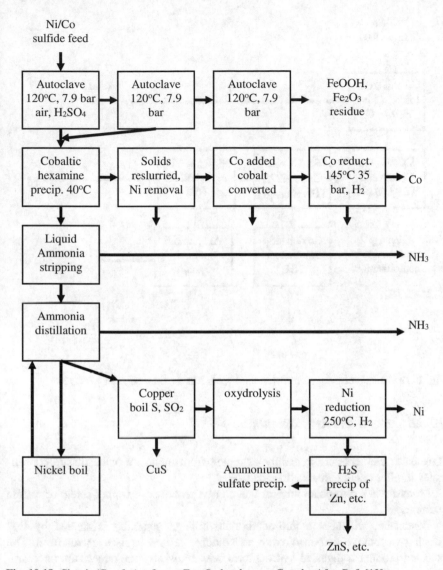

Fig. 10.18 Sherrit (Corefco) refinery, Fort Saskatchewan, Canada. After Ref. [13]

platinum salt is then heated to form platinum sponge that is later purified. Palladium is removed from the remaining PGM solution using a combination of ammonium hydroxide (pH 4–5) and ammonium chloride [14]. The palladium salt is redissolved in HCl (pH 1), then precipitates as diamino-palladous dichloride [14]. In some cases solvent extraction or ion exchange is used to selectively remove PGMs from the PGM rich solution that follows the initial leaching process. Other PGM metals are recovered by a variety of methods. Some specific flow sheets are shown in Figs. 10.20, 10.21, 10.22 and 10.23.

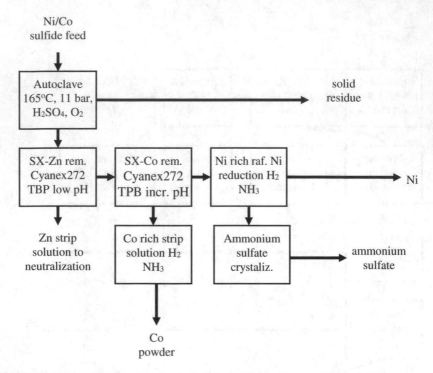

Fig. 10.19 Basic processing elements in the Murrin Murrin refinery. After Ref. [13]

10.6.6 Rare Earth Elements

Rare earth elements are generally extracted from monazite or bastnasite concentrates in either acidic or alkaline media.

Monazite is a phosphate mineral that can be processed by either acidic or caustic solutions.

Bastnasite, which is a fluorocarbonate mineral generally is treated by first roasting to remove carbon dioxide and oxidize cerium to its trivalent form. The calcined product is digested hydrochloric acid to dissolve the noncerium rare earth elements. The residue is high in cerium oxide. The cerium oxide residue, which contains many rare earth fluorides is treated with sodium hydroxide to convert the fluorides to hydroxides, which are then digested in hydrochloric acid before being separated by solvent extraction.

The separation of rare earth elements from leaching solutions generally involves multiple solvent extraction steps.

Examples of rare earth element extraction flow sheets are shown in Figs. 10.24 and 10.25.

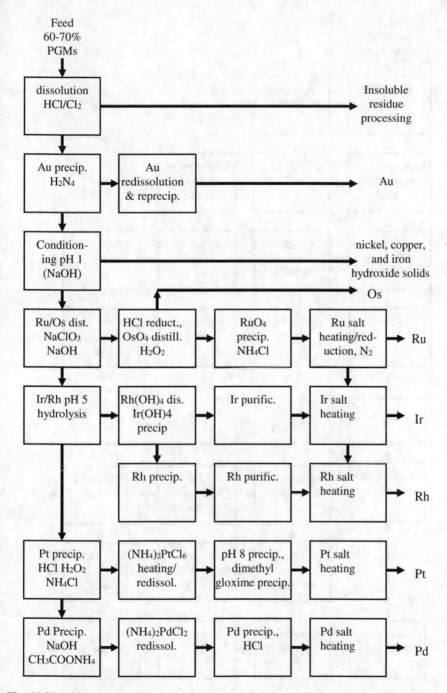

Fig. 10.20 PGMs refining process used at Lonmin's Western Platinum Refinery in Brakpan, South Africa. After Ref. [15]

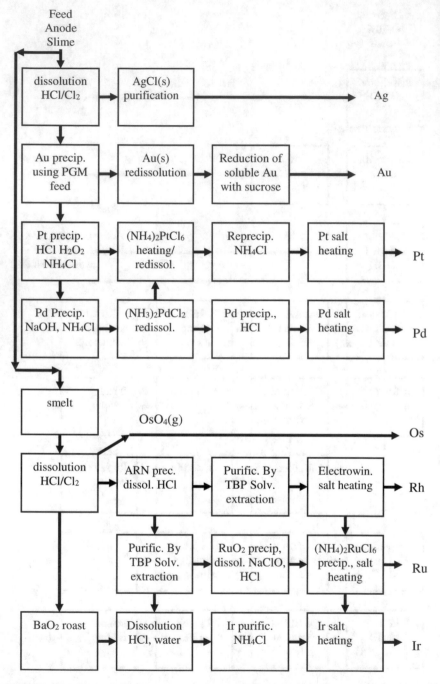

Fig. 10.21 Krastsvetmer (Krasnoyarsk, Russia) refinery flow sheet. Adapted from Ref. [16]. ARN is ammonium rhodium nitrite, and TPB is tri-n-butyl phosphate

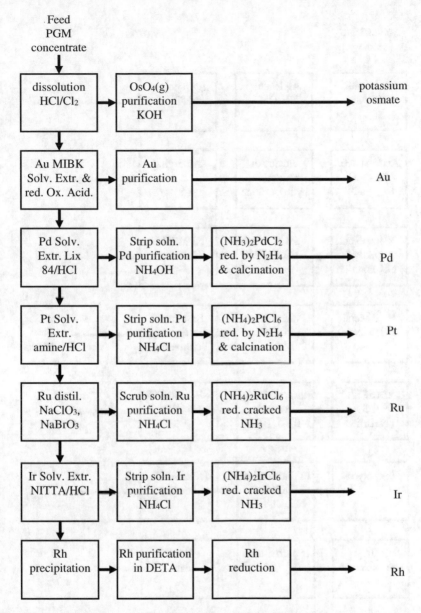

Fig. 10.22 Anglo American Platinum refinery processing in Rustenburg, South Africa. Adapted from Ref. [17]. MIBK is methyl isobutyl ketone, NITTA is n-iso-tridecyltri-decanamide, and DETA is diethylinetriamine

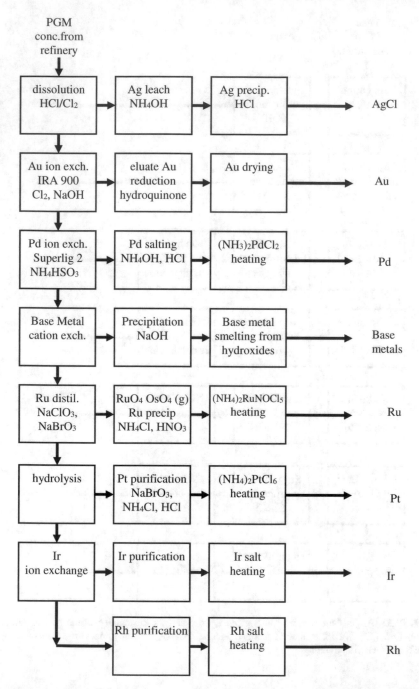

Fig. 10.23 Impala's Springs, South Africa precious metals refinery processing. Adapted from Ref. [18]

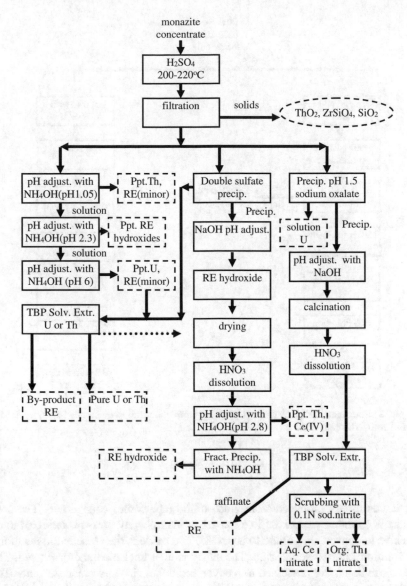

Fig. 10.24 Rare earth (RE) element processing from monazite concentrate in acidic media. Adapted from Ref. [19]

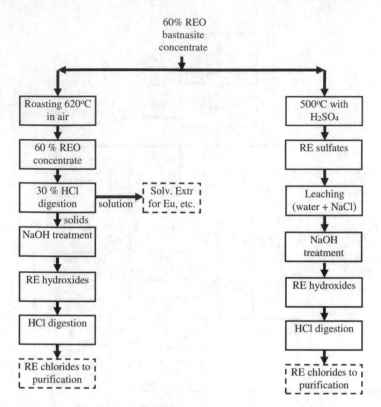

Fig. 10.25 Rare earth (RE) element processing from monazite concentrate in acidic media. Adapted from Ref. [20]

10.6.7 Zinc

Zinc is commonly produced from zinc sulfide (sphalerite) concentrate. The concentrate is commonly oxidized in a roaster or alternatively pressure oxidized in an autoclave to convert the sulfide to an oxide. The oxide is then leached in a sulfuric acid solution to dissolve the zinc. The slurry is then thickened and/or filtered. The tailings can be further processed to recover additional zinc and other elements. The solution from the sulfuric acid based solution is treated with air or oxygen and partially neutralized. Additional precipitates are removed by thickeners and/or filters. The zinc-rich solution is then purified by mixing it with a zinc dust slurry in one or more stages. The zinc dust becomes a sacrificial anode for impurities such as

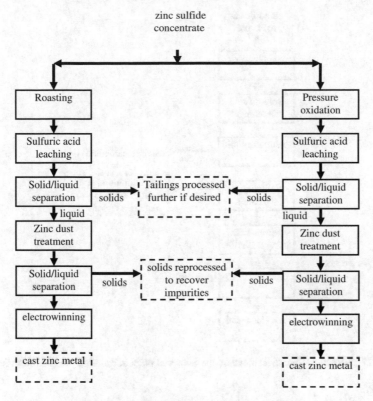

Fig. 10.26 Simplified flow sheet for typical zinc processing from zinc sulfide concentrate. Adapted from Ref. [21]

cadmium and cobalt that are precipitated as metals in exchange for zinc that dissolves. The purified, zinc-rich solution is then sent to electrowinning cells for recovery as cathode zinc. The cathode zinc is generally melted into ingots before it is sold [21]. A flow sheet illustrating the process flow sheet options is presented in Fig. 10.26.

Alternatively, zinc silicate/oxide ores can be processed using sulfuric acid leaching followed by neutralization, solvent extraction, and electrowinning to produce super high quality zinc such as has been practiced at the Skorpion plant in Namibia. A simplified flow sheet for the Skorpion facility is presented in Fig. 10.27.

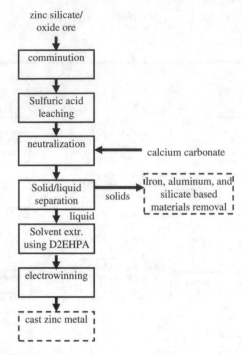

Fig. 10.27 Simplified flow sheet based on the Skorpion plant in Namibia. Adapted from Ref. [22]

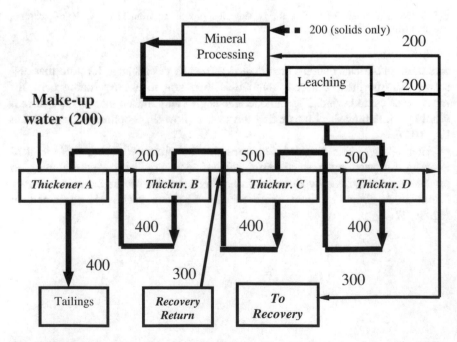

Fig. 10.28 Counter-current decantation circuit for Problem 9.1. Numeric values represent flow rates in tons per hour

References

1. D.M. Muir, E.J. Grimsey, Murdoch University M131 Course Unit Guide (1996), p. 326
2. www.flsmidth.com/~/media/PDF%20Files/.../Alumina1.ashx. Accessed 30 Nov 2012
3. K. Baxter, D. Dreisinger, G. Pratt, The Sepon copper project: developing a flow sheet, in *Hydrometallurgy 2003 Proceedings*, ed. by C.A. Young, A.M. Alfantazi, C.G. Anderson, D. B. Dreisinger, B. Harris, A. James (TMS, Warrendale, 2003), p. 1494
4. J. Marsden, I. House, *The Chemistry of Gold Extraction*, 2nd edn. (SME, Littleton, 2006), p. 523
5. J. Marsden, I. House, *The Chemistry of Gold Extraction*, 2nd edn. (SME, Littleton, 2006), p. 601
6. J. Marsden, I. House, *The Chemistry of Gold Extraction*, 2nd edn. (SME, Littleton, 2006), p. 603
7. J. Marsden, I. House, *The Chemistry of Gold Extraction*, 2nd edn. (SME, Littleton, 2006), p. 575
8. J. Marsden, I. House, *The Chemistry of Gold Extraction*, 2nd edn. (SME, Littleton, 2006), p. 577
9. F.K. Crundwell, M.S. Moats, V. Ramachandran, T.G. Robinson, W.G. Davenport, *Extractive Metallurgy of Nickel, Cobalt, and Platinum-Group Metals* (Elsevier, Amsterdam, 2011), p. 22
10. F.K. Crundwell, M.S. Moats, V. Ramachandran, T.G. Robinson, W.G. Davenport, *Extractive Metallurgy of Nickel, Cobalt, and Platinum-Group Metals* (Elsevier, Amsterdam, 2011), p. 2
11. F.K. Crundwell, M.S. Moats, V. Ramachandran, T.G. Robinson, W.G. Davenport, *Extractive Metallurgy of Nickel, Cobalt, and Platinum-Group Metals* (Elsevier, Amsterdam, 2011), pp. 118–120
12. F.K. Crundwell, M.S. Moats, V. Ramachandran, T.G. Robinson, W.G. Davenport, *Extractive Metallurgy of Nickel, Cobalt, and Platinum-Group Metals* (Elscvicr, Amsterdam, 2011), pp. 471–472
13. F.K. Crundwell, M.S. Moats, V. Ramachandran, T.G. Robinson, W.G. Davenport, *Extractive Metallurgy of Nickel, Cobalt, and Platinum-Group Metals* (Elsevier, Amsterdam, 2011), p. 284
14. F.K. Crundwell, M.S. Moats, V. Ramachandran, T.G. Robinson, W.G. Davenport, *Extractive Metallurgy of Nickel, Cobalt, and Platinum-Group Metals* (Elsevier, Amsterdam, 2011), p. 509
15. F.K. Crundwell, M.S. Moats, V. Ramachandran, T.G. Robinson, W.G. Davenport, *Extractive Metallurgy of Nickel, Cobalt, and Platinum-Group Metals* (Elsevier, Amsterdam, 2011), p. 502
16. F.K. Crundwell, M.S. Moats, V. Ramachandran, T.G. Robinson, W.G. Davenport, *Extractive Metallurgy of Nickel, Cobalt, and Platinum-Group Metals* (Elsevier, Amsterdam, 2011), p. 508
17. F.K. Crundwell, M.S. Moats, V. Ramachandran, T.G. Robinson, W.G. Davenport, *Extractive Metallurgy of Nickel, Cobalt, and Platinum-Group Metals* (Elsevier, Amsterdam, 2011), p. 513
18. F.K. Crundwell, M.S. Moats, V. Ramachandran, T.G. Robinson, W.G. Davenport, *Extractive Metallurgy of Nickel, Cobalt, and Platinum-Group Metals* (Elsevier, Amsterdam, 2011), p 520
19. C. K. Gupta, N. Krishnamurthy, *Extractive Metallurgy of Rare Earths* (CRC Press, Boca Raton, 2005), p. 143
20. C.K. Gupta, N. Krishnamurthy, *Extractive Metallurgy of Rare Earths* (CRC Press, Boca Raton, 2005), pp. 147–148
21. http://www.metsoc.org/virtualtour/processes/zinc-lead/zincflow.asp. Accessed 29 Nov 2012
22. www.mintek.co.za/Mintek75/Proceedings/B04-Sole.pdf. Accessed 29 Nov 2012

Problem

(10.1) For the accompanying counter-current decantation circuit for cyanide leaching (Fig. 10.28), determine the cyanide concentration in the tailings as well as the overall cyanide loss. The cyanide concentration in the leach tanks is 0.7 kg NaCN/metric ton of solution (not slurry). Assume no cyanide losses in the thickeners, and 50% solids in thickener discharge. Also assume 33% solids in leaching vessels. No cyanide is present in the make-up water. Answer: final cyanide concentration = 0.229 kg NaCN/ metric ton of solution, cyanide loss = ?

Chapter 11
General Engineering Economics

> *Engineers need to evaluate the economic value or cost of the processes they design. Companies are in the business of providing goods or services that generate revenue.*

Key Chapter Learning Objectives and Outcomes:
Understand the effect of time and interest on the value of money
Know how to calculate the effects of various cash flow types
Understand how to estimate costs
Understand how to calculate the return on investment
Know how to apply the discounted cash flow method of economic analysis
Be able to quantify the effect of risk on value

11.1 The Effects of Time and Interest

Money is an important commodity in modern society. Money is a measure of value. Items that are purchased at a store require effort and resources to produce them. Employees demand compensation for their time and effort in the form of money. Employees' efforts are rewarded based on the value they provide. The value they provide is related to the time they work and the skills they have acquired.

Money has value associated with time. Money can be used over time to produce more money if it is used appropriately. For example, money can allow an individual or entity to purchase a company that earns money as a function of time and production rate. The ability to use money to make more money over time cements the relationship between time and money. Thus, those that have this resource generally use it to make more money directly. Alternatively, they lend it to others who are willing to pay extra money in the future for use of the money in the present.

Interest is the rate of increasing value of money with time. It is a tool by which the change in value of money over time is measured. The opportunities that money offers to make more money over time leads to the use of interest as a means of measuring the value of money over time. Those who have money have the opportunity to earn interest. Those who borrow money must pay interest. Interest is synonymous with rate-of-return for an investor.

© The Minerals, Metals & Materials Society 2022
M. L. Free, *Hydrometallurgy*, The Minerals, Metals & Materials Series,
https://doi.org/10.1007/978-3-030-88087-3_11

11.1.1 Simple Interest

The effect of interest and time for one period of interest accrual is known as simple interest, and it can be expressed mathematically as:

$$F = P + P(i) = P(1 + i) \tag{11.1}$$

F is the future value of the money. P is the initial or present value of the money. i is the fractional interest rate that applies for the period of interest accrual. Thus, if one invested \$1.00 at 8% ($i = 0.08$) annual interest for one year, the total value at the end of the year would be \$1.08. However, it should be noted that this approach assumes that the interest for each period is paid at the end of the period. All other derivations in this chapter utilize the same assumption, which is referred to as end-of-period payout.

> *Engineering investments must provide reasonable returns on investments when evaluated for the appropriate time scale.*

11.1.2 Normal (Discrete) Compound Interest

The effect of interest over several periods of accrual has a multiplying effect on interest. If an investment is left in an interest bearing account over multiple accrual periods, interest will be earned on the interest that previously accrued in addition to the interest for the initial investment, thereby compounding the effect of the interest. Therefore, the term compound interest refers to interest that is compounded over more than one accrual period. Thus, the total value of the investment at the end of each accrual period is given as:

$$F_1 = P(1 + i) \tag{11.2}$$

$$F_2 = [P(1 + i)](1 + i) \tag{11.3}$$

$$F_3 = [P(1 + i)(1 + i)](1 + i) \tag{11.4}$$

$$F_n = P(1 + i)^n \tag{11.5}$$

Leading to the general compound-interest-based formula for future value after "n" periods of interest accrual [1–9]:

$$F = P(1 + i)^n \tag{11.6}$$

The effect of compound interest is powerful as demonstrated in Fig. 11.1. Figure 11.1 is based on an initial investment of \$7000 that accumulates interest at a rate of 7% for 10 years.

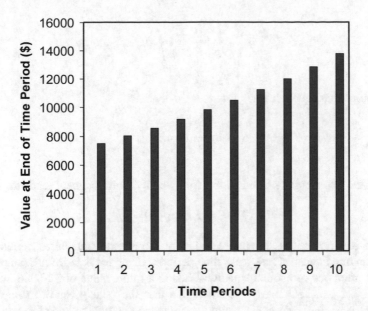

Fig. 11.1. Comparison of value and time for an initial investment of $7,000 that is invested at 7% interest for 10 years earning compounded interest. Note the y-axis represents the value at the end of the time period

Example 11.1 Calculate the future value of a $1,000 investment that earns 10% interest each year for 30 years.

$$F = \$1,000(1 + 0.1)^{30} = \$17,449$$

11.1.3 Continuous Compound Interest

Although nearly all interest is determined on the basis of discrete interest payments occurring at the end of discrete time periods such as months or years, interest can be calculated on a continuous compounding basis using a simple formula. The rate of growth of an investment can be expressed as:

$$\frac{dP}{dt} = Pi \tag{11.7}$$

P is the present value. t is time. i is the interest rate expressed on a fractional basis for the time scale used. Rearrangement of this expression into an integral form leads to:

$$\int_P^F \frac{dP}{P} = i \int_0^t dt \qquad (11.8)$$

Solution of this equation leads to:

$$\ln\left(\frac{F}{P}\right) = i(t) \qquad (11.9)$$

Taking the exponential of both sides, followed by rearrangement leads to:

$$F = P\exp(i(t)) \qquad (11.10)$$

This equation gives the future value of a present value after continuously compounding interest at rate, i, for time period t. Using the data from Example 11.1 in the continuous compound equation leads to a future value of \$20,086, which is 15.1% higher than the value given in Example 11.1 using the discrete compound interest formula. Thus, it should be clear than continuous compounding results in more rapid growth than end-of-period, discrete compounding.

> *The future value of money of an investment includes the effect of time and interest.*

11.1.4   Future Value for a Series of Discrete Uniform Addition (or Annuity) Values

In many settings money flows into or out of an investment in uniform amounts on a regular basis. This type of mone flow occurs results from specified periodic deposit requirements, partial payments for a purchased product, taxes, etc. The future value of such uniform additions can be calculated by summing the effects of each addition with its accompanying interest for each period. The term addition will be used in place of the term annuity, which literally means a series of equal payments made at equal time intervals and is commonly used to describe this type of value. The reason for this use of addition is to avoid confusion with the more common association of annuity with retirement accounts, rather than uniform payment or revenue series. The future value after the first addition, A_1, which occurs at the end of the period "1" is given as:

$$F_1 = A_1 \qquad (11.11)$$

The future value at the end of the second period includes the first addition, A_1, plus interest from A_1, plus A_2. A_2 is added only at the end of the second period. F_2 is given as:

$$F_2 = A_1(1+i) + A_2 \tag{11.12}$$

The future values for the third and fourth periods of time are given as:

$$F_3 = A_1(1+i)(1+i) + A_2(1+i) + A_3 \tag{11.13}$$

$$F_4 = A_1(1+i)(1+i)(1+i) + A_2(1+i)(1+i) + A_3(1+i) + A_4 \tag{11.14}$$

Assuming that the addition values are all equal to each other $(A_1 = A_2 = A_3 \cdots = A_n)$ leads to the following expression for future value for a series of uniform addition values:

$$F = A\left[(1+i)^{n-1} + (1+i)^{n-2} + (1+i)^{n-3} + \cdots + 1\right] \tag{11.15}$$

Multiplication by $(1+i)$ leads to:

$$(1+i)F = A\left[(1+i)^n + (1+i)^{n-1} + (1+i)^{n-2} + \cdots + (1+i)\right] \tag{11.16}$$

Subtraction of the previous expression (before multiplication) from this expression leads to:

$$(1+i)F - F = A[(1+i)^n - 1] \tag{11.17}$$

Rearrangement leads to:

$$F = A\left[\frac{(1+i)^n - 1}{i}\right] \tag{11.18}$$

which is a very useful future value formula in which A represents the uniform addition value that is added at the end of each period. This formula can be used to simultaneously determine the effect of accumulating deposits and associated interest over specified periods of time. An example of the future value of $200 additions over seven periods of time at 12% interest per period is shown in Fig. 11.2.

Example 11.2   Calculate the total value of earnings from a project if $1,000,000 is earned from sales each year for 25 years if the earnings are deposited at the end of each year into an investment portfolio earning 12% annual interest.

For this example, the addition value, A, is equal to $1,000,000, i is 0.12, and n is 25. Using the formula for future value determination from discrete additions leads to:

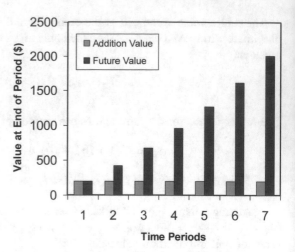

Fig. 11.2 Relationship between value and time periods for cash flow additions of $200 made at the end of each period for seven time periods with an interest rate of 12% per period

$$F = A\left[\frac{(1+i)^n - 1}{i}\right] = \$1,000,000\left[\frac{(1+0.12)^{25} - 1}{0.12}\right] = \$133,333,870$$

Thus, the final future value of the accumulating additions and interest is equal to $133,333,870.

11.1.5 Future Value for a Series of Discrete Addition Values that Increase with an Arithmetic Gradient

Many types of addition values increase with time in even increments according to an arithmetic gradient. For example, consider a scenario in which the addition value may begin with $400 the first year and increment by $100 each year thereafter. The basic formula is obtained by treating the accumulating increment ($100 each year in this scenario) separately from the base value ($400 in this scenario) as illustrated in Fig. 11.3. The base value can be accommodated by the previous future value expression for uniform additions. Therefore, only the series of increments needs to be treated in this section. The derivation begins with a series of future values that include only the increasing increments or gradient value, G. G is not added until the end of the second period since the first period addition is the base value. These assumptions result in:

$$F_{1ag} = 0 \tag{11.19}$$

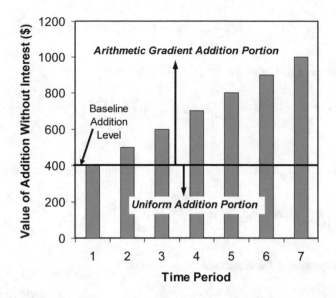

Fig. 11.3 Comparison of addition value (*without interest accrual*) and time periods for an arithmetic gradient addition that begins at $400 and is incremented by $100 each time period

F_{1ag} is the future value of the gradient portion of the arithmetic gradient addition portion only. The future value at the end of the second period includes the first gradient value, G. G is added at the end of the second period, leading to:

$$F_{2ag} = G \tag{11.20}$$

The future value at the end of the third period includes the first gradient value, G. It also includes interest from G. In addition, it includes the second increment of value $2G$. $2G$ is added only at the end of the third period. Thus, F_{3ag} is given as:

$$F_{3ag} = G(1+i) + 2G \tag{11.21}$$

The future values for the fourth and fifth periods of time are given as:

$$F_{4ag} = G(1+i)(1+i) + 2G(1+i) + 3G \tag{11.22}$$

$$F_{5ag} = G(1+i)(1+i)(1+i) + 2G(1+i)(1+i) + 3G(1+i) + 4G \tag{11.23}$$

Thus, a general expression for future value associated with the gradient portion of an arithmetic gradient addition is:

$$F_{ag} = G\left[(1+i)^{n-2} + 2(1+i)^{n-3} + 3(1+i)^{n-4} + \cdots + (n-1)\right] \qquad (11.24)$$

Multiplication of this equation by $(1 + i)$ leads to:

$$F_{ag}(1+i) = G\left[(1+i)^{n-1} + 2(1+i)^{n-2} + 3(1+i)^{n-3} + \cdots + (n-1)(1+i)\right]$$

$$(11.25)$$

Subtraction of the previous equation (before multiplication) from this equation leads to:

$$F_{ag}i = G\left[(1+i)^{n-1} + (1+i)^{n-2} + (1+i)^{n-3} + \cdots + (1+i) + 1\right] - nG \quad (11.26)$$

Previous expressions for F included the formula:

$$\left[(1+i)^{n-1} + (1+i)^{n-2} + (1+i)^{n-3} + \cdots + (1+i) + 1\right] = \frac{(1+i)^n - 1}{i} \quad (11.27)$$

This allows for an additional substitution that leads to:

$$F_{ag}i = G\left[\frac{(1+i)^n - 1}{i}\right] - nG \qquad (11.28)$$

Rearrangement to a more general form leads to:

$$F_{ag} = \frac{G}{i}\left[\frac{(1+i)^n - 1}{i} - n\right] \qquad (11.29)$$

Thus, this expression can be used to calculate the future value of an arithmetic addition gradient. In other words the future value of an addition that increases by a set amount, G, each period after the first period (beginning with the second period). An example of an arithmetic series gradient addition future value chart is presented in Fig. 11.4. Figure 11.4 is established for a base addition of $400 at the end of each period with at $100 per period gradient over a seven year period at 13% interest.

Example 11.3 Calculate the future value of an arithmetic series gradient addition that begins with $700 in the first year and increments by $150 each year thereafter for a total of 10 years. The associated interest rate is 13%.

This problem needs to be separated into two parts. The first part is the base value part ($700 per year), which is accommodated using the uniform series future value formula for additions. The second part is the gradient portion ($100 of increment

Fig. 11.4 Comparison of value versus time for a base addition of $400 at the end of each period with at $100 per period gradient over a seven year period at 13% interest

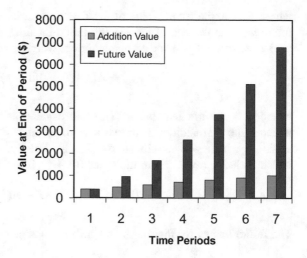

each year after the first year), which is accommodated using the future value formula for arithmetic gradient series.

$$F_{base} = A\left[\frac{(1+i)^n - 1}{i}\right] = \$700\left[\frac{(1+0.13)^{10} - 1}{0.13}\right] = \$12,894$$

$$F_{gradient} = \frac{G}{i}\left[\frac{(1+i)^n - 1}{i} - n\right] = \frac{\$150}{0.13}\left[\frac{(1+0.13)^n - 1}{0.13} - 10\right] = \$9,715$$

$$F_{Total} = F_{base} + F_{gradient} = \$12,894 + \$9715 = \$22,609.$$

11.1.6 Future Value for a Series of Discrete Addition Values that Increase with a Geometric Gradient

One common feature associated with additions is the tendency for additions to increase with time by a specified rate. For example, revenue from a rental may increment by 4% each year, or costs associated with health care may increase by 7% per year. Therefore, it is useful to have a future value equation that appropriately considers such addition scenarios that involve a geometric gradient. Derivation of such an expression can be made by beginning with an expression for future value for the first year:

$$F_{1gg} = A \qquad (11.30)$$

where F_{1gg} is the future value of the first geometric gradient series addition. The future value at the end of the second period includes the first addition, A, with its associated interest plus the second addition, which is equal to $A(1 + g)$, leading to:

$$F_{2gg} = A(1+i) + A(1+g) \tag{11.31}$$

where g is the gradient factor. The future value at the end of the third period includes the first addition with two periods of interest. It also includes the second addition, which was multiplied by $(1 + g)$. Finally, it includes a third addition which includes two gradient multiples $[(1 + g)(1 + g)]$:

$$F_{3gg} = A(1+i)(1+i) + A(1+g)(1+i) + A(1+g)(1+g) \tag{11.32}$$

The future values for the fourth and fifth periods of time are given as:

$$F_{4gg} = A(1+i)^3 + A(1+g)(1+i)^2 + A(1+g)^2(1+i) + A(1+g)^3 \tag{11.33}$$

$$F_{5gg} = A(1+i)^4 + A(1+g)(1+i)^3 + A(1+g)^2(1+i)^2 \\ + A(1+g)^3(1+i) + A(1+g)^4 \tag{11.34}$$

Thus, a general expression for future value associated with the gradient portion of an arithmetic gradient addition is:

$$F_{gg} = A\Big[(1+i)^{n-1} + (1+g)(1+i)^{n-2} \\ + (1+g)^2(1+i)^{n-3} + \cdots + (1+g)^{n-1}\Big] \tag{11.35}$$

Multiplication of this equation by $(1 + i)/(1 + g)$ leads to:

$$F_{gg}\left(\frac{1+i}{1+g}\right) = A\left[\frac{(1+i)^n}{(1+g)} + (1+i)^{n-1} + (1+g)(1+i)^{n-2} \\ + \cdots + (1+g)^{n-2}(1+i)\right] \tag{11.36}$$

Subtraction of the previous expression (prior to multiplication) from this expression leads to:

$$F_{gg}\left(\frac{1+i}{1+g}\right) - F_{gg} = A\left[\frac{(1+i)^n}{(1+g)} - (1+g)^{n-1}\right] \tag{11.37}$$

Further rearrangement leads to:

$$F_{gg} = A \left[\frac{\frac{(1+i)^n}{(1+g)} - (1+g)^{n-1}}{\left(\frac{1+i}{1+g}\right) - 1} \right] \tag{11.38}$$

Additional manipulation leads to:

$$F_{gg} = A \left[\frac{\frac{(1+i)^n}{(1+g)} - (1+g)^{n-1}}{\left(\frac{1+i}{1+g}\right) - \frac{1+g}{1+g}} \right] = A \left(\frac{\frac{(1+i)^n}{(1+g)} - (1+g)^{n-1}\left(\frac{1+g}{1+g}\right)}{\frac{i-g}{1+g}} \right) \tag{11.39}$$

Simplification leads to:

$$F_{gg} = A \left[\frac{(1+i)^n - (1+g)^n}{i - g} \right]; \quad (\text{where}, i \neq g) \tag{11.40}$$

Thus, this expression can be used to calculate the future value of a geometric addition gradient. In other words the future value of an addition that increases by a geometric gradient factor, g. An example of a plot of data with a geometric gradient addition for future health care premium costs (negative values) is shown in Fig. 11.5.

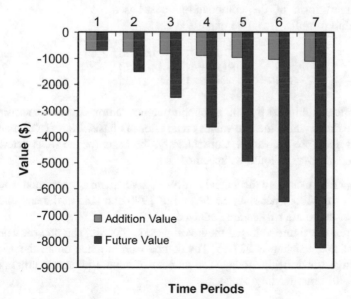

Fig. 11.5 Comparison of value versus time for a geometric series addition for a health care premium that begins with a value of -$700 at the end of the first period and increments by a gradient rate of 8% for seven periods while earning 10% interest each period

Example 11.4 Calculate the future value of sales that grow at a rate of 7% from an initial value of $400,000 per year assuming an annual interest rate of 12%. It is assumed the sales and associated interest accumulate for a total of 15 years.

This is a simple problem using the equation for future values based upon geometric gradients, where g = 0.07, i = 0.12, n = 15, and A = $400,000.

$$F_{gg} = A\left[\frac{(1+i)^n - (1+g)^n}{i - g}\right]$$

$$= \$400,000\left[\frac{(1+0.12)^{15} - (1+0.07)^{15}}{0.12 - 0.07}\right] = \$21,716,274$$

11.2   Return on Investment (ROI)

Most investors characterize the success of an investment based on the return. The return is the additional value they receive in addition to their principal investment. The return is often referred to as the rate of return (ROR), although in some specific contexts it is also known as the return on investment (ROI). It is equivalent to the effective interest rate applied to the investment. The rate-of-return can be calculated using a rearranged form of the compound-interest-based future value formula. It can be expressed as:

> *The return on investment (ROI) is the effective rate of interest earned by an investment.*

$$\text{ROR} = \left[\left(\frac{F}{P}\right)^{1/n} - 1\right] \tag{11.41}$$

It is often multiplied by 100 to obtain percent return. The most accurate ROR values are those which include values on an after-tax basis. Many ROR assessments are based upon pre-tax values. Consequently, the basis for the ROR determination (pre-tax or after-tax) should be specified.

Example 11.5 Calculate the annual return-on-investment for common stock that is purchased for $27.86 per share at the end of 1990 and sold at the end of 1999 for $88.50 per share after taxes are deducted.

The after-tax future value of the investment is $88.50. The present value in the context of this problem is $27.86. The comparison is made under the assumption that the investment begins today. The number of annual interest accrual periods is (1999–1990) nine, leading to:

$$ROR = \left[\left(\frac{88.50}{27.86} \right)^{1/9} - 1 \right] = 0.137$$

or a calculated ROR of 0.137 or 13.7%.

11.3 Cost Estimation

Costs for engineering projects are generally estimated using data from recent similar engineering projects. An accurate and detailed plant design is paramount to successful and realistic cost estimation. When design information is lacking, the accuracy of the cost estimation suffers. Project costs are most accurate when obtained from actual costs for recent similar projects. Often this information can be obtained from vendors and consulting companies. When this information is not available, other information sources, such as handbooks and tables, are used.

Cost estimates are generally based on past known costs of the same or similar items.

11.3.1 Cost Indexes

Most often, the effect of inflation is accounted for through cost indices. The most general and common cost index is the consumer price index (CPI). However, for commercial plant estimations the most common index is the Marshall and Swift Index (often referred to as M&S). Other indexes are also developed for specific industries. These indexes usually consist of a baseline year index value. The baseline index is set at an arbitrary baseline reference value such as 100. The value of the index changes relative to the baseline value. A mathematical formula to calculate value based on cost index values is:

$$C_{new} = C_{past} \frac{Index_{new}}{Index_{past}} \tag{11.42}$$

C_{new} is the new cost for the item. C_{past} is the cost for the item in the past. $Index_{new}$ is the new index value. $Index_{past}$ is the index value at the time period for which the past cost was obtained.

Example 11.6 Calculate the estimated cost of an equipment item in 2002 that cost $790,000 in 1980 based upon the Marshall and Swift Index.

This problem is solved using the values of the Marshall and Swift Index, which is found in the journal Chemical Engineering, in 1980 (675) and 2002 (1104.2) and the cost index equation:

$$C_{new} = C_{past}\frac{Index_{new}}{Index_{past}} = \$790,000\frac{1,104.2}{675} = \$1,292,323$$

11.3.2   General Inflation Estimation

Because cost index data is not always readily available for necessary items, it is useful to estimate the effect of inflation in a general way. Inflation corrections can be made by multiplying the cost estimates from an earlier time by the inflation adjustment factor, which can be calculated assuming an average inflation rate. The calculated average inflation adjustment factor is given as:

$$I_{AF} = (1+i)^n \tag{11.43}$$

I_{AF} is the adjustment factor. i is the fractional rate of inflation per time period (usually years). n is the number of time periods (usually years) between the present time and the time in which the cost information was generated.

11.3.3   Equipment Size Cost Adjustment

It is common to find cost information for production facilities using the same process as is desired for a new project. The cost information obtained for the existing similar facility can be easily adjusted for the effect of inflation as discussed previously. Often, cost information for other facilities is based upon a different level of production or facility size. Thus, it

> *The capital equipment cost to produce an item generally decreases as the number of items produced increases. This is often referred to as the economy of scale.*

is frequently necessary to adjust for size differences. A general formula that is in common use is [4].

$$C_{new} = C_{existing}\left(\frac{X_{new}}{X_{existing}}\right)^y \tag{11.44}$$

C_{new} is the new facility cost. $C_{existing}$ is the existing facility cost. X_{new} is the new facility capacity. $X_{existing}$ is the existing facility capacity. y is the cost ratio exponent, which is between 0.4 and 0.8 and is often close to 0.6 for most facilities.

11.3.4   Working Capital

All new projects need some financial resource in the form of working capital in order to run the operation. Working capital is essentially the amount of money needed to operate during lost revenue periods. It pays temporarily for items such as supply costs, payroll expenditures, utility bills, and maintenance bills. Although the project will result in a stream of sales revenue to replenish the cash outflow due to expenses, there will always be a need to have enough cash on hand to sustain the operation for a reasonable period of time. Because this monetary buffer is not available for investment, it is accounted for as a one time working capital cost. It is often around 20% of the total capital investment [10].

11.3.5   Start-Up Costs

Any new operation has costs associated with starting a project. Start-up items include training new personnel and adjusting equipment settings to achieve the desired production levels. Each of these items incurs costs that are unique to starting up a new process. These costs are usually small but significant one time costs. Start-up costs are usually 3–5% of annual operating costs [11].

11.3.6   Indirect Costs (Overhead Costs)

All projects need indirect support. This support can be in the form of property and liability insurance, plant security, etc. A reasonable estimate for such costs is 4% of the fixed capital investment cost each year [12].

11.4   Discounted Cash Flow Economic Analysis

In most design situations, several technically viable options are evaluated for economic viability. Projected net revenue (and earned interest) from the project is compared to the potential value of the project investment based upon a minimum acceptable rate-of-return. In other words, the financial evaluator determines if the overall anticipated net revenue for a potential project will be greater than an equivalent investment in an interest-bearing venue with an acceptable rate-of-return. Usually, the comparison is made to a hypothetical or comparative reference

> *The discounted cash flow method of economic analysis discounts the effects of time and interest to allow for comparisons of costs incurred at different times in an investment cycle.*

investment. It is based on a relatively high reference interest rate of between 10 and 15% due to the risk associated with corporate projects. If the value of the projected net revenue (or net cash flow plus interest) exceeds the value of the money in a comparable or hypothetical reference investment at the minimum acceptable reference interest rate (or minimum acceptable rate-of-return) for a determined period of time, the project is economically acceptable. The project is acceptable under these conditions. Under these conditions the investor will earn more money from the project than in a comparative reference investment.

A more conventional way of economical assessment is to determine if the project will result in a net income. This analyis is based on cash inflow and outflow. Interest is added at the specified minimum acceptable rate-of-return. If money is expended the investment plus interest accrual is negative. Thus, money invested in a project represents an initial opportunity cost. To the investor the initial investment is a loss, because the money is spent. It is possible that this money will not be recovered. In contrast, money that comes in as a result of the project is positive. Consequently, an investor desires to appropriately compare the value of expected income to the initial investment cost. Financial comparisons require proper consideration of interest at an acceptable rate-of-return. However, this comparison cannot be made without discounting the effect of time and interest. Each value must be appropriately converted to a common reference basis. The most common comparative basis is present value.

11.4.1 Equivalency Conversions

Values such as future values, present values, and addition values have equivalencies to each other. Equivalency conversions require discounting for the effect of time and interest. The equivalencies are generated by manipulating the equations that have already been shown. Future values can be converted to present values using the equation:

Future values can be converted to an equivalent present value using the appropriate formula.

$$P = \frac{F}{(1+i)^n} \tag{11.45}$$

This is simply a rearrangement of the initial expression for future values using compound interest. Future values can also be converted to present values by multiplying by the factor (P/F):

$$P = F\left(\frac{P}{F}\right) \tag{11.46}$$

(P/F) is known as the single payment compound factor. It is given by simple rearrangement of previous equations:

$$\frac{P}{F} = \frac{1}{(1+i)^n} = (1+i)^{-n} \tag{11.47}$$

When such conversion factors are used they are often included in a set of parentheses. Their citation includes the applicable interest rate and time period. The following example illustrates the conversion with the citation format.

Example 11.7 A company expects to sell a car to a junk yard for $500 after driving it for 20 years. Determine the equivalent present value of the anticipated future car sale. The applicable annual interest rate is 7%.

In this example the future value of the car is $500. The interest rate is 7%. The number of time periods is 20. Using the single payment compound factor:

$$P = F\left(\frac{P}{F}, 7\%, 20\right) = F\left(\frac{1}{(1+i)^n}\right) = \$500\left(\frac{1}{(1+0.07)^{20}}\right) = \$129.21$$

Thus, the salvage value of the car in 20 years is equivalent to a present value of $129.21.

Similarly, uniform addition values can be converted to present values by multiplying their respective future values by (P/F):

$$P = F\left(\frac{P}{F}\right) = \left(A\left[\frac{(1+i)^n - 1}{i}\right]\right)\left(\frac{1}{(1+i)^n}\right) \tag{11.48}$$

Alternatively, they can be converted to present value equivalency by multiplying by the factor P/A. This factor can be found by rearrangement of the previous equation to:

$$\frac{P}{A} = \left[\frac{(1+i)^n - 1}{i(1+i)^n}\right] \tag{11.49}$$

The conversion of a uniform addition value to a present value is demonstrated in the following example.

Example 11.8 Determine the equivalent investment cost of an anticipated maintenance cost of $5,000 per year over a 15-year period at an interest rate of 12%.

This problem can be solved by multiplying the given maintenance cost, which is a uniform addition cost, by the uniform addition P/A factor:

$$P = A\left(\frac{P}{A}, 12\%, 15\right) = -\$5000\left[\frac{(1+0.12)^{15} - 1}{0.12(1+0.12)^{15}}\right] = -\$24,054$$

> *The net present value is the sum or net of all values converted to their equivalent present values.*

The same approach can be used to determine other necessary conversion factors that are given in Table 11.1. In other words, it is feasible to compare present values to equivalent uniform additions or future values. It is also possible to convert all values to a future value equivalent reference. Similarly, comparisons can be made on the basis of equivalent addition values. The most common reference is the present value.

11.4.2   Net Present Value Analysis

Project financial viability is commonly made based on net present value. All addition and future values are discounted for time and interest by converting them to equivalent present values. The sum of the equivalent present values is the net present value.

The value of a project over time is illustrated in Fig. 11.6. The net value of a project begins with the negative value of the initial investment. After some time it is anticipated that project income will compensate for the investment. Gradually, the net present value will become less negative. Eventually, the net present value will become positive as illustrated in Fig. 11.6. It is apparent from Fig. 11.6 that the project needs to complete the sixth time period. This time is shown as "b" in the plot. This time is needed for the project to produce a net present value of zero. This point equals a "break-even" value. It is also known as the payout period.

Table 11.1 Economic equivalency conversion factors for normal interest compounding

Factor name	Formula
Single payment present value factor	$\frac{P}{F} = \frac{1}{(1+i)^n} = (1+i)^{-n}$ (11.50)
Single payment factor	$\frac{F}{P} = (1+i)^n$ (11.51)
Uniform series present value factor	$\frac{P}{A} = \left[\frac{(1+i)^n-1}{i(1+i)^n}\right]$ (11.52)
Capital-recovery factor	$\frac{A}{P} = \left[\frac{i(1+i)^n}{(1+i)^n-1}\right]$ (11.53)
Uniform series factor	$\frac{F}{A} = \left[\frac{(1+i)^n-1}{i}\right]$ (11.54)
Sinking-fund deposit factor	$\frac{A}{F} = \left[\frac{i}{(1+i)^n-1}\right]$ (11.55)
F/G factor for arithmetic series	$\frac{F}{G} = \frac{1}{i}\left[\frac{(1+i)^n-1}{i} - n\right]$ (11.56)
F/A_g factor for geometric series	$\frac{F}{A_g} = \left[\frac{(1+i)^n-(1+g)^n}{i-g}\right]$; (where, $i \neq g$) (11.57)
P/G factor for arithmetic series	$\frac{P}{G} = \left[\frac{(1+i)^n-in-1}{i^2(1+i)^n}\right]$ (11.58)
P/A_g factor for geometric series	$\frac{P}{A_g} = \left[\frac{(1+i)^n-(1+g)^n}{(i-g)(1+i)^n}\right]$; (where, $i \neq g$) (11.59)

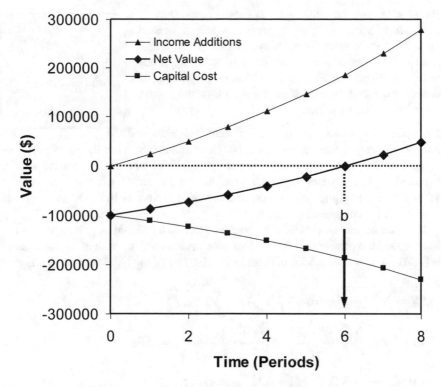

Fig. 11.6 Illustration of net value and investment cost and cash flow addition as a function of time. (11% interest, additions of $23,500, and an initial investment of −$100,000). The mark at "*b*" represents the payback period

where

P   present value
F   future value
A   addition value
G   arithmetic gradient portion of addition value
g   geometric gradient (fractional rate per period)
i   interest (fractional rate per period)
n   number of periods of interest accrual.

The net present value (NPV) can be expressed mathematically as the sum of all of the relevant values, multiplied by the appropriate equivalency conversion factors:

$$\text{NPV} = \sum P + \left(\frac{P}{A}\right) \sum A + \left(\frac{P}{A}\right) \sum F \qquad (11.60)$$

P represents present values such as capital costs. (*P/A*) is the uniform series addition to present value conversion factor. *A* represents addition values or income on a uniform series addition basis (gradients must be appropriately converted). (*P/F*) is the future to present value conversion factor or single payment compound factor. *F* represents future values.

> *A positive net present value indicates the investment will provide more income than an equivalent investment at the specified ROI.*

Example 11.9 Calculate the net present value of a project investment that requires a capital expenditure of $1,700,000 to purchase equipment that is expected to generate $450,000 of net annual income that is expected to last 7 years at which time the equipment is expected to sell for $500,000 as used equipment. The applicable rate of return is 10%. Determine if the investment is economically viable.

This is a simple net present value problem in which the sum of present values is - $1,700,000 (the negative sign is added because this is a cost), the sum of addition values is $450,000, *n* is 7, *F* is $500,000, and i is 0.10. Use of the NPV equation leads to:

$$\text{NPV} = \sum P + \left(\frac{P}{A}, 10\%, 7\right) \sum A + \left(\frac{P}{F}, 10\%, 7\right) \sum F$$

$$\frac{P}{A}, 10\%, 7 = \left[\frac{(1+i)^n - 1}{i(1+i)^n}\right] = \left[\frac{(1+0.1)^7 - 1}{0.1(1+0.1)^7}\right] = 4.868$$

$$\frac{P}{F}, 10\%, 7 = (1+i)^{-n} = (1+0.1)^{-7} = 0.5132$$

$$\text{NPV} = -\$1,700,000 + (4.868)(\$450,000) + (0.5132)(\$500,000) = \$747,200$$

The positive NPV of $747,200 indicates that over the 7 years of the project, the project will be worth the present equivalent of $727,200 more than the initial investment. In other words, this project will generate more than the minimum rate of return. The rate-of-return generated by this project can be determined by calculating the interest rate that gives an NPV equal to zero. Alternatively it can be calculated using future values and the ROR formula (Eq. 11.41).

11.4.2.1   Cash Flow Sources of Additions

Positive and negative cash flows arise from a wide variety of sources. Sources can include sales, maintenance, operation, taxes, tax deductions, supplies, overhead, etc. The future value of any source is:

$$F = \sum_{k=1}^{n} A_k (1+i)^{n-k} \tag{11.61}$$

where k is the specific time period from 1 to n, and A_k is the cash addition made during period "k". In other words the future value of a given cash flow is the sum of the individual additions that occur each period plus the associated interest. The value is based on the time they are added through the final period "n".

A classic example of changing additions is depreciation tax deduction. Because most items lose more value during the initial periods of use than they do near the end of their useful life, they are often depreciated in a nonlinear fashion. Therefore, the tax deduction changes each year. One common way of determining the depreciated future value of an item is the double declining balance method. This method gives the future depreciated value of an item as:

$$F = P\left(1 - \frac{2}{N}\right)^n \tag{11.62}$$

where

N number of periods of useful life
P present value (also the initial purchase price of the item, and
n number of accrual periods.

The method of determining the future value of the depreciating item is dependent upon current tax laws. Modern tax laws in the U. S. allow a double declining balance approach. This approach is known as the modified accelerated cost recovery system (MACRS). MARCS allows the depreciation to occur only for six months of the first year. Depreciation extends to a specified year during the middle of the depreciation life. The decline is linearized after the middle of the depreciation life [13].

The resulting depreciation tax deductions are determined each period. The calculation is made by multiplying the decrease in future book value during that period by the applicable corporate tax rate. Mathematically, this is expressed as:

$$T_{D,k} = (F_{k-1} - F_k)t \tag{11.63}$$

where

$T_{D,k}$ tax deduction for period "k".
F_{k-1} future book value of the item at the time prior to "k" F_k future book value of the item at time "k".

Often, tax deductions due to equipment depreciation are small. Some simplifications are commonly made, including the assumption that the depreciation is linear. Another useful quantity for simplified tax deduction analysis is the salvage value. The salvage value represents the lowest possible resale value of an item if it is sold as scrap material. The simplified periodic tax deduction for simple declining depreciation is given as:

$$T_{D(\text{lin.depr.})} = \frac{1}{N}(C - S)t \tag{11.64}$$

N = number of years over which the item is depreciated. C = capital cost of the item, which has a negative present value. S = salvage value of the item. t = the corporate tax rate.

Example 11.10 Determine the uniform addition value of the tax deduction associated with depreciation for a pipeline that is depreciated linearly over its 15-year life from an initial cost of \$30,000 to a salvage scrap value of \$5000 if the applicable rate of return is 12% and the associated corporate tax rate (federal, state, and local taxes combined) is 48%.

$$T_{D(\text{lin.depr.})} = \frac{1}{N}(C - S)t = \frac{1}{15}(\$30,000 - \$5,000)0.48 = \$800$$

The general form for determining the future value of the depreciation tax deduction is:

$$F = \sum_{k=1}^{n} T_{D,k}(1 + i)^{n-k} \tag{11.65}$$

This is essentially the same as Eq. 11.61 where the addition or annuity value, A, in this case is the tax deduction for each period "k". The present value of the deduction is positive, and for a typical linear deduction it can be expressed as:

$$P = \frac{\frac{1}{N}(C - S)t[(1 + i)^n - 1]}{i(1 + i)^n} \tag{11.66}$$

Because corporate taxes are based on profit rather than income, taxes are paid only on net profits after all expenses and tax deductions have been subtracted from income. However, taxes represent an additional cost that is subtracted from the pretax profit. The net effect of taxes is obtained by multiplying the sum of all non depreciable values by $(1 - t)$. In other words it is one minus the corporate tax rate fraction. The present value of fixed (depreciable) capital expenses does not need to be multiplied by $(1 - t)$. Fixed capital costs are distributed as equivalent addition values (costs) each year. For a typical, simplified analysis in which the depreciation of all depreciable item is linear the following expression can be used:

$$\text{NPV} = \sum P_{\text{Fixed}} + \sum \frac{P_N}{A}t\left(\frac{1}{N}\right)(|P_{\text{Fixed}}| - S)$$
$$+ (1 - t)\left[\sum P_{\text{Other}} + \frac{P}{A}\sum A\right] + \frac{P}{F}\sum S \tag{11.67}$$

ΣP_{Fixed} is the sum of all fixed or depreciable capital costs. ΣP_{Other} is the sum of all other or nondepreciable capital costs. P_N/A is the present value to addition value conversion factor for the number of years of depreciation of the fixed capital items. P/A is the present value to addition value conversion factor for the entire project. P_{fixed} is the present value of fixed capital investment cost (items such as equipment for which depreciation can be made). ΣA is the sum of the addition values (not including tax deductions for depreciation) t is the tax rate. N is the number of depreciation periods. S is salvage value. P/F is the present value to future value conversion factor for the entire project life.

Example 11.1 Calculate the after-tax and pre-tax, net present value of a new project that has the following financial data:

Fixed capital investment (land, equipment, constr. etc.	$15,000,000
Working capital	$ 3,000,000
Maintenance costs per anum	$ 1,050,000
Indirect costs per anum	$ 600,000
Annual sales revenue	$ 7,500,000
Other annual operating costs (supplies/raw materials)	$ 1,300,000
Salvage value (at the end of the project)	$ 1,500,000
Minimum annual rate of return	12%
Tax rate	45% (35% fed., 10% local)
Depreciation (straight line or linear with an average 15 year life) Project Life	15 years

The solution to this problem involves the expression for NPV that includes linear depreciation:

$$\text{NPV} = \sum P_{\text{Fixed}} + \sum \frac{P_N}{A} t \left(\frac{1}{N}\right) (|P_{\text{Fixed}}| - S)$$

$$+ (1 - t)\left[\sum P_{\text{Other}} + \frac{P}{A}\sum A\right] + \frac{P}{F}\sum S$$

$$\frac{P_N}{A} = \frac{P}{A}, 12\%, 15 = \left[\frac{(1+i)^n - 1}{i(1+i)^n}\right]$$

$$= \left[\frac{(1+0.12)^{15} - 1}{0.12(1+0.12)^{15}}\right] = 6.811$$

$$\frac{P}{F}, 12\%, 15 = (1+i)^{-n} = (1+0.12)^{-15} = 0.1827$$

The sum of addition values (not including depreciation claims for tax purposes) is:

$$\sum A = \$7,500,000 - \$1,050,000 - \$600,000$$
$$- \$1,300,000 = \$4,550,000$$

$$\sum P_{\text{Fixed}} = -\$15,000,000$$

$$\sum P_{\text{Other}} = -\$3,000,000$$

$$\text{NPV} = -15,000,000 + 6.811 \left[0.45 \left(\frac{1}{15} \right) (13,500,000) \right.$$
$$+ (1 - 0.45)[-3,000,00 + 6.811(4,550,000)]$$
$$+ 1,500,000(0.1827)$$

After tax NPV = \$3,427,033.

The pretax formula is:

$$\text{NPV} = \sum P_{\text{Fixed}} + \left[\sum P_{\text{Other}} + \frac{P}{A} \sum A + \frac{P}{F} \sum S \right]$$
$$\text{NPV} = -18,000,000 + 6.811(4,550,000)$$
$$+ 1,500,000(0.1827) = \$13,264,100$$

This example illustrates the importance of including taxes in any analysis. The equivalent rates-of-return that result at NPV = 0 are 15.6% (after-tax) and 24.4% (pre-tax) in this example. In this type of analysis the NPV equation is solved iteratively to find the rate of return, i, value that leads to a zero NPV value. However, it should also be noted that other investments that are used for comparison, such as stocks, also have associated taxes that must be properly considered for fair comparisons. Consequently, in many analyses, a pre-tax NPV may be appropriate when compared to other investments on a pre-tax basis.

All of the previous examples have been based on a common time element. All values were converted to equivalencies based upon a common project life number. In many cases the available data set will involve pieces that pertain to different time elements. For example, a tank or pipeline may need to be replaced two or three times during the life of a project in high wear situations. Thus, the time element for such items will be different than those of the overall project. Such differences in time can be easily accommodated by using the previous expressions appropriately as demonstrated in the following example.

Example 11.12 Calculate the after-tax, net present value of a new project that has the following financial data:

Fixed capital investment (depreciable equipment)	$15,000,000
Equipment replacement cost (once every 10 years) (treat as non-capital maintenance expense)	$1,500,000
Working/other capital	$3,000,000
Maintenance costs per anum	$1,050,000
Indirect costs per anum	$600,000
Annual sales revenue	$7,900,000
Other annual operating costs (supplies/raw materials)	$1,500,000
Salvage value (10% of capital cost)	$1,500,000
Minimum annual rate of return	12%
Tax rate	45% (35% fed., 10% local)
Depreciation (straight line or linear with an average 15 year life) project life	25 years

In order to accommodate the replacement costs every ten years, it is useful to convert such values to future values at the end of the project and then convert them back to equivalent addition values for the entire project life before incorporating them into the overall expression. (They cannot be used as equivalent future values unless proper consideration of taxes are considered. Because taxes are determined on an annual basis, the numbers need to be in terms of annual or addition values in order to appropriately consider applicable addition values based taxes.)

Replacement 1: At the end of year ten the replacement is made. This cost leads to a final end-of-project (15 years from year 10 to year 25) future value given by $(F = P(F/P))$

$$\frac{F}{P}, 12\%, 15 = (1+i)^n = (1+0.12)^{15} - 5.474$$

$$F_{\text{repl1}} = P\left(\frac{F}{P}, 12\%, 15\right) = -\$1,500,000(5.474)$$

$$= -\$8,211,000$$

The tax deduction portion, which is an addition value, is found by similar analysis:

$$\frac{F}{A}, 12\%, 15 = \frac{(1+i)^n - 1}{i} = \frac{(1+0.12)^{15} - 1}{0.12} = 37.28$$

$$F_{TD,\text{repl1}} = \left[t\frac{1}{N}(|P_{\text{Fixed}}| - S)\right]\frac{F}{A}$$

$$= 0.45\left(\frac{1}{15}\right)(1,500,000 - 150,000)(37.28)$$

$$= \$1,509,840$$

Conversion to an equivalent addition value for the entire project is made by multiplication by A/F over the entire project life:

$$\left(\frac{A}{F}, 12, 25\right) = \left[\frac{i}{(1+i)^n - 1}\right] = 0.0075$$

$$A_{\text{repl1equiv}} = F\left(\frac{A}{F}, 12, 25\right) = (\$8,211,000 + 1,509,840)[0.0075]$$

$$= \$50,259$$

Replacement 2: at the end of year twenty the replacement is made. This cost leads to a final end-of-project (5 years from year 20 to year 25) future value given by $(F = P(F/P))$

$$\frac{F}{P}, 12\%, 5 = (1+i)^n = (1+0.12)^5 = 1.7623$$

$$F_{\text{repl2}} = P\left(\frac{F}{P}, 12\%, 5\right) = -\$1,500,000(1.7623)$$

$$= -\$2,644,000$$

The resulting tax deduction is found as:

$$\frac{F}{A}, 12\%, 5 = \frac{(1+i)^n - 1}{i} = \frac{(1+0.12)^5 - 1}{0.12} = 6.353$$

$$F_{\text{TD,repl2}} = \left[t\frac{1}{N}(|P_{\text{Fixed}}| - S)\right]\frac{F}{A}$$

$$= 0.45\left(\frac{1}{15}\right)(1,500,000 - 150,000)(6.353)$$

$$= \$257,290$$

Conversion to an equivalent addition value for the entire project is made by multiplication by A/F:

$$\left(\frac{A}{F}, 12, 25\right) = \left[\frac{i}{(1+i)^n - 1}\right] = 0.0075$$

$$A_{\text{repl2equiv}} = F\left(\frac{A}{F}, 12, 25\right)$$

$$= (-\$2,644,000 + 257,290)[0.0075] = -\$17,900$$

The remainder of the solution to this problem involves the expression for NPV that includes linear depreciation:

$$\frac{P}{A}, 12\%, 25 = \left[\frac{(1+i)^n - 1}{i(1+i)^n}\right] = \left[\frac{(1+0.12)^{25} - 1}{0.12(1+0.12)^{25}}\right] = 7.843$$

$$\frac{P_N}{A}, 12\%, 15 = \left[\frac{(1+i)^n - 1}{i(1+i)^n}\right] = \left[\frac{(1+0.12)^{15} - 1}{0.12(1+0.12)^{15}}\right] = 6.811$$

$$\frac{P}{F}, 12\%, 25 = (1+i)^{-n} = (1+0.12)^{-25} = 0.05882$$

The sum of addition values (not including depreciation claims for tax purposes) is:

$$\sum A = 7,900,000 - 1,050,000 - 600,000$$
$$- 1,300,000 - 50,259 - 17,900 = \$4,881,841$$
$$\sum F = \$1,500,000 + \$150,000 + \$150,000 = \$1,800,000$$
$$\sum P = -\$15,000,000 - \$3,000,000 = -\$18,000,000$$
$$\text{NPV} = -15,000,000 + 6.811\left(0.45\left(\frac{1}{15}\right)\right)(13,500,000)$$
$$+ (1 - 0.45)[-3,000,000 + 7.843(4,881,841)]$$
$$+ 1,800,000(0.05882) = \$5,991,646$$

After tax NPV = $5,991,646.

Thus, this project is worth $5,991,646 more to the investor than could be achieved by investment in another venue at the specified rate-of-return based on an after tax NPV.

11.5   Evaluating Financial Effects of Risk

Another important element in economic evaluations is risk. All projects involve risk. Therefore, it is imperative that the effects of risk on project economics are assessed in an appropriate way. Risk is most often viewed in terms of probability and associated expectations.

> *The economic effect of risk can be quantified based on reasonable assumptions.*

11.5.1   Expected Values

11.5.1.1   Simple Probability Estimation

The principle of probability based estimation can be described using a gambling scenario. Assume an individual went to a gambling casino. The gambler placed a quarter into a slot machine. The slot machine had a designed payout of $100 when three of the same item were randomly selected. The payout event occurs approximately once every 729 uses. The expected value of patronizing such a machine with one quarter can be expressed as:

$$E = \sum \rho P$$
$$E = \left(\frac{1}{729}\right)(\$100) + (1)(-\$0.25) = -\$0.1128 \tag{11.68}$$

E is the expected value. ρ is the probability of the outcome. P is the present value of the outcome. Thus, the expected value of using such a slot machine is a cost of about 11 cents per use. It is true that just one attempt may result in the successful outcome. However, the reality is that the long-term average use of the slot machine will result in the loss of 11 cents. The 11 cent loss to the patron remains with the casino. Consequently, the casino is guaranteed long-term income. The patron is guaranteed losses if such a machine is used a sufficient number of times. This approach of determining expected values can also be used to estimate the potential value associated with investment risks. The following example evaluates investment risk.

Example 11.13   A fuel company that is seeking to augment its supply of natural gas is proposing to drill more wells. Determine the expected value of each well if the drilling cost is $750,000. The success rate for each well is 35%. The average income on a successful well is $1,800,000 assuming that the income is recovered in one year.

In this case there is a 100% chance that a well will cost $750,000, and a 35% chance of successfully realizing an income of $1,800,000. Thus,

$$E = \sum \rho P (0.35)(\$1,800,000) + (1)(-\$750,000) = -\$120,000$$

Consequently, this type of venture is not profitable under these conditions.

11.5.1.2   Probability Tree Estimation

Probability trees are another method for determining expected values. These trees are designed to help illustrate and determine expected values from potential investments that have several different possible outcomes. For example, consider a company that desires to determine the expected value of a research program that is designed to produce a successful modified version of an existing process for which the following probability tree has been constructed:

Using the data from the decision tree Table 11.2 can be constructed:

Thus, the expected value of this project is expressed as:

$$
\begin{aligned}
E = {} & (1.00)(-\$200,000) + 0.10(-\$150,000) \\
& + 0.06(-\$750,000) + 0.012(0) + 0.009(\$900,000) \\
& + 0.033(\$4,500,000) + 0.006(\$9,500,000) = -\$46,400
\end{aligned}
$$

Therefore, despite the potential for high values at the end of the research, the overall expected value is not favorable due to the low probability of achieving one of the high value outcomes. Another example is presented in Example 11.13.

Example 11.14 Calculate the expected value of this year's maintenance costs for a seven-year-old automobile if the annual probabilities of transmission, engine, miscellaneous failures are 0.15, 0.07, and 0.25, respectively and the costs of repair vary depending upon dealer and extent of damage with their respective probabilities as shown in the diagram (see also Table 11.3).

$$
\begin{aligned}
E = {} & 0.09(-\$1800) + 0.06(-\$1500) \\
& + 0.049(-\$1200) + 0.021(-\$2600) \\
& + 0.1625(-\$700) + 0.0875(-\$1500) = -\$610.40
\end{aligned}
$$

Table. 11.2 Decision tree data from Fig. 11.7

Event	Probability	Values
Exploratory testing	(1 00)(1.0) − 1.00	−$200,000
Additional lab testing	(0.10)(1.0) = 0.10	−$150,000
Pilot-scale testing	(0.10)(0.6) = 0.06	−$750,000
Outcome 1	(0.10)(0.6) (0.20) = 0.012	$0
Outcome 2	(0.10)(0.6) (0.15) = 0.009	$900,000
Outcome 3	(0.10)(0.6) (0.55) = 0.033	$4,500,000
Outcome 4	(0.10)(0.6) (0.10) = 0.006	$9,500,000

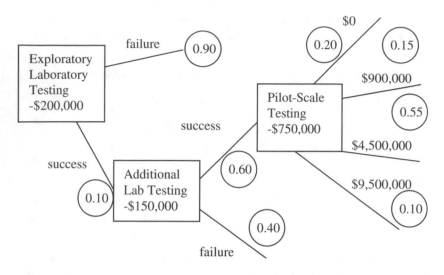

Fig. 11.7 Probability tree for a research program with associated probabilities and projected values for the associated outcomes

Table. 11.3 Decision tree data from Fig. 11.8

Event	Probability	Values
Transm. outcome 1	(0.15)(0.60) = 0.09	−$1800
Transm. outcome 2	(0.15)(0.40) = 0.06	−$1500
Engine outcome 1	(0.07)(0.70) = 0.049	−$1200
Engine outcome 2	(0.07)(0.30) = 0.021	−$2600
Other outcome 1	(0.25)(0.65) = 0.1625	−$ 700
Other outcome 2	(0.25)(0.35) = 0.0875	−$1500

11.5.1.3   Monte Carlo Simulation-Based Estimation

The maintenance cost scenario presented in Example 11.10 illustrates that costs can vary for an individual item under consideration. The variance in cost for one item is likely to be different than for other items. Although the probability tree approach to addressing elements of uncertainty or risk in economic assessment is useful and relatively simple, it can only accommodate a few possibilities. A more accurate assessment can be made using statistical functions with associated distributions of values.

An example of a more realistic distribution in value is presented in Fig. 11.9 for annual income or value from $100,000 of investment based upon a sampling of 25 investors. The results show that on average the investors earned $15,760. However, there is considerable variation in the data. Some earned little and others earned more than the average. The line in Fig. 11.9 represents the predicted distribution of values based upon a normal distribution function. The data seem to fit the normal

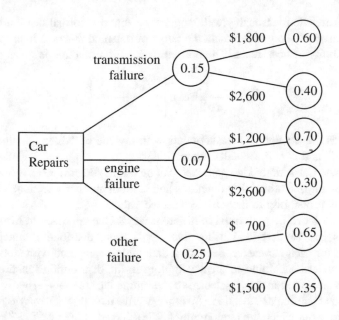

Fig. 11.8 Probability tree for annual car repairs given in Example 11.10

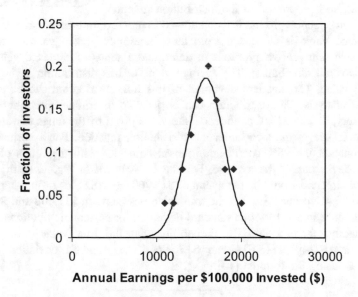

Fig. 11.9 Comparison of fraction of investors versus annual earnings per $100,000 for a hypothetical group of 25 investors. The diamonds represent actual data points, and the solid line represents a best-fit normal distribution function.

distribution function reasonably well, suggesting that the normal distribution function is a reasonable way to represent the range of potential outcome in proportion to their probability of occurrence. The normal distribution function is given as [14].

$$f(x) = \frac{1}{\sigma\sqrt{2\pi}} \exp\left[\frac{-(x - x_{\text{ave}})^2}{2\sigma^2}\right]$$ (11.69)

where $f(x)$ is the fraction of measurements with a value of "x", x_{ave} is the average value of "x", and σ is the standard deviation.

The Monte Carlo simulation approach can be used to determine expected values. This method utilizes probability functions and random number selection. Values are weighted by probability to determine expected values.

This method can be performed in different ways. One approach to Monte Carlo simulations can be described using five steps. First, distribution functions are determined for each parameter. Second, a data table is prepared with slots that are filled with values according to their weighting in the distribution function. Third, slots in the table are randomly selected to determine the value for that particular selection. In other words the number pool from which the random number is selected is weighted. The weighting a higher proportion near the average and lower proportions far from the average. The weighting is in proportion to the distribution function.

> *Statistical functions can be used to help quantify the effect of risk as well as the uncertainty associated with costs.*

The random selection process is then allowed to proceed for each parameter. Fourth an expected value is calculated from these randomly picked values. Fifth, the random selection process proceeds to determine a series of expected values that form an overall distribution. The average value of this distribution is the overall expected value. The standard deviation of this final distribution can be used for statistical analysis. The distribution can be used to determine appropriate confidence intervals. The 95.4% confidence interval is given by the range of values that comprise 95.4% of the area under the distribution function. For a normal distribution function the 95% confidence interval coincides with the limits of ± two standard deviations of the average. For the data shown in Fig. 11.9, the 95.4% confidence interval would be the average ($15,760) ± 2($2146) or in other words, there is a 95.4% probability that the true value lies between $11,468 and $20,052. The confidence interval for one standard deviation for normal distributions is 68%. Consequently, there is a 68% probability that the true value lies between $15,760 ± 1($2146) [$13,614 and $17,906]. The standard deviation can be determined using the formula [15].

$$\sigma = \sqrt{\frac{1}{N-1}\sum(x_j - x_{\text{ave}})^2}$$ (11.70)

σ is the standard deviation of x. N is the number of measurements. x_j is the value of x at position j. x_{ave} is the average value of x. In other words the standard deviation is a measure of the variation of each value of a given parameter in a collection. Important values for confidence intervals as they relate to standard deviations and normal distributions can be determined. The mean value $\pm0.57\sigma$ is used for 50% confidence. The mean value $\pm1.28\ \sigma$ is used for the 80% confidence interval. The mean value $\pm2.38\sigma$ gives the 98% confidence interval. The mean value $\pm3.4\sigma$ is used for the 99.9% confidence interval. The mean value $\pm4.0\ \sigma$ gives the 99.99% confidence interval [16].

11.5.2   Contingency Factors

An important additional element of cost estimation is the provision or contingency for unforeseen costs. These unforeseen costs may be associated with economic risk that arises due to project delays, large increases in equipment prices, changes in design, etc. Such contingency adjustments to the estimated project costs are usually made after all other costs have been established. However, they are made prior to a final economic analysis. Contingencies are designed to minimize financial risks associated with new projects. The more clearly a project is defined, the lower the contingency becomes. For preliminary evaluations, the overall contingency may be 35 or 40%. As the project becomes well defined and reliable cost estimates are obtained from recent similar projects, the overall contingency may drop to 15%. In other words, if a project is expected to cost $1,000,000, a complete economic assessment may be based upon the expected cost of $1,000,000 plus a contingency value of 15% of the total cost ($150,000) for a total estimated project cost of $1,150,000.

11.5.3   Sensitivity Analysis

Another approach to assessing risk is a sensitivity analysis. This consists of changing input variables such as capital cost, operating cost, and projected revenue and measuring the sensitivity of the response variable. A common response variable is rate-of-return. Thus, the change in response is measured relative to a change in input variables. The resulting changes in the response variable will be most sensitive to changes in the most important input variables. Thus, this approach is designed to reveal the most important input variables. The most important input variables require the most accurate information. An example of a simple sensitivity analysis is presented in Tables 11.4 and 11.5.

In Table 11.4 it is clear that the changes in the initial investment cost have a very large impact on the change in the rate-of-return. Thus, the rate-of-return is very sensitive to the initial investment cost.

In Table 11.5 it is clear that the changes in the operating addition values have a significant impact on the change in the rate-of-return. However, the influence of

Table. 11.4 Analysis of sensitivity of rates-of-return to capital cost value changes

Investment value	% change in NPV	ROR (%)	% Change in ROR
−$200,000	−50	60	200
−$300,000	−25	40	100
−$400,000	0	20	0
−$500,000	25	00	−100
−$600,000	50	−20	−200

Table. 11.5 Analysis of sensititvity of rates-of-return to operating addition value changes

Operating addition	% change in A	ROR (%)	% Change in ROR
−$40,000	−50	37	85
−$60,000	−25	28	40
−$80,000	0	20	0
−$100,000	25	12	−40
−$120,000	50	2	−90

operating addition values on the rate-of-return is not as great as the initial investment. Thus, the rate-of-return is less sensitive to the operating costs than it is to the
initial investment cost.

Engineering Economics Terminology

Amortization—Payment plan to fulfill a financial obligation.
Annuity—A series of equal payments made at equal time periods. Also commonly
associated with specific retirement account distributions.
Assets—Everything owned by or owed to a corporation.
Bond—A financial instrument that is similar to a promissory note. The bond holder
is a creditor to whom interest payments are usually made periodically and the
principal is repaid at maturity.
Book value—Original cost minus accumulated depreciation.
Capital—Monetary resource used for purchases that are depreciated, depleted, or
amortized over time such as equipment and buildings.
Capital recovery—The process of recovering capital through periodic payments.
Cash flow—Money that goes into or out of a financial venture.
Common stock—An instrument representing financial ownership.
Current assets—All potential values that are reasonably available within a business
cycle. These include receivables, cash, bonds, and inventories.
Current liabilities—Company debt that is payable within the business cycle (e.g.
1 year)
Debenture—A promissory note such as bond that is issued by a company.
Deferred annuity—A series of payments that are deferred to another time period.
Depletion—Decline in available resource due to extraction.
Depreciation—Decline in value of an asset.

Direct cost—Costs attributable to specific products, operations, or services.

Discounted cash flow—Monetary value that has been discounted for the effect of time and interest to a common reference such as present value.

Equity—Portion of capital owned by an investor.

Expected value—Value expected from an investment after consideration of risk.

Fixed cost—Cost that is unaffected by production level.

Fixed assets—Assets that consist of depreciable, depletable, or amortizable items such as buildings, equipment, mineral deposits, etc.

Goodwill—Value associated with intangibles such as public or customer satisfaction

Indirect cost—Costs not directly traceable to production, operations, or services. Often referred to as overhead costs.

Intangible assets—Assets that are not physical in nature such as goodwill and trademarks

Liabilities—All claims such as wages, dividends, bonds, accounts payable, etc against a corporation.

Marginal cost—The cost of one additional production or service unit.

MACRS Modified accelerated cost recovery system—A method of property depreciation that uses specific double-declining depreciation schedules that differ depending upon specific items and their useful life.

Opportunity cost—Cost associated with resource commitment that precludes other options.

Overhead cost—See indirect cost.

Payback period—Time necessary to achieve a net present value of zero from an investment.

Preferred stock—An instrument of capital ownership with fixed dividends that are paid before common stock dividends, no voting rights, and priority asset recovery over common stock in the event of asset liquidation.

Rate of return—Rate of value change for an investment.

Recovery period—See payback period.

Retained earnings—Profit that is reinvested into the institution of origin rather than distributed to common stock holders as dividends.

Salvage value—Value of a used item that is recovered when it is sold.

Working capital—Investment funds needed to finance general operations such as supply costs, utility bills, and payroll obligations. It is effectively the money in the corporate bank account in excess of current liabilities that is used to run the operation.

References

1. D.G. Newnan, J.P. Lavelle, T.G. Eschenbach, in *Essentials of Engineering Economic Analysis*, 2nd edn (Oxford Press, New York, 2002), p. 86
2. T.G. Eschenbach, *Engineering Economy: Applying Theory to Practice*, 2nd edn (Oxford Press, New York, 2003), p. 26

3. M.S. Bowman, in *Applied Economic Analysis for Technologiests, Engineers, and Managers* (Prentice-Hall, Upper Saddle River, 1999), p. 130
4. J.R. Couper, in *Process Engineering Economics* (Marcel-Dekker, New York, 2003), p. 155
5. F.J. Stermole, in *Economic Evaluation and Investment Decision Methods*, 3rd edn (Investment Evaluations Corporation, Golden, 1980), p. 13
6. G.J. Thuesen, W.J. Fabrycky, in *Engineering Economy*, 7th edn (Prentice-Hall, Englewood Cliffs, 1989), p. 40
7. J.L. Riggs, D.D. Bedworth, S.U. Randhawa, in *Engineering Economics*, 4th edn (McGraw-Hill, New York, 1996), p. 29
8. H. Levy, M. Sarnat, in *Capital Investment and Financial Decisions*, 5th edn (Prentice-Hall, New York, 1994), p. 36
9. R.A. Brealey, S.C. Myers, in *Capital Investment and Valuation* (McGraw-Hill, New York, 2003), p. 39
10. J.R. Couper, *Process Engineering Economics* (Marcel-Dekker, New York, 2003), pp. 110–111
11. J.R. Couper, in *Process Engineering Economics* (Marcel-Dekker, New York, 2003), p. 115
12. J.R. Couper, in *Process Engineering Economics* (Marcel-Dekker, New York, 2003), p. 136
13. Jones, in *Principles and Prevention of Corrosion*, 2nd edn (Prentice-Hall, 1996), pp. 538–548
14. J.R. Taylor, in *An Introduction to Error Analysis* (Oxford University Press, New York, 1982), p. 111
15. J.R. Taylor, in *An Introduction to Error Analysis* (Oxford University Press, New York, 1982), p. 87
16. J.R. Taylor, in *An Introduction to Error Analysis* (Oxford University Press, New York, 1982), p. 116

Problems

(11-1) Calculate the future value of an equipment item at the end of 12 years if it is purchased today for $15,000 based upon an accepted annual rate of return of 15%. (Answer = $80,254)

(11-2) Calculate the future value of an annual cash flow of $1,800,000 over 20 years if the applicable annual interest rate is 12%.

(11-3) Calculate the future value of an operating expense that begins at the end of the first year at $10,000 per year and increases by $1,000 per year for each of the next 14 years to a final level of $24,000 per year at the end of the final 15th year of the project. Assume that the accepted annual rate of return is 13%.

(11-4) Calculate the future value over 20 years for an annual health care premium that begins at the end of the first year at $3,000 per year and increases annually at a 7% rate. Assume that the accepted rate of return is 12%. (Answer = -$346,597)

(11-5) Calculate the rate of return for an investment that sold for an after-tax equivalent of $800,000 at the end of 20 years if the initial investment was $150,000.

(11-6) Estimate the current cost of an item that could be purchased for $500 20 years ago if the current relevant cost index for the item is 925 and the previous cost index value (20 years ago) was 438.

(11-7) A company expects to sell property for $800,000 in 10 years. Determine the equivalent present value of the anticipated future property sale if the applicable annual interest rate is 6.5%.

(11-8) Determine the equivalent current cost of an anticipated expense of $89,000 per year over a 25-year period assuming an annual interest rate of 10%. (Answer = $807,857)

(11-9) Determine the equivalent present value of accumulating health care costs that begin at $7,000 per year and increase by 8% each year for 24 additional years (25 years of total cost accumulation) if the applicable interest rate is 11%. (5 points)

(11-10) Calculate the net present value of a company expansion that requires $5,300,000 to purchase equipment that is expected to generate $1,300,000 of net annual income that is expected to last 12 years. The value of the equipment at the end of the project is expected to be $900,000. The applicable rate of return is 12%. Determine if the investment is economically viable given the required rate of return. (Answer = $2,983,230)

(11-11) Calculate the after-tax, net present value of a new project that has the following financial data:

Fixed capital investment (land, constr., equipment, etc.)	$3,750,000
Working capital	$750,000
Maintenance costs per anum yr 1 (increase by 8%/year)	$575,000
Indirect costs per anum	$100,000
Annual sales revenue (year 1) (increase by 10% /year)	$1,700,000
Other annual operating costs (supplies/raw materials)	$300,000
Salvage value (at the end of the project)	$400,000
Minimum annual rate of return	11%
Tax rate	35%
Depreciation (straight line or linear with an average 15-year life) project life	15 years

(11-12) Calculated the expected value (per nickel) when playing slot machines with nickels when the average payout of $10 occurs an average of once per 350 nickels entered.

(11-13) An energy company that is seeking to produce more oil is proposing to drill more wells. Determine the expected value of each well if the average exploration cost associated with each well is $400,000, the average drilling cost in the proposed exploration area will be $900,000 per well, and the average success rate for each well is 30%, and the average income on a successful well is $3,500,000 assuming that the income is recovered in one year.

Chapter 12
General Engineering Statistics

> *Statistics provide important tools that can help engineers to make objective evaluations and decisions.*

Key Chapter Learning Objectives and Outcomes:
Know how to evaluate measurement error and uncertainty
Understand how to use the correct number significant figures
Know how to calculate basic statistical terms
Know how to obtain basic statistical probabilities and confidence intervals
Know how to calculate and evaluate statistical data
Understand how to apply statistical design of experiments and evaluate its data

12.1   Uncertainty

All measurements have some degree of uncertainty. Some uncertainty arises from the level of accuracy of the device used to make the measurements. Additional uncertainty accompanies human error associated with the recording and utilization of the information. Uncertainty is reported with an associated measurement in the form [1]:

$$x = x_{best} \pm \delta x \tag{12.1}$$

where x is the measurement quantity, x_{best} is the best estimate for a measurement of x, and δx is the uncertainty or error associated with the measurement. The uncertainty term represents the quantity by which a measurement is likely to differ above or below the best estimate of the measurement. The uncertainty for an instrument reflects the accuracy of the instrument. For example, if a digital multimeter is accurate to within 0.001 V of the correct value, a measurement of a voltage source of exactly 10.034 V would be reported as 10.034 ± 0.001 V. This, of course, assumes accurate calibration. However, because the measurement requires the connection of wire clips to both the multimeter and the voltage source, which can affect the measured value along with electrical noise associated with overhead lights

© The Minerals, Metals & Materials Society 2022
M. L. Free, *Hydrometallurgy*, The Minerals, Metals & Materials Series,
https://doi.org/10.1007/978-3-030-88087-3_12

and adjacent power sources, the real uncertainty is likely much larger. If the combination of all potential uncertainties is 0.005 V, the measured value of the exact voltage source of 10.034 V should be reported as 10.034 ± 0.005 V.

12.1.1   Significant Figures

It is easy to inappropriately state numbers with more figures than can be reasonably justified. If a measured distance was measured to be 10.25 ± 0.04 m, it is likely that the instrument gave more figures than were relevant. Perhaps the meter reported 10.2532 m, and the uncertainty of the measurement was calculated to be 0.0374 m. It would not be appropriate to report 10.2532 ± 0.0374 m because the figures to the right of 10.25 are well below the level of uncertainty. In other words, the uncertainty is greater than 0.01, so any figure that is less than 0.01 in the measured value is uncertain and, therefore, insignificant. Consequently, values are rounded to the order of magnitude (or same left-most decimal position) of the uncertainty. In this case the uncertainty is of the order of magnitude of 0.01. Thus, the corresponding measurement of 10.2532 is rounded to the nearest 0.01 value or 10.25. Moreover, the most appropriate statement of the measurement is 10.25 ± 0.04 m. Thus, there are only four significant figures in the measured value—even though there were six figures given by the instrument.

> *Measurement uncertainties can be used to more accurately report and evaluate processes.*

If values are multiplied, divided, or added to, the number of significant figures in the final computed value should not be greater than the value with the lowest number of significant figures.

Example 12.1   A laboratory technician weighs 0.0058 g of sodium chloride (Mw 58.443) to prepare a 1-L solution of 0.0001 M NaCl. Determine the resulting concentration and the correct number of significant figures. Neglect uncertainties.

The calculated value of the concentration is calculated to be (0.0058/58.443) 0.000099242 M. However, it should be noted that the weighed value has only two significant figures; therefore, the final concentration should be reported as 0.000099 M.

12.1.2   Systematic or Dependent (Maximum) Uncertainties in Sums and Differences

The uncertainties in sums and differences are additive. If the uncertainty in the weight of particle A is 5 g and the uncertainty or error in the weight of particle B is 3 g, the uncertainty associated with the combined weights of A and B would be approximately 8 g. The general formula for systematic or dependent uncertainties, which represents a maximum uncertainty, in sums and differences, δq, for the combined measurement, q, of x, y, z, ($q = (x + y + z)$ or $q = (x - y - z)$) is [2]:

$$\delta q \approx \delta x + \delta y + \delta z \qquad (12.2)$$

δq is the uncertainty in the measurement of the sum or difference, q. δx is the uncertainty in the measurement of x. δy is the uncertainty in the measurement of y. δz is the uncertainty in the measurement of z.

12.1.3 Systematic or Dependent (Maximum) Uncertainties in Products and Quotients

The uncertainties in products and quotients are also additive when the values are utilized on a fractional uncertainty basis. The general formula for systematic or dependent uncertainties, which represents maximum uncertainty, associated with products and quotients is [3]:

$$\frac{\delta q}{|q|} \approx \frac{\delta x}{|x|} + \frac{\delta y}{|y|} + \frac{\delta z}{|z|} \qquad (12.3)$$

12.1.4 Systematic or Dependent (Maximum) Uncertainties in Powers

The uncertainties for variables with powers that have systematic or dependent uncertainties are determined by multiplying the fractional uncertainty by the power. The general formula for the maximum systematic uncertainty in q, δq, where $q = x^n$ is [4]:

$$\frac{\delta q}{|q|} \approx n \frac{\delta x}{|x|} \qquad (12.4)$$

where δq is the uncertainty in the measurement of q, n is the power, and δx is the uncertainty in the measurement of x.

12.1.5 Random and Independent Uncertainties in Sums and Differences

The propagation of errors or uncertainty in values that are independently measured and have random uncertainties is determined by addition in quadrature. The general formula for the random uncertainties in sums and differences, δq, for the combined measurement, q, which is a function of x, y, z, ($q = (x + y + z)$ or $q = (x - y - z)$) is [5]:

$$\delta q = \sqrt{(\delta x)^2 + (\delta y)^2 + (\delta z)^2} \tag{12.5}$$

This expression always gives a reduced estimate of the uncertainty relative to the systematic or dependent uncertainty case (see Eq. (12.2)).

12.1.6 Random and Independent Uncertainties in Products and Quotients

Random uncertainties in products and quotients are also additive in quadrature when the values are utilized on a fractional uncertainty basis. The general formula for random or independent uncertainties associated with products and quotients is [6]:

$$\frac{\delta q}{q} = \sqrt{\left(\frac{\delta x}{x}\right)^2 + \left(\frac{\delta y}{y}\right)^2 + \left(\frac{\delta z}{z}\right)^2} \tag{12.6}$$

Example 12.2 Calculate the volume and associated uncertainty of a heap leaching lift if the length (l), width (w), and height (h) are respectively 750 ± 20 m, 150 ± 10 m, 9 ± 2 m.

The first step is to determine the fractional uncertainty in the volume calculation:

$$\frac{\delta V}{|V|} = \sqrt{\left(\frac{\delta l}{l}\right)^2 + \left(\frac{\delta w}{w}\right)^2 + \left(\frac{\delta h}{h}\right)^2} = \sqrt{\left(\frac{20}{750}\right)^2 + \left(\frac{10}{150}\right)^2 + \left(\frac{2}{9}\right)^2} = 0.2335$$

The next step is to determine the best estimate of the volume:

$$V_{\text{best}} = lwh = (750\,\text{m})(150\,\text{m})(9\,\text{m}) = 1{,}013{,}000\,\text{m}^3$$

Next, the fractional uncertainty in volume is converted into a direct uncertainty:

$$\delta V = \frac{\delta V}{|V|}|V| = 0.2335(1{,}013{,}000\,\text{m}^3) = 236{,}000\,\text{m}^3$$

Finally, the numbers are combined into a value with the correct number of significant figures:

$$V = V_{\text{best}} \pm \delta V = 1{,}000{,}000 \pm 200{,}000\,\text{m}^3$$

12.1.7 Random and Independent Uncertainties in Functions of Several Variables

Random uncertainties in functions of several variables are also additive in quadrature. The partial derivatives of the variables in the function are multiplied by the uncertainty in the variable measurement. The general formula for random or independent uncertainties associated with functions of several variables is [7]:

$$\delta q = \sqrt{\left(\frac{\partial q}{\partial x}\delta x\right)^2 + \left(\frac{\partial q}{\partial y}\delta y\right)^2 + \left(\frac{\partial q}{\partial z}\delta z\right)^2} \tag{12.7}$$

Example 12.3 Calculate the reaction constant and associated uncertainty for a reaction system where $R = kC_A C_B{}^2$ if the reaction rate (R), concentration of A (C_A), and concentration of B (C_B), are respectively 0.1 ± 0.003 mol/hr, 0.01 ± 0.0005 mol/l, 0.01 ± 0.0003 mol/l.

The first step is to determine the functional relationship between the variables:

$$k = \frac{R}{C_A C_B^2}$$

The next step is to determine the uncertainty in the rate constant:

$$\delta k = \sqrt{\left(\frac{\partial k}{\partial R}\delta R\right)^2 + \left(\frac{\partial k}{\partial C_A}\delta C_A\right)^2 + \left(\frac{\partial k}{\partial C_B}\delta C_B\right)^2}$$

$$\delta k = \sqrt{\left(\frac{\delta R}{C_A^1 C_B^2}\right)^2 + \left(-\frac{R}{C_B^2}\delta C_A\right)^2 + \left(-\frac{R}{C_A}\delta C_B\right)^2}$$

$$\delta k = \sqrt{\left(\frac{0.003}{(0.01)(0.01)^2}\right)^2 + \left(-\frac{0.1}{(0.01)^2}0.0005\right)^2 + \left(-2\left(\frac{0.1}{0.01}\right)0.0003\right)^2} = 3000$$

The next step is to determine the best estimate of the reaction constant:

$$k = \frac{R}{C_A C_B^2} = \frac{0.1\,\text{mol/hr}}{0.01\,\text{mol/l}(0.01\,\text{mol/l})^2} = 100{,}000\,\text{mol}^{-2}\text{l}^{-3}\text{hr}^{-1}$$

Finally, the numbers are combined into a value with the correct number of significant figures:

$$k = k_{\text{best}} \pm \delta k = 100{,}000 \pm 3000\,\text{mol}^{-2}\text{l}^{-3}\text{hr}^{-1}$$

12.2   Basic Statistical Terms and Concepts

The correct application of statistics requires an ppropriate understanding of basic statistical terms and concepts. The term *population* signifies a collection of all individual items of a specific type or system such as the population of inhabitants within an entire city. *Sample* is defined as a randomly selected group or subset of items within a population. Examples of samples include a set of opinions obtained through a survey given to a small portion of a city's population or a group of measurements of the quantity of a chemical in a series of sample vials taken from the same solution source. As the number of data pieces in a sample becomes large, it approximates a population. The *sample mode* is the most commonly occurring value within a sample. The *sample median* is the value that is in the middle of a sample. In other words the sample median value is larger than or equal to half of the values and less than or equal to half of the values in a sample. The *sample mean* is the numerical average of all of the values contained within a sample. Thus, for the sample of values 1, 4, 4, 4, 7, 8, 8, 9, and 9, the sample mode is 4, the sample median is 7, and the sample mean is 6. Mathematically, the sample mean is defined as [8]:

$$\bar{x} = \sum_{i=1}^{n} \frac{x_i}{n} = \frac{x_1 + x_2 + \ldots + x_{n+}}{n} \tag{12.8}$$

$\bar{x}$ is the sample mean value. x_i is the individual value of x. Thus, $x_1, x_2, \ldots x_n$ are respective values of the first, second, and nth x values. n is the number of values in the sample. Sample variance is a measure of data scatter relative to the sample mean value that is based upon the squared difference between individual values and the sample mean as well as the

> *Mean values are key statistical data points.*

number of degrees of freedom, which will be explained later. Sample variance is defined mathematically as [9]:

$$s^2 = \sum_{i=1}^{n} \frac{(x_i - \bar{x})^2}{n-1} = \frac{n \sum_{i=1}^{n} x_i^2 - \left(\sum_{i=1}^{n} x_i\right)^2}{n(n-1)} \tag{12.9}$$

s^2 is the sample variance. $\bar{x}$ is the sample mean value. x_i is the individual value of x, where $x_1, x_2, \ldots x_n$ are respective values of the first, second, and nth x values. n is the number of values in the sample. Sample standard deviation is a measure of data deviation from the sample mean value that is based upon addition of individual deviations added in quadrature, then normalized by the number of degrees of freedom. The sample standard deviation is equivalent to the square root of the sample variance. The mathematical formula for sample standard deviation is [10]:

$$s = \sqrt{\sum_{i=1}^{n} \frac{(x_i - \bar{x})^2}{n-1}} = \sqrt{\frac{n \sum_{i=1}^{n} x_i^2 - \left(\sum_{i=1}^{n} x_i\right)^2}{n(n-1)}} \tag{12.10}$$

> **The standard deviation provides an indication of variability in measurements.**

The number of *degrees of freedom*, v, is defined as the number of independent data pieces. In other words, it is the number of pieces of information needed to complete a set of information. The reliability of an estimate increases with increasing degrees of freedom. Note that if a sample contains five values and the sample mean value is known, only four of the values are needed, together with the sample mean, to determine the fifth and final value. Consequently, there are $n - 1$ or four degrees of freedom for this example. Moreover, the number of degrees of freedom for sample variance and sample standard deviation is always $n - 1$. This is because the sample mean must already be known. However, if an entire population, which is assumed to be large, is used to determine these values, alternative definitions are used. This is partially due to the fact that $n - 1$ and n are nearly the same when n is large. Thus, for entire populations, the *population variance* is defined as [11]:

$$\sigma^2 = \sum_{i=1}^{n} \frac{(x_i - \mu)^2}{n} \tag{12.11}$$

σ^2 is the population variance. μ is the population mean. It should also be noted that the sample mean and population mean are the same. However, the population mean is often referred to by the symbol, μ. The corresponding *population standard deviation* is defined as:

$$\sigma = \sqrt{\sum\nolimits_{i=1}^{n} \frac{(x_i - \mu)^2}{n}} \qquad (12.12)$$

Most often, statistical analyses are based on samples rather than populations. Therefore, the corresponding sample-based formulas involving $\bar{x}$, s^2, and s are generally used.

12.3   The Normal Distribution

Many repetitive natural phenomena such as weather as well as many physical measurements have values that are distributed in a common pattern around an average value. This type of distribution of values is known as a *normal distribution*. Mathematically, the normal distribution can be represented in terms of the density of entities at a particular value as a normal distribution density function, which is expressed as [12]:

$$f(x) = \frac{1}{\sigma\sqrt{2\pi}} e^{-\left(\frac{1}{2}\left[\frac{(x-\mu)}{\sigma}\right]^2\right)} \qquad (12.13)$$

where $f(x)$ is the density or frequency of occurrence. An example of a normal distribution where the mean is 50 and the standard deviation is 15 is presented in Fig. 12.1.

12.4   Probability and Confidence

The probability of an occurrence or of obtaining a measurement within a specific range of values is obtained by integrating under the distribution function curve to find the associated area. The area under the entire curve is equal to one. Moreover, the probability of finding a value within a large enough range from the mean value is always one. Mathematically, the probability of a given event occurring within a specified range of values, x_1 and x_2, is expressed as:

> *The degree to which a conclusion can be made based on data is quantified by the probability and confidence interval.*

$$P(x_1 < x < x_2) = \int_{x_1}^{x_2} f(x)dx \qquad (12.14)$$

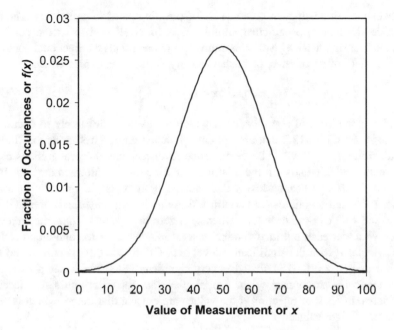

Fig. 12.1 Normal distribution plot of $f(x)$ versus the variable x with a population mean value of 50 and a population standard deviation of 15

The term $f(x)$ can represent any one of several distribution functions that will be given later. For cases in which the normal distribution function applies, this expression becomes:

$$P(x_1 < x < x_2) = \int_{x_1}^{x_2} \frac{1}{\sigma\sqrt{2\pi}} e^{-\left(\frac{1}{2}\left[\frac{(x-\mu)}{\sigma}\right]^2\right)} dx \qquad (12.15)$$

For the special case of the range of values within one standard deviation of the mean, the probability of occurrence is 0.68 or 68%. As illustrated in Fig. 12.2, the area under the curve within one standard deviation (15) of the mean (50), which is equivalent to the associated probability of event occurrence, is 0.68. Thus, it can be stated that there is 68% confidence in a randomly selected value from the distribution occurring within one standard deviation of the mean value, or in other words between the values of 35 and 65 for the data presented in Fig. 12.2. Thus, confidence describes the likelihood of a particular outcome and is directly related to probability. The range of values leading to a particular confidence is often referred to as the confidence interval. The significance level, α, is equivalent to one minus the fractional confidence or probability of occurrence, or mathematically written it is expressed as:

$$P(x_1 < x < x_2) = 1 - \alpha \qquad (12.16)$$

where α is the level of significance at which the event is not likely to occur.

The areas in Fig. 12.2 that are not cross-hatched are referred to as the "tails" of the distribution as shown. The significance level contains the probabilities associated with the tail ends of the distribution function if intermediate values are selected. Such significance levels that include both "tails" of the distribution are associated with two-tail tests. In contrast, those including only one of the "tails" are associated with one-tail tests (e.g. when x_1 is zero or x_2 is ∞). Thus, the area under the curve under each tail in two-tailed tests is $\alpha/2$, whereas for one-tailed tests the area associated with the significance level is α. Often, $\alpha = 0.05$ is considered to be significant, and $\alpha = 0.01$ is treated as very significant. Statistics-based decisions are often made for acceptance when a particular data set falls within the necessary confidence interval or rejection if the values are outside that range, which is within the level of significance.

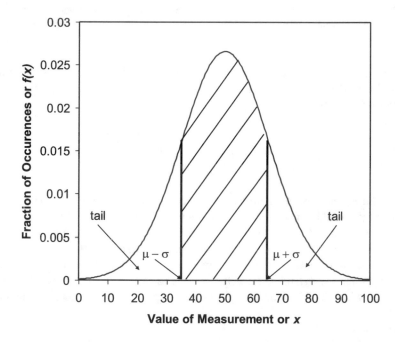

Fig. 12.2 Normal distribution plot of $f(x)$ versus the variable x with a population mean value of 50 and a population standard deviation of 15. The cross-hatched middle section represents the area of 0.68 under the curve within one standard deviation of the mean

Example 12.3 A company desires to ensure that at least 99% of their 500 g products weigh at least 500 g. To accomplish this task a lower limit of the weight measuring device that determines whether or not the product is rejected must be set properly. What lower weight setting must be used as the rejection criteria if the if the population mean value is 510 g and the population standard deviation is 5 g if the population is known to follow a normal distribution?

This is a one-tailed test because the range under consideration is above a specified limit rather than between certain finite limits. This problem can be solved by numerical integration (using a spreadsheet with small x step sizes of Δx to simulate dx) of the normal distribution density function to find the lower limit of weight that insures a probability of 99% or 0.99:

$$P(x_1 < x < \infty) = f(x)$$

For a normal distribution, the density distribution function, $f(x)$, is given as:

$$f(x) = \int_{x_1}^{\infty} \frac{1}{\sigma\sqrt{2\pi}} e^{-\left(\frac{1}{2}\left[\frac{(x-\mu)}{\sigma}\right]^2\right)} dx$$

Consequently,

$$P(x_1 < x < \infty) = 0.99 = \int_{x_1}^{\infty} \frac{1}{\sigma\sqrt{2\pi}} e^{-\left(\frac{1}{2}\left[\frac{(x-\mu)}{\sigma}\right]^2\right)} dx$$

The associated numerical integration leads to an x_1 value of 498.36.

The previous example shows that confidence interval problems can be solved rather easily using numerical integration. However, numerical integration can be rather inconvenient using a calculator. Data tables are also not practical for such an analysis because there are an infinite number of possible means and standard deviations. Another difficulty with practical statistical analyses is the fact that the population standard deviation is often unknown. Consequently, alternative approaches to address these challenges have been established. Most notably, t and z variables and associated functions are utilized to analyze data that originate from normal population distributions. These will be discussed in connection with hypothesis testing.

12.4.1   z-based Determinations of Probabilities of Continuous Function Values Within Normal Populations

The z-based assessments are used to evaluate samples with large numbers of data pieces (<30) that follow normal distributions. This approach utilizes the variable z as the basis for the assessment. The variable z is called the standard normal variable. It is used to convert normally distributed data to a standard normal distribution for which the mean is zero and the standard deviation is one. The variable z consists of the ratio of the measured departure of a measurement value from the stated

> *z is the standard normal variable. It is used to convert normally distributed data to a standard normal distribution for which the mean is zero and the standard deviation is one.*

population mean. It is relative to the population standard deviation. Consequently, for very large numbers of data pieces in a sample, the transformation of x from a normal population distribution to z will lead to a population mean value of zero and a population standard deviation of one. Furthermore, the use of z converts values from wide ranging applications and magnitudes to a common version of a standard normal distribution function for which a common table can be constructed. The value of z for large populations with continuous normal distributions is [13]:

$$z = \frac{(x - \mu)}{\sigma} \quad \text{(for continuous normal}$$

$$\text{distributions and large populations)} \tag{12.17}$$

Because z is a transformation variable that applies to several statistical scenarios, the user should ensure the correct form is used for the desired situation. Substitution of z and dz ($dz = dx/\sigma$) into the normal distribution equation in place of x and dx leads to:

$$P(z_1 < z < z_2) = \int_{z_1}^{z_2} \frac{1}{\sqrt{2\pi}} e^{-\left(\frac{z^2}{2}\right)} dz \tag{12.18}$$

Thus, the probability of a value of z between z_1 and z_2 or the area under the normal distribution curve between these values can be determined using this integral. Figure 12.3 shows the probability or area under the curve associated with the interval for z between approximately -2 and 2, which is approximately 0.95. The accompanying Table (12.1a and b) give accurate values of the probability of obtaining a value of z that is less than z_1. Thus, the determination of the probability between z_1 and z_2 is found as the difference in the associated areas or probability values in the appropriate table(s).

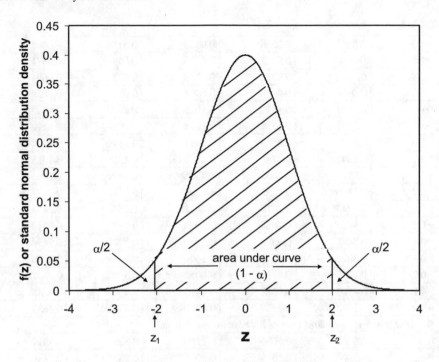

Fig. 12.3 Comparison of the standard normal distribution density value to the corresponding value of z. The range between z_1 and z_2 gives an area or probability equal to $1 - \alpha$ and the distribution tail areas below z_1 and above z_2 have areas or probabilities of $\alpha/2$

Table 12.1 a Area under the standard normal distribution density function from $-\infty$ to z, **b** Area under the standard normal distribution curve from $-\infty$ to z

(a) *Area under the standard normal distribution density function from $-\infty$ to z*				
z	0.00	0.02	0.04	0.06	0.08
−2.9	0.0019	0.0017	0.0016	0.0015	0.0014
−2.8	0.0026	0.0024	0.0023	0.0021	0.0020
−2.7	0.0035	0.0033	0.0031	0.0029	0.0027
−2.6	0.0047	0.0044	0.0041	0.0039	0.0037
−2.5	0.0062	0.0059	0.0055	0.0052	0.0049
−2.4	0.0082	0.0078	0.0073	0.0069	0.0066
−2.3	0.0107	0.0102	0.0096	0.0091	0.0087
−2.2	0.0139	0.0132	0.0125	0.019	0.0133
−2.1	0.0179	0.0170	0.0162	0.0154	0.0146
−2.0	0.0228	0.0217	0.0207	0.0197	0.0188
−1.9	0.0287	0.0274	0.0262	0.0250	0.0239
−1.8	0.0359	0.0344	0.0329	0.0314	0.0301
−1.7	0.0446	0.0427	0.0409	0.0392	0.0375

(continued)

Table 12.1 (continued)

−1.6	0.0548	0.0526	0.0505	0.0485	0.0465
−1.5	0.0668	0.0643	0.0618	0.0594	0.0571
−1.4	0.0808	0.0778	0.0749	0.0722	0.0694
−1.3	0.0968	0.0934	0.0901	0.0869	0.0838
−1.2	0.1151	0.1112	0.1075	0.1038	0.1003
−1.1	0.1357	0.1314	0.1271	0.1230	0.1190
−1.0	0.1587	0.1539	0.1492	0.1446	0.1401
−0.9	0.1841	0.1788	0.1736	0.1685	0.1635
−0.8	0.2119	0.2061	0.2005	0.1949	0.1894
−0.7	0.2420	0.2358	0.2296	0.2236	0.2177
−0.6	0.2743	0.2676	0.2611	0.2546	0.2483
−0.5	0.3085	0.3015	0.2946	0.2877	0.2810
−0.4	0.3446	0.3372	0.3300	0.3228	0.3156
−0.3	0.3821	0.3745	0.3669	0.3594	0.3520
−0.2	0.4207	0.4129	0.4052	0.3974	0.3897
−0.1	0.4602	0.4522	0.4443	0.4364	0.4286
−0.0	0.5000	0.4920	0.4840	0.4761	0.4681

(**b**) *Area under the standard normal distribution curve from* $-\infty$ *to z*

0.0	0.5000	0.5080	0.5160	0.5239	0.5319
0.1	0.5398	0.5478	0.5557	0.5636	0.5714
0.2	0.5793	0.5871	0.5948	0.6026	0.6103
0.3	0.6179	0.6255	0.6331	0.6406	0.6480
0.4	0.6554	0.6628	0.6700	0.6772	0.6844
0.5	0.6915	0.6985	0.7054	0.7123	0.7190
0.6	0.7257	0.7324	0.7389	0.7454	0.7517
0.7	0.7580	0.7642	0.7704	0.7764	0.7823
0.8	0.7881	0.7939	0.7995	0.8051	0.8106
0.9	0.8159	0.8212	0.8264	0.8315	0.8365
1.0	0.8413	0.8461	0.8508	0.8554	0.8599
1.1	0.8643	0.8686	0.8729	0.8770	0.8810
1.2	0.8849	0.8888	0.8925	0.8962	0.8997
1.3	0.9032	0.9066	0.9099	0.9131	0.9162
1.4	0.9192	0.9222	0.9251	0.9278	0.9306
1.5	0.9332	0.9357	0.9382	0.9406	0.9429
1.6	0.9452	0.9474	0.9495	0.9515	0.9535
1.7	0.9554	0.9573	0.9591	0.9608	0.9625
1.8	0.9641	0.9656	0.9671	0.9686	0.9699
1.9	0.9713	9.9726	0.9738	0.9750	0.9761
2.0	0.9772	0.9783	0.9793	0.9803	0.9812
2.1	0.9821	0.9830	0.9838	0.9846	0.9854
2.2	0.9861	0.9868	0.9875	0.9881	0.9887
2.3	0.9893	0.9898	0.9904	0.9909	0.9913

(continued)

Table 12.1 (continued)

2.4	0.9918	0.9922	0.9927	0.9931	0.9934
2.5	0.9938	0.9941	0.9945	0.9948	0.9951
2.6	0.9953	0.9956	0.9959	0.9961	0.9963
2.7	0.9965	0.9967	0.9969	0.9971	0.9973
2.8	0.9974	0.9976	0.9977	0.9979	0.9980
2.9	0.9981	0.9982	0.9984	0.9985	0.9986

Example 12.4 Using the accompanying z tables, calculate the probability of obtaining a value of reaction rate between 10 and 20 mol/l/hr, assuming that the reaction rates follow a continuous normal distribution and the population mean rate is 15 mol/l/hr and the population standard deviation is 3 mol/l/hr.

The first step to accomplishing this two-tailed evaluation is to convert the rate values to corresponding z values (in this case for a continuous normal distribution function):

$$z_1 = \frac{(x_{low} - \mu)}{\sigma} = \frac{(10 - 15)}{3} = -1.67$$

$$z_2 = \frac{(x_{high} - \mu)}{\sigma} = \frac{(20 - 15)}{3} = 1.67$$

Using the z tables:

$$P(z < z_1 \text{ or } -1.67) = 0.0475 \text{ (interpolated)}$$
$$P(z < z_2 \text{ or } 1.67) = 0.9525 \text{ (interpolated)}$$

Thus, the probability of finding z between z_1 and z_2 is found by taking the difference between the probability of finding z below z_2 and the probability of finding z below z_1:

$$P(z_1 < z < z_2) = P(z < z_2) - P(z < z_l) = 0.9525$$
$$- 0.0475 = 0.9050 \text{ or } 90.5\%$$

12.4.2 z-based Probability Analysis of Distributions of Means from Continuous Normal Distributions

If a series of samples of size n are taken from a large normally distributed population, the resulting distribution of mean values as well as the sum of the values within each sample will follow a normal distribution. The standard deviation of the resulting distribution of sample means can be computed as [14]:

$$\sigma_{\bar{x}} = \frac{\sigma}{\sqrt{n}} \qquad\qquad (12.19)$$

where $\sigma\,\bar{x}$ is equal to the standard deviation of the means of a large population from which samples of size n are taken and the population standard deviation is σ. Consequently, if an analysis of mean values from a series of samples is desired when the standard deviation of the population is known, the z-based approach can be utilized by substituting the term for the standard deviation of the mean and the sample mean into the previous expression for z to give:

$$z = \frac{(\bar{x} - \mu)}{\frac{\sigma_{\bar{x}}}{\sqrt{n}}} \qquad\qquad (12.20)$$

This equation can then be used to evaluate the probability of observing sample mean values by applying the same methodology demonstrated with the previous use of z.

12.4.3   z-based Estimate of Population Mean Based Upon Several Sample Means

The use of the z-table can also be reversed and used to determine an estimate of either the population mean or the population standard deviation, provided that one of these two values is known along with the value of the population mean and the sample size. The process begins with the establishment of the desired confidence interval, 100% $(1 - \alpha)$, and the formulation of the statement:

$$P(z_{\alpha/2} < z < z_{1-\alpha/2}) = 1 - \alpha = P\left(z_{\alpha/2} < \frac{\bar{x} - \mu}{\frac{\sigma_{\bar{x}}}{\sqrt{n}}} < z_{1-\alpha/2} \right) \qquad (12.21)$$

which is then followed by rearrangement to give the estimate of μ for the confidence interval:

$$\bar{x} - z_{\alpha/2}\frac{\sigma_{\bar{x}}}{\sqrt{n}} < \mu < \bar{x} + z_{1-\alpha/2}\frac{\sigma_{\bar{x}}}{\sqrt{n}} \qquad\qquad (12.22)$$

Example 12.5 Estimate the population mean value with 95% confidence based upon 10 samples with a sample size of 8 if the mean value of the various sample means is 10.2 and the standard deviation of the sample means is 1.7.

The values for z are:

$$z_{\alpha/2} = z_{1-\alpha/2} = 1.96$$

The basic formula is:

$$\bar{x} - z_{\alpha/2}\frac{\sigma_{\bar{x}}}{\sqrt{n}} < \mu < \bar{x} + z_{1-\alpha/2}\frac{\sigma_{\bar{x}}}{\sqrt{n}}$$

Substitution of the values into the equation leads to:

$$10.2 - (1.96)\frac{1.7}{\sqrt{8}} < \mu < 10.2 + (1.96)\frac{1.7}{\sqrt{8}} = (9.02 < \mu < 11.38)$$

This same approach can be used to estimate values that are unknown for other statistical scenarios using different statistical variables.

12.4.4   z-based Probability Determination of Discrete Values by Approximation of Binomial Distributions

Individual experiments are often evaluated on the basis of the result that is classified as either success or failure. The collective data within a sample that is obtained from multiple experiments that are classified as either success or failure often follows a binomial distribution. When the number of experiments or data pieces within a sample is large, the resulting binomial distribution can be approximated by the normal distribution. Furthermore, the transformation variable, z, for a large-size sample (>30) that follows a binomial distribution, which is a discrete rather than continuous distribution (Discrete functions can only have integer values. In contrast, continuous functions are not restricted to integer values) is given as:

$$z = \frac{x - np}{\sqrt{np(1-p)}} \quad \text{(for discrete binomial}$$

$$\text{distribution with } n > 30)$$

(12.23)

x is the number of times a successful outcome was achieved. n is the sample size. p is the probability of the successful outcome. The discrete nature of binomial distributions forces the width of the area elements to be one. The value of the element is centered half way between the outer boundaries of the neighboring integer values. Consequently, if the area under the curve associated with a discrete integer value of x is desired, the associated values $x - 0.5$ and $x + 0.5$ need to be used as the limits. Moreover, the value of x will always be offset by 0.5 above or below the desired centered value of the element. This ensures the entire area associated with that element is measured. For example, if the probability of observing 5 positive outcomes is desired, the associated area is found between 4.5 and 5.5. These are the outer boundaries of the element centered at 5 that has a width of one.

Example 12.6 Through extensive testing over a 5-year period a company has established that 29% of one of its products will fail by the end of 6 years of service. If a random sample of 39 units of this product are tested for failure after exactly 6 years of service, determine the probability that between 12 and 16 of these units will have failed.

The probability of success for 6 year service is 71% or 0.71. The value that is being tested is the probability that between 23 and 27 parts out of a sample size of 39 will succeed. Consequently, the desired probability will be determined between a lower limit of x of 22.5 and an upper limit of x of 27.5. The corresponding z values are:

$$z_1 = \frac{x_1 - np}{\sqrt{np(1-p)}} = \frac{22.5 - (39)(0.71)}{\sqrt{(39)(0.71)(1-0.71)}} = -1.83$$

$$z_2 = \frac{x_2 - np}{\sqrt{np(1-p)}} = \frac{27.5 - (39)(0.71)}{\sqrt{(39)(0.71)(1-0.71)}} = -0.07$$

Thus, the corresponding probabilities or areas associated with these values are:

$$P(z < z_1) = 0.0336$$
$$P(z < z_2) = 0.4721$$

Consequently, the probability of finding z between z_1 and z_2, which correspond to probability of finding between 12 and 16 failed parts out of 39, is found as the difference between the z_1 and z_2 probabilities:

$$P(z_1 < z < z_2) = 0.4721 - 0.0336 = 0.4385$$

12.4.5   z-based Evaluation of the Sample Mean When Population Mean and Standard Deviation Are Known

Mathematicians have shown that sample mean values from any distribution are normally distributed with respect to population means provided that the sample size is large. This observation is an important core tool of statisticians. It is known as the Central Limit Theorem. Consequently, a value of z can be determined that will transform sample mean distributions to the standard normal distribution. The value of the transformation variable z when evaluating the likelihood of measuring a given mean value from a sample of size n when the population mean and standard deviation are known is determined using the formula:

$$z = \frac{\bar{x} - \mu}{\sigma/\sqrt{n}} \quad \text{(for evaluation of sample}$$

$$\text{mean using a large sample size } n > 30)$$

(12.24)

Therefore, this equation allows the user to determine the probability of measuring a sample mean. It requires a population with a known mean and standard deviation as a function of sample size. The transformation variable, z, and the associated data tables are used in the assessment.

12.4.6   z-based Comparison of Population Means with Known Population Variances and Sample Means

It is sometimes useful to compare population means using sample means and known population variances. These circumstances are often applicable to random variables that have normal distributions. They often apply for reasonably large sample sizes. This is a somewhat unusual case for which the value of z for populations 1 and 2 from which samples 1 and 2 are obtained is:

$$z = \frac{(\bar{x_1} - \bar{x_2}) - (\mu_1 - \mu_2)}{\sqrt{\frac{\sigma_1^2}{n_1} + \frac{\sigma_2^2}{n_2}}}$$

(12.25)

Note that if the variances for the two populations are equivalent, the equation can be simplified.

12.4.7   z-based Determination of the Sample Size to Achieve a Desired Confidence Level in Sample Mean

Rearrangement of the expression for z used to compare sample and population means for a specified confidence level of $1 - \alpha$ leads to:

$$n = \left[\frac{z_{\alpha/2}\sigma}{\bar{x} - \mu}\right]^2 \quad \text{(for determination of sample}$$

$$\text{size for normally distributed means)}$$

Thus, when a quality control scenario requires the sample mean to be within a specified range of the known sample mean within a specified confidence limit, this equation can be used to determine the sample size needed to achieve the desired confidence level.

Example 12.7 Determine the solution sample size needed to ensure at the 95% confidence level that a randomly selected sample mean will be within 1 ppm of the known solution population mean concentration value of 100 ppm if the population standard deviation is 3 ppm.

The 95% confidence interval corresponds to $1 - \alpha = 0.95$ or $\alpha = 0.05$ and $\alpha/2 = 0.025$. The value of z corresponding to $\alpha/2 = 0.025$ is -1.96 based upon the z data tables. Thus,

$$
n = \left[\frac{z_{\alpha/2}\sigma}{\bar{x} - \mu}\right]^2 = \left[\frac{-1.96(3)}{101 - 100}\right]^2 = 34.6 \,(\text{or } 35)
$$

12.4.8   t-based Evaluation of a Sample Mean to Known Population Mean and Sample Deviation

As demonstrated, the z-based approach to probability evaluations has great utility. However, the z-based approach requires large samples and a knowledge of the population mean and standard deviation. Often the population standard deviation is not known. Consequently, the z-based methodology cannot be directly applied to many data sets. Thus, a slightly modified transformation is appropriate.

Because the sample mean value and the standard deviation of the associated distribution of means follow a normal distribution, the approximate substitution of sample standard deviation for standard deviation can be made to yield [15]:

$$
s_{\bar{x}} = \frac{s}{\sqrt{n}} \tag{12.26}
$$

This modified transformation is made by substituting the sample standard deviation for population standard deviation. The modified transformation variable is t.

The t-distribution, sometimes known as the student t-distribution, is very similar to the standard normal distribution. It approaches the standard normal distribution as the sample size becomes large. The variable, t, is a transformation variable that is very similar to z, since t is used to transform data into a representative density function

> *The variable, t, is used to transform data into a representative density function with a mean value of zero.*

with a mean value of zero. Thus, the t variable is identical to z when the number of data points in a sample approaches infinity. For practical purposes, t and z are very similar when the sample contains more than 30 data points. However, the *t* distribution function more closely approximates normally distributed real-world data from limited sample sizes than the *z* distribution function. Consequently, a valuable feature of the *t*-based analysis is the absence of a known population standard deviation. The population standard deviation is usually more difficult to obtain than the sample standard deviation. The variable *t* is computed on the basis of $n - 1$ degrees of freedom using the formula [15]:

$$t = \frac{\bar{x} - \mu}{\frac{s}{\sqrt{n}}} \tag{12.27}$$

This is analogous to the formula for *z* using the Central Limit Theorem. Because *t* is not the same as *z*, it has a different distribution density function that is given as [16]:

$$f(t) = \frac{\Gamma[(v+1)(0.5)]}{\Gamma[(0.5)(v)(\sqrt{\pi v})]} \left(1 + \frac{t^2}{v}\right)^{-[(v+1)/2]} \tag{12.28}$$

The gamma function, Γ, is a constant for a given number of degrees of freedom. The gamma function is defined as [17]:

$$\Gamma(y) = \int_0^\infty x^{(y-1)} e^{-x} dx \tag{12.29}$$

The normal distribution is only dependent upon population mean and standard deviation. The *t* distribution is dependent upon degrees of freedom in addition to sample values. The large difference between the *t* and *z* distributions is associated with the sample size, *n* is found in the denominator of the expression for *t*. As with the application of the *z*-based assessments, the *t* information is used by determining the probability of occurrence of *t*. Because the density function needed to compute the probability of *t* involves a rather complex mathematical function, tabulated data such as those shown in Table 12.2 are used. In order to find the appropriate probability in the *t* table, the number of data points in the sample or sample size must be known. The *t* distribution is affected strongly by the sample size or the degrees of freedom, which is equivalent to $n - 1$ as shown in Fig. 12.4. In other words, $v = n - 1$ for a typical set of data.

Example 12.8   A company plant operator needs to be sure that the average level of iron in a process flow stream is less than 3000 ppm in order to prevent precipitation in downstream processes. Consequently, a technician is sent to gather 9 random solution samples from the process stream for iron analysis. The results from the solution analysis are 2710, 2890, 2660, 2780, 2930, 2920, 2840, 2770, and 2670 ppm iron. Can the operator have at least 99% confidence that the mean iron level is less than 3000 ppm?

Table 12.2 Values of the *t*-distribution function (based on one-tailed significance levels)

V	α 0.4	α 0.25	α 0.1	α 0.05	α 0.025	α 0.01	α 0.005
1	0.325	1.000	3.078	6.314	12.71	31.82	63.66
2	0.289	0.816	1.886	2.920	4.303	6.965	9.925
3	0.277	0.765	1.638	2.353	3.182	4.541	5.841
4	0.271	0.741	1.533	2.132	2.776	3.747	4.604
5	0.267	0.727	1.476	2.015	2.571	3.365	4.032
6	0.265	0.718	1.440	1.943	2.447	3.143	3.707
7	0.263	0.711	1.415	1.895	2.365	2.998	3.499
8	0.262	0.706	1.397	1.860	2.306	2.896	3.355
9	0.261	0.703	1.383	1.833	2.262	2.821	3.250
10	0.260	0.700	1.372	1.812	2.228	2.764	3.169
11	0.260	0.697	1.363	1.796	2.201	2.718	3.106
12	0.259	0.695	1.356	1.782	2.179	2.681	3.055
13	0.259	0.694	1.350	1.771	2.160	2.650	3.012
14	0.258	0.692	1.345	1.761	2.145	2.624	2.977
15	0.258	0.691	1.341	1.753	2.131	2.602	2.947
16	0.258	0.690	1.337	1.746	2.120	2.583	2.921
17	0.257	0.689	1.333	1.740	2.110	2.567	2.898
18	0.257	0.688	1.330	1.734	2.101	2.552	2.878
19	0.257	0.688	1.328	1.729	2.093	2.539	2.861
20	0.257	0.687	1.325	1.725	2.086	2.528	2.845
21	0.257	0.686	1.323	1.721	2.080	2.518	2.831
22	0.256	0.686	1.321	1.717	2.074	2.508	2.819
23	0.256	0.685	1.319	1.714	2.069	2.500	2.807
24	0.256	0.685	1.318	1.711	2.064	2.492	2.797
25	0.256	0.684	1.316	1.708	2.060	2.485	2.787
26	0.256	0.684	1.315	1.706	2.056	2.479	2.779
27	0.256	0.684	1.314	1.703	2.052	2.473	2.771
28	0.256	0.683	1.313	1.701	2.048	2.467	2.763
29	0.256	0.683	1.311	1.699	2.045	2.462	2.756
30	0.256	0.683	1.310	1.697	2.042	2.457	2.750

$$\bar{x} = \sum_{i=1}^{n} \frac{x_i}{n} = \frac{x_1 + x_2 + \ldots + x_{n+}}{n} = 2797$$

$$s = \sqrt{\frac{n \sum_{i=1}^{n} x_i^2 - \left(\sum_{i=1}^{n} x_i\right)^2}{n(n-1)}}$$

$$= \sqrt{\frac{9(70{,}478{,}900) - 633{,}528{,}900}{9(8)}} = 104$$

Fig. 12.4 Comparison of the distribution density value as a function of t for the t distribution density function for different degrees of freedom ($v = 1$, 4, and 30)

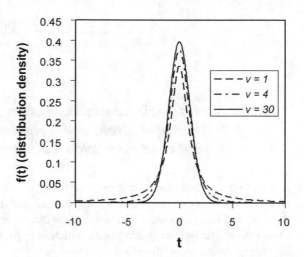

These calculations are more easily performed using a spreadsheet with statistical functions. In this case the mean of interest is 3000 ppm, so the value for t is:

$$t = \frac{\bar{x} - \mu}{\frac{s}{\sqrt{n}}} = \frac{2797 - 3000}{\frac{104}{\sqrt{9}}} = -5.86$$

Because the stated goal is to determine if the sample mean is less than 3000 ppm, only one tail of the distribution is to be assessed. The value of t from the table that corresponds to a significance level of $\alpha = 0.01$ for $n - 1$ ($9 - 1 = 8$) degrees of freedom is 2.896, which is much smaller in magnitude than the 5.86 calculated. (Note the negative sign would correspond to the opposite end of the t distribution table, but because the distribution is symmetric it can be applied to positive or negative values) . Consequently, the conclusion is that the operator can have greater than 99% confidence that the iron level is below 3000 ppm.

12.4.9   t-based Evaluation of Population Means from Sample Means and Variances and the Assumption of Equivalent but Unknown Population Variances

It is common to compare sample means to evaluate one method relative to another. If the sample means and variances are known and the population variances are equivalent, then the t distribution can be used to evaluate the population means using the following form of t:

$$t = \frac{(n_1 + n_2 - 2)[(\overline{x_1} - \overline{x_2}) - (\mu_1 - \mu_2)]}{[s_1^2(n_1 - 1) + s_2^2(n_2 - 1)]\sqrt{\frac{1}{n_1} + \frac{1}{n_2}}} \qquad (12.30)$$

12.4.10 t-based Evaluation of Population Means from Sample Means and Variances and Unknown but Different Population Variances

It is common to compare sample means to evaluate one method relative to another. If the sample means and variances are known and the population variances are different and unknown, then the t distribution can be used to evaluate the population means using the following form of t', which is approximately equal to t when the degrees of freedom are given as:

$$v = \frac{[(s_1^2/n_1) + (s_2^2/n_2)]^2}{\frac{(s_1^2/n_1)^2}{n_1 - 1} + \frac{(s_2^2/n_2)^2}{n_2 - 1}} \qquad (12.31)$$

and t' is obtained using:

$$t' = \frac{[(\overline{x_1} - \overline{x_2}) - (\mu_1 - \mu_2)]}{\sqrt{\frac{s_1^2}{n_1} + \frac{s_2^2}{n_2}}} \qquad (12.32)$$

The value of t' is then used as if it was t to determine the appropriate probability.

12.4.11 F-distribution (Used for Analysis of Sample Variance and Comparing More than Two Means)

In many statistical analyses it is useful to compare the measured variance between different samples. The ratio of variances provides a useful metric for evaluating variance differences, and this ratio is known as the F value and it is given as:

$$F = \frac{s_1^2 \sigma_2^2}{s_2^2 \sigma_1^2} = \frac{\frac{x}{v_x}}{\frac{y}{v_y}} \qquad (12.33)$$

Note that the population standard deviations cancel each other if sample data for the same population distribution are compared. The associated F distribution density function is given as [18]:

$$f(F) = \frac{\left(\frac{v_1}{v_2}\right)^{0.5v_1} \Gamma[(v_1 + v_2)(0.5)]}{(0.5v_2)\Gamma[(0.5)(v)(\sqrt{\pi v})]} \left(\frac{F^{(0.5v_1 - 1)}}{\left(1 + \frac{v_1 F}{v_2}\right)^{0.5(v_1 + v_2)}}\right)^{-[(v+1)/2]} \tag{12.34}$$

A plot of the F distribution density function is presented in Fig. 12.5, and tabulated values are presented in Table 12.3. Use of the F-distribution

> *The F value is a measure of the relative variances of two samples.*

will be demonstrated in connection with the analysis of variance section of this chapter.

12.4.12 χ^2 (Chi-Squared) Evaluation of Independence

In many statistical comparisons it is desirable to compare the distribution of measured data relative to a known distribution of a population. Thus, the ratio of sample and population variance, multiplied by the number of degrees of freedom, determines

> *Chi-squared is a comparison of the ratio of sample and population variance multiplied by the degrees of freedom.*

the value of chi-squared. The value of chi-squared can be calculated using the equation:

$$\chi^2 = \frac{(n-1)s^2}{\sigma^2} = \sum_{i=1}^{n} \frac{(x_i - \bar{x})^2}{\sigma^2} \tag{12.35}$$

Fig. 12.5 Comparison of the distribution density value as a function of F for the F distribution density function for different degrees of freedom as indicated

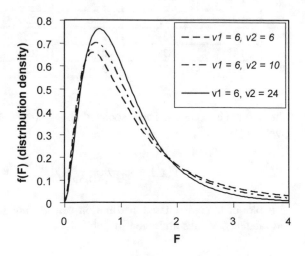

Table 12.3 Values of the F-distribution function (one-tailed $\alpha = 0.05$ significance level)

v_2	v_1, $\alpha = 0.05$	v_1, $\alpha = 0.05$	v_1, $\alpha = 0.05$	v_1, $\alpha = 0.05$	v_1, $\alpha = 0.05$	v_1 $\alpha = 0.05$
	1	2	3	5	10	15
1	161.5	199.5	215.7	230.2	241.9	246.0
2	18.51	19.00	19.16	19.30	19.40	19.43
3	10.13	9.55	9.28	9.01	8.79	8.70
4	7.71	6.94	6.59	6.26	5.96	5.86
5	6.61	5.79	5.41	5.05	4.75	4.62
6	5.99	5.14	4.76	4.39	4.06	3.94
7	5.59	4.74	4.35	3.97	3.64	3.51
8	5.32	4.46	4.07	3.69	3.35	3.22
9	5.12	4.26	3.86	3.48	3.14	3.01
10	4.96	4.10	3.71	3.33	2.98	2.85
11	4.84	3.98	3.59	3.20	2.85	2.72
12	4.75	3.89	3.49	3.11	2.75	2.62
13	4.67	3.81	3.41	3.03	2.67	2.53
14	4.60	3.74	3.34	2.96	2.60	2.46
15	4.54	3.68	3.29	2.90	2.54	2.40
16	4.49	3.63	3.24	2.85	2.49	2.35
17	4.45	3.59	3.20	2.81	2.56	2.31
18	4.41	3.55	3.16	2.77	2.41	2.27
19	4.38	3.52	3.13	2.74	2.38	2.23
20	4.35	3.49	3.10	2.71	2.35	2.20
21	4.32	3.47	3.07	2.68	2.32	2.18
22	4.30	3.44	3.05	2.66	2.30	2.15
23	4.28	3.42	3.03	2.64	2.27	2.13
24	4.26	3.40	3.01	2.62	2.25	2.11
25	4.24	3.39	2.99	2.60	2.24	2.09
26	4.23	3.37	2.98	2.59	2.22	2.07
27	4.21	3.35	2.96	2.57	2.20	2.06
28	4.20	3.34	2.95	2.56	2.19	2.04
29	4.18	3.33	2.93	2.55	2.18	2.03
30	4.17	3.32	2.92	2.53	2.16	2.01

where χ^2 is chi-squared. The associated chi-squared distribution density function is given as [19]:

$$f(\chi^2) = \frac{1}{2^{0.5v}\Gamma(0.5v)} (\chi^2)^{(0.5v-1)} e^{-0.5(\chi^2)} \qquad (12.36)$$

A plot of the chi-squared distribution density function is presented in Fig. 12.6, and tabulated values are given in Table 12.4.

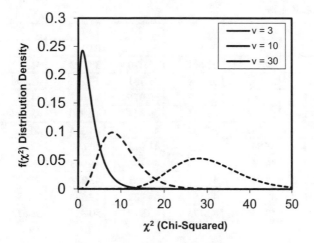

Fig. 12.6 Comparison of the distribution density value as a function of chi-squared for the chi-squared distribution density function for different degrees of freedom as indicated

Table 12.4 Values of the χ^2 (chi-squared)—distribution function

V	α 0.25	α 0.10	α 0.05	α 0.25	α 0.01	α 0.005	α 0.001
1	1.323	2.706	3.841	5.024	6.635	7.879	10.827
2	2.773	4.605	5.991	7.378	9.210	10.597	13.815
3	4.108	6.251	7.815	9.348	11.345	12.838	16.268
4	5.385	7.779	9.488	11.143	13.277	14.860	18.465
5	6.626	9.236	11.070	12.832	15.086	16.750	20.517
6	7.841	10.645	12.592	14.449	16.812	18.548	22.457
7	9.037	12.017	14.067	16.013	18.475	20.278	24.322
8	10.219	13.362	15.507	17.535	20.090	21.955	26.125
9	11.389	14.684	16.919	19.023	21.666	23.589	27.877
10	12.549	15.987	18.307	20.483	23.209	25.188	29.588
11	13.801	17.275	19.675	21.920	24.725	26.757	31.264
12	14.845	18.549	21.026	23.337	26.217	28.300	32.909
13	15.984	19.812	22.362	24.736	27.688	29.819	34.528
14	17.117	21.065	23.685	26.119	29.141	31.319	36.123
15	18.245	22.307	24.996	27.488	30.578	32.801	37.697
16	19.369	23.542	26.296	28.845	32.000	34.267	39.252
17	20.489	24.769	27.587	30.191	33.409	35.718	40.790
18	21.605	25.989	28.869	31.526	34.805	37.156	42.312
19	22.718	27.204	30.144	32.852	36.191	38.582	43.820
20	23.828	28.412	31.410	34.170	37.566	39.997	45.315

(continued)

Table 12.4 (continued)

	α	α	α	α	α	α	α
V	0.25	0.10	0.05	0.25	0.01	0.005	0.001
21	24.935	29.615	32.671	35.479	38.923	41.401	46.797
22	26.039	30.813	33.924	36.781	40.289	42.796	48.268
23	27.141	32.007	35.172	38.076	41.638	44.181	49.728
24	28.241	33.196	36.415	39.364	42.980	45.558	51.179
25	29.339	34.382	37.652	40.646	44.314	46.928	52.620
26	30.434	35.563	38.885	41.923	45.642	48.290	54.052
27	31.528	36.741	40.113	43.194	46.963	49.645	55.476
28	32.620	37.916	41.337	44.461	48.278	50.993	56.893
29	33.711	39.087	42.557	45.722	49.588	52.336	58.302
30	34.800	40.256	43.773	46.979	50.892	53.672	59.703

12.4.13   Chi-Squared Evaluations of the Fit of Data to Distribution Functions (Goodness-of-Fit Evaluations)

The chi-squared variable used for goodness-of-fit evaluations is given as:

$$\chi^2 = \sum_{i=1}^{n} \frac{(x_i - x_{expected})^2}{x_{expected}} \qquad (12.37)$$

where $x_{expected}$ is the expected value of x or the expected frequency of measuring a value based upon the chosen distribution function and x is the observed value or the frequency of measuring a value.

The number of degrees of freedom for chi-squared tests is determined by multiplying the number of columns in an assessment minus one times the number of rows minus one (#columns—1)(#rows—1), where the number of rows is equal to the number of classifications and the number of columns is the number of categories.

Example 12.9 A random solution measurement system is established to monitor solution discharge concentrations from five different locations in the same process stream. Determine if the random measurement system is truly random according to a normal distribution function at the 0.01 significance level if a total of 1000 measurements are taken and the number of measurements taken from each location is 180, 200, 195, 215, and 210 from locations A, B, C, D, and E, respectively.

The first step in this assessment is to determine the expected values. Because 1000 measurements were taken from the five locations, each location is expected to have 1/5th of the total or 200 measurements. Thus the associated chi-squared value is:

$$\chi^2 = \sum_{i=1}^{n} \frac{(x_i - x_{\text{expected}})^2}{x_{\text{expected}}}$$

$$\chi^2 = \frac{(180 - 200)^2}{200} + \frac{(200 - 200)^2}{200} + \frac{(195 - 200)^2}{200} + \frac{(215 - 200)^2}{200}$$

$$+ \frac{(210 - 200)^2}{200} = 3.80$$

This problem has two categories (expected and measured) and five classifications, so there are $(2-1)(5-1)$ or 4 degrees of freedom. Consequently, the chi-squared value from the chi-squared data table (or computer output) at the 0.01 significance level is 13.277. Because the measured value is less than the table or computer derived value at the 0.01 significance level, there is no reason to suspect the random sampling system doesn't function properly.

12.5 Linear Regression and Correlation

Variables are commonly related to each other. Often, such relationships are linear. A typical equation of a line is given as:

$$y_i = \alpha + \beta x_i \tag{12.38}$$

where α and β are constants. Determination of estimates for α and β is often made using the method of least squares, which involves the minimization of the sum of squares between the estimated and measured y values. The formula for determining the least squares estimates a and b for the coefficients α and β are [20]:

$$b = \frac{n \sum_{i=1}^{n} x_i y_i - \left(\sum_{i=1}^{n} x_i\right)\left(\sum_{i=1}^{n} y_i\right)}{n \sum_{i=1}^{n} x_i^2 - \left(\sum_{i=1}^{n} x_i\right)^2} = \frac{\sum_{i=1}^{n} (x_i - \bar{x})(y_i - \bar{y})}{\sum_{i=1}^{n} (x_i - \bar{x})^2} \tag{12.39}$$

$$a = \frac{\sum_{i=1}^{n} y_i - b \sum_{i=1}^{n} x_i}{n} = \bar{y} - b\bar{x} \tag{12.40}$$

If the values for the function y are normally distributed, the confidence intervals for a and b can be evaluated using the t transformation variable for $n - 2$ degrees of freedom where the t value appropriate for b is:

$$t = \frac{b - \beta}{\frac{s}{\sqrt{\sum_{i=1}^{n} (x_i - \bar{x})^2}}} \tag{12.41}$$

whereas the corresponding t value for a for $n - 2$ degrees of freedom is:

$$t = \frac{a - \alpha}{s \sqrt{\sum_{i=1}^{n} \frac{x_i^2}{n \sum_{i=1}^{n} (x_i - \bar{x})^2}}} \tag{12.42}$$

Another statistical analysis that has great utility is correlation. Many variables are linearly correlated with others, and it is useful to quantify the degree of linearity in the relationship between sample variables using the sample correlation coefficient. The sample correlation coefficient is calculated using the formula [21]:

$$r = \frac{s_{xy}}{\sqrt{s_x s_y}} = \frac{n \sum_{i=1}^{n} x_i y_i - \left(\sum_{i=1}^{n} x_i\right) \left(\sum_{i=1}^{n} y_i\right)}{\sqrt{n \sum_{i=1}^{n} x_i^2 - \left(\sum_{i=1}^{n} x_i\right)^2} \sqrt{n \sum_{i=1}^{n} y_i^2 - \left(\sum_{i=1}^{n} y_i\right)^2}} \tag{12.43}$$

If the sample correlation coefficient is near one, the correlation is nearly perfect. If the correlation is near zero, there is little or no correlation between x and y. If the correlation coefficient is near negative 1, the correlation is opposite of the measured correlation or the variable y changes in the opposite direction rela-

> *The correlation coefficient quantifies the correlation between variables. If x and y are perfectly correlated, it is 1. It is zero if they are uncorrelated. It is negative one if they are negatively correlated.*

tive to x. However, the quantitative value of the correlation coefficient does not reflect the degree of linearity. Instead, it is appropriate to compare the degree of linearity through the coefficient of determination, which is equivalent to r^2 or:

$$r^2 = \frac{s_{xy}^2}{s_x s_y} = \frac{\left[n \sum_{i=1}^{n} x_i y_i - \left(\sum_{i=1}^{n} x_i\right) \left(\sum_{i=1}^{n} y_i\right)\right]^2}{\left[n \sum_{i=1}^{n} x_i^2 - \left(\sum_{i=1}^{n} x_i\right)^2\right] \left[n \sum_{i=1}^{n} y_i^2 - \left(\sum_{i=1}^{n} y_i\right)^2\right]} \tag{12.44}$$

   The coefficient of determination gives the fraction of the total variation in variable y that is accounted for through a linear relationship with x. Thus, if r^2 is 0.96, 96% of the variation in y values can be accounted for by the linear relationship with the associated x values. When the variables x and y are normally distributed, the distribution of r values follows the t-distribution for $n - 2$ degrees of freedom if the value of t is calculated using the formula [22]:

$$t = r \frac{\sqrt{n - 2}}{\sqrt{1 - r^2}} \tag{12.45}$$

   Using the resulting t value statistics the probability of a linear correlation between x and y can be determined.

12.6   Selecting Appropriate Statistical Functions

The vast majority of common statistical analyses are performed on the basis of the normal distribution density function, since most of the statistics using the z-based approaches, t-based methods, F-tests, and chi-squared tests all have some connection to the normal distribution density function. However, many types of data fit functions other than the normal distribution density function. Thus, it useful to know what other functions are available and to verify that the data fit the function that is used in the analysis. Consequently, this section will begin with an introduction of other distribution functions that will be followed by a method to evaluate the applicability of the function to the desired data set.

12.6.1   Log-Normal Distribution Function

The log-normal distribution density function is given as [23]:

$$f(x) = \frac{1}{x\sigma\sqrt{2\pi}} e^{-\left(\frac{1}{2}\left[\frac{(\ln(x)-\mu)}{\sigma}\right]^2\right)} \tag{12.46}$$

12.6.2   Weibull Distribution Density Function

The Weibull distribution density function is written mathematically as [24]:

$$f(x) = y\beta x^{\beta-1} e^{-yx^{\beta}} \tag{12.47}$$

where y, and β are parameters with values greater than one, and the value of x must also be positive.

12.6.3   Exponential Distribution Density Function

The exponential distribution density function is written mathematically as [25]:

$$f(x) = \frac{1}{\beta} e^{-x/\beta} \tag{12.48}$$

where β is a parameter with value greater than one. The value of x must also be positive.

12.6.4   Double Exponential Distribution Density Function

The exponential distribution density function is written mathematically as:

$$f(x) = \frac{1}{\alpha} \exp\left[\frac{-(x-\lambda)}{\alpha} - \exp\left(\frac{(x-\lambda)}{\alpha} \right) \right] \tag{12.49}$$

where β and α are parameters with values greater than one. The value of x must also be positive.

12.6.5   Determination of the Most Appropriate Distribution Function

One common approach to determining the appropriate statistical function is to convert the data into the cumulative distribution function format to facilitate the evaluation. In other words, the data are converted to cumulative probabilities, and the converted cumulative probability data are compared with the integrals of the density functions. The cumulative probabilities are determined by first ranking the values. The ranking is performed by creating associated bins with boundary limits, and it is followed by dividing the frequency of occurrence of a bin category divided by $n + 1$ (the addition of one is needed due to the implicit boundaries of 0 and ∞ that would otherwise create an unaccounted for bin category or element). The associated cumulative distribution functions to which the data can be compared are given in the accompanying section. That section also contains a linearized form of each function that facilitates graphical evaluations and linear correlation coefficient comparisons.

12.6.6   *Weibull Cumulative Distribution Function*

The Weibull cumulative distribution function is written mathematically as:

$$F(x) = 1 - e^{-\alpha x^{\beta}} \tag{12.50}$$

and the associated linearized form is:

$$\ln\left[\ln\left(\frac{1}{1 - F(x)}\right)\right] = \ln(\alpha) + \beta \ln(x) \tag{12.51}$$

Thus, an appropriate plot of $\ln(\ln(1/(1-F(x))))$ versus $\ln(x)$ will allow the fit of the data to be analyzed using correlation coefficient data.

12.6.7   *Exponential* **Distribution Density Function**

The exponential cumulative distribution function is written mathematically as:

$$F(x) = 1 - e^{-\beta x} \tag{12.52}$$

The associated linearized form is:

$$\ln[1 - F(x)] = \beta x \tag{12.53}$$

Thus, an appropriate plot of $\ln(1-F(x))$ versus x will allow the fit of the data to be analyzed using correlation coefficient data.

12.6.8   Double Exponential Distribution Density Function

The exponential distribution density function is written mathematically as:

$$F(x) = \exp\left[-\exp\left(\frac{-(x-\lambda)}{\alpha}\right)\right]$$ (12.54)

and the associated linearized form is:

$$\ln[-\ln(F(x))] = -\frac{(x-\lambda)}{\alpha}$$ (12.55)

Thus, an appropriate plot of $\ln(-\ln(F(x)))$ versus x will allow the fit of the data to be analyzed using correlation coefficient data.

12.6.9   Normal and Log-Normal Distribution Data Evaluations

The normal cumulative distribution and log-normal cumulative distribution functions cannot be linearized. Thus, the assessment of the data relative to these functions is often made by plotting the data using appropriate graphs as provided in Figs. 12.7 and 12.8. A summary table of statistical formulas that are rooted in analysis of normal distributions is presented in Table 12.5.

Example 12.10 Determine the most appropriate statistical distribution density function that fits the following data for corrosion pit penetration in micrometers: 30, 50, 70, 80, 100, 120, 140, 150, 170, 180, 200, 220, 240, 260, 280, 310, 320, 330, 350, 380, 420, 450, 500, 550.

There are 24 data points in this set to be ranked. Thus, $f(x)$ for each data point is equivalent to $1/(24 + 1) = 0.04$. The cumulative distribution function is determined by the cumulative sum of $f(x)$. These numbers are tabulated in the accompanying table along with the associated frequency of occurrence, the distribution density and cumulative distribution function values.

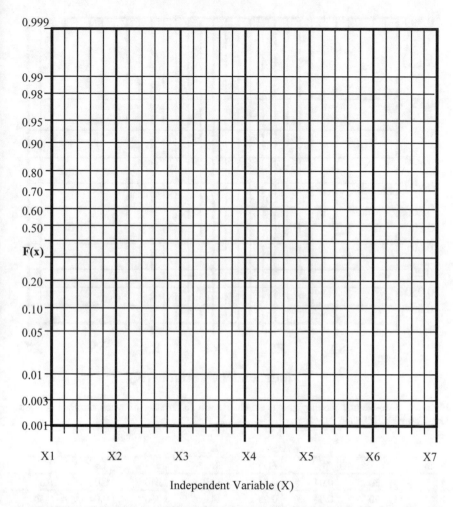

Fig. 12.7 Normal plotting graph, which gives a linear data fit when the data follows a Normal distribution

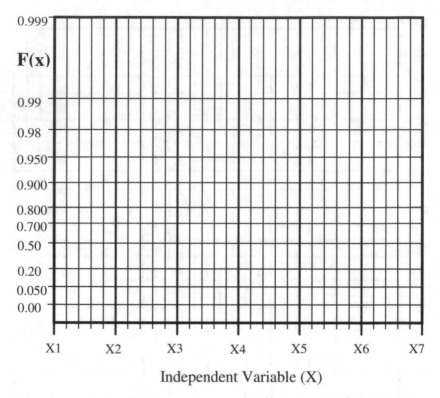

Fig. 12.8 Log-Normal plotting graph, which gives a linear data fit when the data follows a Log-Normal distribution

Rank	x	$f(x)$	$F(x)$	Rank	x	$f(x)$	$F(x)$
1	30	0.04	0.04	13	240	0.04	0.52
2	50	0.04	0.08	14	260	0.04	0.56
3	70	0.04	0.12	15	280	0.04	0.60
4	80	0.04	0.16	16	310	0.04	0.64
5	100	0.04	0.20	17	320	0.04	0.68
6	120	0.04	0.24	18	330	0.04	0.72
7	140	0.04	0.28	19	350	0.04	0.76
8	150	0.04	0.32	20	380	0.04	0.80
9	170	0.04	0.36	21	420	0.04	0.84
10	180	0.04	0.40	22	450	0.04	0.88
11	200	0.04	0.44	23	500	0.04	0.92
12	220	0.04	0.48	24	550	0.04	0.96

Table 12.5 Summary of selected statistical analysis variable formulas and utilization for normally distributed variables and near normally distributed variables

Goal	Deg Free-dom	n	Distribution Variable and Other Relevant Formulas (Calculated distribution values z, t, F, and χ^2 are used to locate associated probabilities in applicable tables)
Estimate probability of value $x_1 < x < x_2$	n	1	$z = \frac{(x-\mu)}{\sigma}$
Estimate probability of value $x_1 < x < x_2$ for binomial variables	n	>30	$z = \frac{x-np}{\sqrt{np(1-p)}}$
Estimate probability of mean sample value $P\,(\bar{x}_1 < \bar{x} < \bar{x}_2)$	n	> 30	$z = \frac{(\bar{x}-\mu)}{\frac{\sigma}{\sqrt{n}}}$
Estimate value of mean	n	>30	$\bar{x} - z_{\alpha/2}\frac{\sigma}{\sqrt{n}} < \mu < \bar{x} + z_{\alpha/2}\frac{\sigma}{\sqrt{n}}$
Estimate probability of value P $(x_1 < x < x_2)$	$n-1$	<30	$t = \frac{\bar{x}-\mu}{\frac{s}{\sqrt{n}}}$
Estimate value of mean	$n-1$	<30	$\bar{x} - t_{\alpha/2}\frac{s}{\sqrt{n}} < \mu < \bar{x} + t_{\alpha/2}\frac{s}{\sqrt{n}}$
Estimate probability of difference between population means $P[x_1 < (\mu_1 - \mu_2) < x_2]$	N	>30	$z = \frac{(\bar{x}_1-\bar{x}_2)-(\mu_1-\mu_2)}{\sqrt{\frac{\sigma_1^2}{n_1}+\frac{\sigma_2^2}{n_2}}}$
Estimate probability of difference between population means $P[x_1 < (\mu_1 - \mu_2) < x_2]$	$n-1$	< 30	$t = \frac{(n_1+n_2-2)[(\bar{x}_1-\bar{x}_2)-(\mu_1-\mu_2)]}{[s_1^2(n_1-1)+s_2^2(n_2-1)]\sqrt{\frac{1}{n_1}+\frac{1}{n_2}}}$ if $\sigma_1 = \sigma = \sigma_2$
Estimate probability of difference between population means $P[x_1 < (\mu_1 - \mu_2) < x_2]$	V See Formula	<30	$t \approx \frac{[(\bar{x}_1-\bar{x}_2)-(\mu_1-\mu_2)]}{\sqrt{(s_1^2/n_1)+(s_2^2/n_2)}}$ $v = \frac{[(s_1^2/n_1)+(s_2^2/n_2)]^2}{\frac{(s_1^2/n_1)^2}{n_1-1}+\frac{(s_2^2/n_2)^2}{n_2-1}}$
Estimate probability of variances not different	$n-1$		$F = \frac{s_1^2\sigma_2^2}{s_2^2\sigma_1^2}$
Estimate probability of sample and pop. variances not diff	$n-1$		$\chi^2 = \frac{(n-1)s^2}{\sigma^2} = \sum_{i-1}^{n}\frac{(x_i-\bar{x})^2}{\sigma^2}$
Estimate the variance σ^2	$n-1$		$\frac{(n-1)s^2}{\chi_{\alpha/2}^2} < \sigma^2 < \frac{(n-1)s^2}{\chi_{1-\alpha/2}^2}$

Plotting the data leads to the following assessment data:

Function to evaluate	Plot	r^2
Weibull	$\ln(\ln(1/(1-F(x))))$ versus $\ln(x)$	0.996
Exponential	$\ln(1-F(x))$ versus x	0.875
Double exponential	$\ln(-\ln(F(x)))$ versus x	0.990

The fit of the data to a normal and log-normal distribution were performed by plotting the data using the appropriate plot grid (see Figs. 12.7 and 12.8). The resulting plots of the data did not appear to be linear. Because the fit for the Weibull distribution was the best, the Weibull distribution function is the most appropriate for the associated data.

12.7   Hypothesis Testing

It is common to utilize statistics to make decisions regarding the validity of con-clusions that are made based upon quantitative data. One very common approach to statistics-based decision making involves the establishment of hypotheses. This process begins by establishing a null hypothesis (H_o), which is based upon the theory that is believed to be incorrect, and an alternative hypothesis (H_1), which is based upon the theory that is believed to be correct. The next step is the selection of the desired significance level, α, and the type of test (one-tailed or two-tailed test). Assuming the random sample data is already available and that the null hypothesis is true (the objective is to prove it should be rejected, so the proof begins by assuming it is true), the next step is to determine the appropriate statistical values. Next, a decision is made to accept or reject the null hypothesis. Finally, based upon the decision regarding the null hypothesis, a conclusion is made to accept or not accept the alternative hypothesis. If the null hypothesis is rejected, the alternative hypothesis is accepted. If the null hypothesis is not rejected, the alternative hypothesis is not accepted. This approach applies to all types of general statistical function tests that are discussed in this chapter.

Example 12.11   Determine if corrosion inhibitor A is better than inhibitor B at the 0.05 significance level given the following corrosion rate (mils per year) data: For inhibitor A: 9.5, 10.5, 11.2, 12.3, 12.5, 12.8, 13.5, 15.2; For inhibitor B: 8.5, 8.9, 9.9, 10.1, 10.3, 10.5, 10.7, and 11.2.

Determine what is known or can be calculated: s_A, s_B, s^2_A, s^2_B $\overline{x_A}$, $\overline{x_B}$, n_A, n_B, ($n_A = n_B < 30$).

Perform a t-test to compare the mean values of inhibitors A and B.

H_1: The performance of inhibitor A is different than the performance of inhibitor B.
H_0: There is no difference between the performances of inhibitors A and B.
$\alpha = 0.05$ (one-tailed because the performance of A must be better than B)

$$n_A = n_B = 8$$

$$\bar{x}_A = \sum_{i=1}^{n} \frac{x_i}{n} = \frac{9.5 + 10.5 + 11.2 + 12.3 + 12.5 + 12.8 + 13.5 + 15.2}{8} = 12.19$$

$$\bar{x}_B = \sum_{i=1}^{n} \frac{x_i}{n} = \frac{8.5 + 8.9 + 9.9 + 10.1 + 10.3 + 10.5 + 10.7 + 11.2}{8} = 10.01$$

$$s_A = \sqrt{\sum_{i=1}^{n} \frac{(x_i - \bar{x})^2}{n-1}} = \sqrt{\frac{n \sum_{i=1}^{n} x_i^2 - \left(\sum_{i=1}^{n} x_i\right)^2}{n(n-1)}} = \sqrt{\frac{8(1211) - 9510}{8(7)}} = 1.78$$

$$s_B = \sqrt{\sum_{i=1}^{n} \frac{(x_i - \bar{x})^2}{n-1}} = \sqrt{\frac{n \sum_{i=1}^{n} x_i^2 - \left(\sum_{i=1}^{n} x_i\right)^2}{n(n-1)}} = \sqrt{\frac{8(808) - 6413}{8(7)}} = 0.94$$

$$t \approx \frac{[(\bar{x}_1 - \bar{x}_2) - (\mu_1 - \mu_2)]}{\sqrt{(s_1^2/n_1) + (s_2^2/n_2)}} = \frac{[(12.19 - 10.01) - (0)]}{\sqrt{(1.78^2/8) + (0.94^2/8)}} = 4.08$$

$$v = \frac{[(s_1^2/n_1) + (s_2^2/n_2)]^2}{\frac{(s_1^2/n_1)^2}{n_1-1} + \frac{(s_2^2/n_2)^2}{n_2-1}} = \frac{[(1.78^2/8) + (0.94^2/8)]^2}{\frac{(1.78^2/8)^2}{7} + \frac{(0.94^2/8)^2}{7}} = 10.6 \approx 11$$

Thus, examining the t-distribution table with 11 degrees of freedom for $a = 0.05$ gives a value of t of 1.796, which is considerably lower than the calculated value of 4.08.

Decision: Reject H_0.

Conclusion: Accept H_1: The performance of inhibitor A is different than the performance of inhibitor B.

12.8 Analysis of Variance (ANOVA)

It is often desirable to determine if two or more different treatments differ in their effect on a specific outcome. One way to approach such a determination is to perform an analysis of variance (ANOVA) of the data sets. The logic for utilizing such an approach is rooted in the understanding that if the variance between the mean outcomes of the treatments is very different from the average variance of the outcomes obtain from within the individual treatment samples, the treatments produce different outcomes. Conversely, if the variance of mean outcomes between the samples is small and the average variance of outcomes within each sample is large, the conclusion is that the treatments do not produce different outcomes. In other words, if the variability between the sample means is larger than expected by random measurements about a

> *ANOVA is used to evaluate effects of variables on specific outcomes.*

common mean then the sample means must be different. The F statistic predicts the probability of obtaining a sample variance from one data set that is significantly different than one obtained from another data set when the respective population variances are known. Thus, the F-distribution is used for such evaluations along with two new terms, $s^2{}_B$ and $s^2{}_W$, that respectively estimate the variance between sample means and the average variance within sample sets. The estimated variance between sample means is:

$$s_B^2 = \frac{SS_{\text{Treatment}}}{k-1} = \frac{\sum_{i=1}^{k} n_i \left(\bar{x}_i - \frac{\sum_{i=1}^{k} \bar{x}_i}{k} \right)^2}{k-1}$$

$$= \frac{\sum_{i=1}^{k} \frac{\left(\sum_{j=1}^{n_i} x_{ij} \right)^2}{n_i} - \frac{\left(\sum_{i=1}^{k} \sum_{i=1}^{n_i} x_{ij} \right)^2}{N}}{k-1}$$

(12.56)

where the subscript B represents "between" samples, N is the total number of measurements made for the entire collection of samples, $SS_{\text{Treatment}}$ is the sum of squares of associated with the treatment or the sum of squares between samples, k is the number of samples and n is the sample size. The estimated average variance within sample sets is presented in Eq. (12.57):

$$s_W^2 = \frac{SSE}{\left(\sum_{i=1}^{k} n_i \right) - k} = \frac{SST - SS_{\text{Treatm.}}}{N-k}$$

$$= \frac{\left[\sum_{i=1}^{k} \sum_{j=1}^{n_i} x_{ij}^2 - \frac{\left(\sum_{i=1}^{k} \sum_{j=1}^{n_i} x_{ij}^2 \right)}{N} \right] - \left[\sum_{i=1}^{k} \frac{\left(\sum_{j=1}^{n_i} x_{ij} \right)^2}{n_i} - \frac{\left(\sum_{i=1}^{k} \sum_{i=1}^{n_i} x_{ij} \right)^2}{N} \right]}{N-k}$$

(12.57)

where the subscript W represents "within" samples, SSE is the error sum of squares or the sum of squares within the samples, and SST is the sum of squares total from the entire set of samples.

One important key to utilizing these equations to determine a corresponding value of F is the assumption that if the treatments are assumed to produce the same result as the standard treatment, then the variance associated with the populations should be the same, leading to:

$$F = \frac{s_B^2 \sigma_W^2}{s_W^2 \sigma_B^2} = \frac{s_B^2}{s_W^2}$$

(12.58)

Note that there are $k - 1$ degrees of freedom for variance between samples and $N - k$ degrees of freedom for the variance within samples, where N is the total sum of sample sizes or the total number of measurements made for all of the samples.

Example 12.12   Determine whether or not treatments a, b, c, and d have the same effect at the 0.05 significance level on a process outcome given the following relative process outcome data for each group: a: 96, 99, 108, 102, 96, 98, 105; b: 121, 115, 114, 122, 109; c: 110, 108, 99, 100, 103, 112, 92; d: 88, 83, 86, 93, 96, 89, 102, 105, 88, 95, 103.

ANOVA analysis table for Example 12.12:

	x_a, x_a^2	x_b, x_b^2	x_c, x_c^2	x_d, x_d^2	
	969,216	12,114,641	11,012,100	887,744	
	999,801	11,513,225	10,811,664	836,889	
	10,811,664	11,412,996	999,801	867,396	
	10,210,404	12,214,884	10,010,000	938,649	
	969,216	10,911,881	10,310,609	969,216	
	989,604		11,212,544	897,921	
	10,511,025		928,464	10,210,404	
				10,511,025	
				887,744	
				959,025	
				10,310,609	***Sum Total***
$\sum x$	704	581	724	1,028	***3037***
$\sum x^2$	70,930	67,627	75,182	96,622	***310,361***
n	7	5	7	11	***30***
$(\sum x)^2$	495,616	337,561	524,176	1,056,784	
$\dfrac{(\sum x)^2}{n}$	70,802	67,512	74,882	96,071	***309,267***
$\bar{x}$	100.6	116.2	103.4	93.5	

Problems of this type are often solved by setting up associated hypotheses and performing an ANOVA on the data using tables to facilitate the data analysis.

H_1: The treatments differ in process outcome effectiveness.
H_0: The treatments do not differ in process outcome effectiveness.

$$SS_{\text{Treatment}} = \sum_{i=1}^{k} \frac{\left(\sum_{j=1}^{n_i} x_{ij}\right)^2}{n_i} - \frac{\left(\sum_{i=1}^{k} \sum_{i=1}^{n_i} x_{ij}\right)^2}{N} = 309,267 - \frac{(3,037)^2}{30} = 1,821$$

$$SST = \sum_{i=1}^{k} \sum_{j=1}^{n_i} x_{ij}^2 - \frac{\left(\sum_{i=1}^{k} \sum_{j=1}^{n_i} x_{ij}^2\right)}{N} = 310,361 - \frac{(3,037)^2}{30} = 2,915$$

$$SSE = SST - SS_{\text{Treatment}} = 2,915 - 1,821 = 1,094$$

$$F = \frac{s_B^2}{s_W^2} = \frac{\frac{SS_{\text{Treatment}}}{k-1}}{\frac{SSE}{N-k}} = \frac{\frac{1,821}{3}}{\frac{1,094}{26}} = 14.43$$

$$F_{(3,26)\alpha=0.05} = 2.98$$

$$F > F_{(3,26)\alpha=0.05}$$

Decision: Reject H_O: The treatments do not differ in process outcome effectiveness

Accept H_1: The treatments differ in process outcome effectiveness.

Conclusion: There is at least 95% confidence that the treatments differ in process outcome effectiveness.

12.9   Factorial Design and Analysis of Experiments

In order to measure the effect of three variables, A, B, and C on a key process output variable, Y, it may be necessary to perform experiments. If A, B, and C are completely independent of each other, it would be necessary to perform tests at a minimum of two levels to determine if the change in the level of the selected variable has an influence on the output variable. Thus, a minimum of six tests would be needed to make an assessment of the three variables. However, most applications require some testing to determine the interdependence or independence amongst the variables. In other words, will the level of A have an influence on the effect of C, or will B have an effect on A, or will A and B together have an effect on the performance of C? A complete evaluation of all possible effects and interdependencies would require a significant number of additional tests if they were to be evaluated as separate test series. However, the factorial design of experiments method allows the user to perform fewer tests and obtain more information regarding interdependencies than traditional methods involving the changing of one variable at a time.

> *Factorial designs are very useful to evaluate the effects of multiple variables on one or more specific outcomes based on a relatively small number of experiments.*

A set of tests that follow the factorial design approach with variables A, B, and C tested at high (+) and low (-) levels is presented in Table 12.6.

A graphical illustration of the relative levels and outcomes of these tests is presented in Fig. 12.9. The relative effects of parameters A, B, and C and their associated interactions can be computed using the following:

$$(A) = \frac{-Y_1 + Y_2 - Y_3 + Y_4 - Y_5 + Y_6 - Y_7 + Y_8}{4} \quad (12.59)$$

$$(B) = \frac{-Y_1 - Y_2 + Y_3 + Y_4 - Y_5 - Y_6 + Y_7 + Y_8}{4} \quad (12.60)$$

$$(C) = \frac{-Y_1 - Y_2 - Y_3 - Y_4 + Y_5 + Y_6 + Y_7 + Y_8}{4} \quad (12.61)$$

$$(AB) = \frac{+Y_1 - Y_2 - Y_3 + Y_4 + Y_5 - Y_6 - Y_7 + Y_8}{4} \quad (12.62)$$

$$(AC) = \frac{+Y_1 - Y_2 + Y_3 - Y_4 - Y_5 + Y_6 - Y_7 + Y_8}{4} \quad (12.63)$$

$$(BC) = \frac{+Y_1 + Y_2 - Y_3 - Y_4 - Y_5 - Y_6 + Y_7 + Y_8}{4} \quad (12.64)$$

$$(ABC) = \frac{-Y_1 + Y_2 + Y_3 - Y_4 + Y_5 - Y_6 - Y_7 + Y_8}{4} \quad (12.65)$$

$$\text{mean} = \frac{Y_1 + Y_2 + Y_3 + Y_4 + Y_5 + Y_6 + Y_7 + Y_8}{8} \quad (12.66)$$

Thus, the value of the effect indicates the influence that parameter or parameter interaction has on the outcome. Thus if the average value of Y is 100 and the C effect is calculated to be 12, then the average change in C from the lower level to the higher level caused a change in Y of 12.

Table 12.6 Three factor design of experiments at high (+) and low (-) levels of A, B, and C

Run	A	B	C	AB	AC	BC	ABC	Observation
1	−	−	−	+	+	+	−	Y_1
2	+	−	−	−	−	+	+	Y_2
3	−	+	−	−	+	−	+	Y_3
4	+	+	−	+	−	−	−	Y_4
5	−	−	+	+	−	−	+	Y_5
6	+	−	+	−	+	−	−	Y_6
7	−	+	+	−	−	+	−	Y_7
8	+	+	+	+	+	+	+	Y_8

Note that the high and low levels for the interactions A with B (AB), A with C (AC), B with C (BC) and A with B and C (ABC) also show high and low levels for the purpose of determining the relative effects of such interactions on the observed values

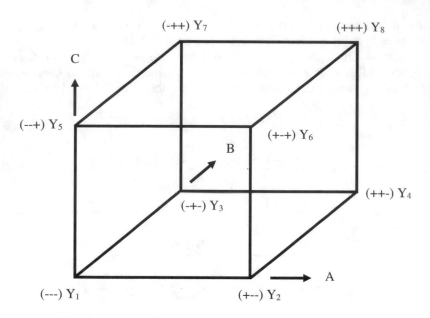

Fig. 12.9 Schematic illustration of 2^3 factorial design of experiments, which includes a three factor set of parameters at two levels

Example 12.13 Using the following data from electrodeposition experiments, determine the most important factor influencing current efficiency. The factors are initial Cu^{2+}/Cu^{+}(C) ratio, initial Fe^{3+}/Fe^{2+}(F) ratio, and HCl(H) concentration. The high and low levels are 0.111 and 9 for the copper and iron and 0.05 and 0.15 m for HCl. The factorial design of experiments and effects is shown in the accompanying table.

Table (Example 12.13): Factorial Design of Experiments Example [Data from P. K. Sarswat, Ph.D. Dissertation, Univ. of Utah, 2010.]

Run	C	F	H	CF	CH	FH	CFH	Curr.Effic. (%)
1	−	−	−	+	+	+	−	54
2	+	−	−	−	−	+	+	51.6
3	−	+	−	−	+	−	+	32
4	+	+	−	+	−	−	−	22
5	−	−	+	+	−	−	+	47
6	+	−	+	−	+	−	−	40
7	−	+	+	−	−	+	−	27
8	+	+	+	+	+	+	+	28

$$(Cu^{2+}/Cu^{+}) = \frac{-54+51.6-32+22-47+40-27+28}{4} = -4.6$$

Thus, the copper ion ratio had a slightly negative effect on current efficiency. Similar calculations can be made for each of the other factors and the related interactions. The calculated values are as follows:

$$(Fe^{3+}/Cu^{2+}) = \frac{-54-51.6+32+22-47-40+27+28}{4} = -20.9$$

$$(HCl) = \frac{-54-51.6-32-22+47+40+27+28}{4} = -4.4$$

$$(Cu/Fe..Interaction) = 0.1$$
$$(Cu/HCl..Interaction) = 1.6$$
$$(Fe/HCl..Interaction) = 4.9$$
$$(Cu/Fe/HCl..Interaction) = 3.9$$

Thus, the ferric/ferrous ion ratio had a very substantial (average of -20.9% decrease) negative effect on current efficiency. Thus, when this ratio is high, the current efficiency is low. This conclusion is logical because the ferric ion presence will lead to unwanted cathodic reduction of ferric ions that will compete with electrons needed for copper reduction.

12.10   Taguchi Method

In the previous section, the benefits and methods of factorial designs of experiments were discussed. If an experimenter needed to evaluate the effect of 6 variables a full-factorial design of experiments would require 2^6 or 64 experiments. Often, such a number of experiments is difficult to justify in industry. Alternative approaches such as fractional factorial and Taguchi methods are often used to reduce the number of experiments. [p. 40–44, Ranjit K. Roy, A primer on the Taguchi Method, 1990, Van Nostrand Reinhold, NY]. Taguchi methods are relatively popular because they utilize a fewer experiments than traditional factorial designs. The Taguchi Method utilizes standard orthogonal arrays. It assumes that variables or factors do not interact.

> *The Taguchi method is a specialized form of factorial design of experiments that uses noise to evaluate parameter effects.*

Genichi Taguchi's main objective for his method is to improve a product's quality. The premise is that improving quality or maintaining high quality is the best approach to benefit society. Quality can be improved or maintained at high levels through proper control of control factors. The control factors are the variables that can be controlled in the desire process and related experiments. Control variable levels are set to achieve quality by being least sensitive to noise variables. Noise variables are the variables that cannot be controlled in the process.

One common orthogonal design for experiments is shown in Table 12.7 that is commonly used for the Taguchi method.

Systems with 4–7 variables can use Table 12.7 by eliminating columns as needed.

The analysis of the data is often made on the basis of the signal to noise ratio. The signal to noise (SN) ratio is effectively a comparison of the response relative to the variation or noise. The SN analysis is made using a formula that is selected based on the analysis objective. Table 12.8 provides the SN formulas.

y_i is the measured value for test i in a series of n tests performed at the same condition, y_{ave} is the average of measured values, σ is the standard deviation, and n is the number of measurements under one specified condition. Often, the data analysis is performed by treating the SN ratio as the response. Corresponding ANOVA (as discussed previously) can then be performed with the SN ratio as the response.

Example 12.14 Compare the SN ratio for treatments a, b, c, and d to assess which treatment has the greatest influence on a process outcome given the following relative process outcome data for each group: a: 96, 99, 108, 102, 96, 98, 105; b:

Table 12.7 L8 Orthogonal design for experiments (7 factors, 2 levels)

Test	Factor	Factor	Factor	Factor	Factor	Factor	Factor
	A	B	C	D	E	F	G
1	−	−	−	−	−	−	−
2	−	−	−	+	+	+	+
3	−	+	+	−	−	+	+
4	−	+	+	+	+	−	−
5	+	−	+	−	+	−	+
6	+	−	+	+	−	+	−
7	+	+	−	−	+	+	−
8	+	+	−	+	−	−	+

Adapted from Roy, p. 51

Table 12.8 SN formulas [26]

Analysis objective	SN ratio
Maximize response	$SN_{max} = -10\log\left(\frac{1}{n}\sum_{i=1}^{n}\frac{1}{y_i^2}\right)$ (12.67)
Achieve target	$SN_{tgt} = 10\log\frac{y_{ave}^2}{\sigma^2}$ (12.68)
Minimize response	$SN_{max} = -10\log\left(\frac{1}{n}\sum_{i=1}^{n}y_i^2\right)$ (12.69)

121, 115, 114, 122, 109; *c*: 110, 108, 99, 100, 103, 112, 92; *d*: 88, 83, 86, 93, 96, 89, 102, 105, 88, 95, 103. Assume the highest outcome is the best. Utilize a Taguchi signal to noise comparison to evaluate the information. Note that the data is the same as in Example 12.12 to facilitate a comparison.

$$SN_{\max(a)} = -10 \log \left(\frac{1}{7} \sum_{i=1}^{n} \frac{1}{96^2} + \frac{1}{99^2} + \frac{1}{108^2} + \frac{1}{102^2} \right.$$
$$\left. + \frac{1}{96^2} + \frac{1}{98^2} + \frac{1}{105^2} \right) = 40.02$$
$$SN_{\max(b)} = 41.3$$
$$SN_{\max(c)} = 40.2$$
$$SN_{\max(d)} = 39.3$$

Generally, the factors that give the highest SN ratios are the most influential parameters. Thus, in this case treatment b has the highest SN ratio.

A more rigorous and effective evaluation of data using the Taguchi method involves multiple measurements for each condition. A SN analysis is performed using the multiple measurements for each condition. Next, the SN values are compared for different levels and conditions. Additionally, an ANOVA is performed and the F values are compared as discussed previously.

In some cases a comparison of SN values and response values are compared to identify the most important factors in simplified comparison. An example of this approach is given in Example 12.15.

Example 12.15 A student performs 18 tests according to a modified orthogonal L18 design for 8 variables [adapted from Roy, p. 224]. This design tests up to 7 variables at 3 levels and one variable at two levels. In this application of the design, the first and last two variables are omitted. The student measures the following data: (Table 12.9).

Analyze the data using a simplified Taguchi SN and response evaluation by constructing SN and response versus factor levels.

The resulting data are plotted in the accompanying Figs. (12.10, 12.11, 12.12, 12.13, 12.14, 12.15, 12.16, 12.17, 12.18 and 12.19). The data in the plots show that Factors C and D are very influential in increasing the rate of extraction. The SN values are based on Eq. 12.67.

Table 12.9 Chalcopyrite leaching data for chloride based leaching with compounds (factors) A–E. The rate of leaching in terms of fraction of copper extracted per hour was performed twice to evaluate the noise level for each compound (factor)

Test	Factor A	Factor B	Factor C	Factor D	Factor E	Rate 1 (fr./hr)	Rate 2 (fr./hr)
1	Low	Low	Low	Low	Low	0.0013	0.0010
2	Low	Med	Med	Med	Med	0.0038	0.0042
3	Low	High	High	High	High	0.0100	0.0108
4	Med	Low	Low	Med	Med	0.0050	0.0046
5	Med	Med	Med	High	High	0.0042	0.0046
6	Med	High	High	Low	Low	0.0050	0.0046
7	High	Low	Med	Low	High	0.0029	0.0027
8	High	Med	High	Med	Low	0.0075	0.0071
9	High	High	Low	High	Med	0.0046	0.0042
10	Low	Low	High	High	Med	0.0113	0.0096
11	Low	Med	Med	Low	High	0.0033	0.0036
12	Low	High	Low	Med	Low	0.0033	0.0029
13	Med	Low	Med	High	Low	0.0075	0.0071
14	Med	Med	High	Low	Med	0.0063	0.0058
15	Med	High	Low	Med	High	0.0042	0.0033
16	High	Low	High	Med	High	0.0075	0.0067
17	High	Low	Low	High	Low	0.0050	0.0043
18	High	High	Med	Low	Med	0.0025	0.0029

Fig. 12.10 Comparison of SN levels for Factor A in Example 12.15

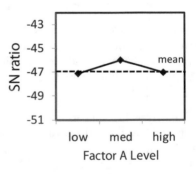

Fig. 12.11 Comparison of rates for Factor A in Example 12.15

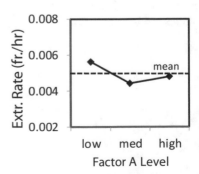

Fig. 12.12 Comparison of
SN levels for Factor B in
Example 12.15

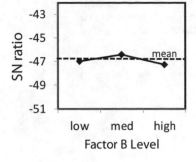

Fig. 12.13 Comparison of
rates for Factor B in Example
12.15

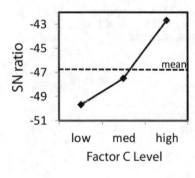

Fig. 12.14 Comparison of
SN levels for Factor C in
Example 12.15

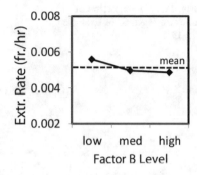

Fig. 12.15 Comparison of
rates for Factor C in Example
12.15

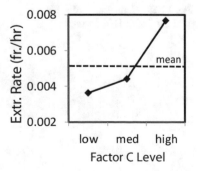

Fig. 12.16 Comparison of
SN levels for Factor D in
Example 12.15

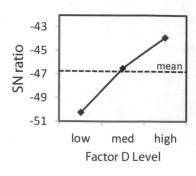

Fig. 12.17 Comparison of
rates for Factor D in Example
12.15

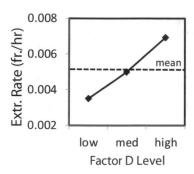

Fig. 12.18 Comparison of
SN levels for Factor E in
Example 12.15

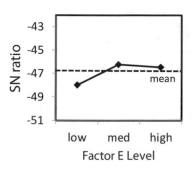

Fig. 12.19 Comparison of
rates for Factor E in Example
12.15

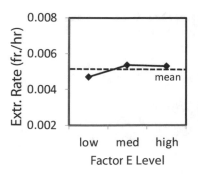

References

1. J.R. Taylor, *An Introduction to Error Analysis* (Oxford University Press, New York, 1982a), p. 15
2. J.R. Taylor, *An Introduction to Error Analysis* (Oxford University Press, New York, 1982b) p. 45
3. J.R. Taylor, *An Introduction to Error Analysis* (Oxford University Press, New York, 1982c) p. 48
4. J.R. Taylor, *An Introduction to Error Analysis* (Oxford University Press, New York, 1982d) p. 51
5. J.R. Taylor, *An Introduction to Error Analysis* (Oxford University Press, New York, 1982e) p. 56
6. J.R. Taylor, *An Introduction to Error Analysis* (Oxford University Press, New York, 1982f) p. 57
7. J.R. Taylor, *An Introduction to Error Analysis* (Oxford University Press, New York, 1982g), p. 73
8. R.E. Walpole, R.H. Meyers, S.L. Meyers, *Probability and Statistics for Engineers and Scientists*, 6th edn. (Simon & Schuster, Upper Saddle River, 1998a), p. 4
9. R.E. Walpole, R.H. Meyers, S.L. Meyers, *Probability and Statistics for Engineers and Scientists*, 6th edn. (Simon & Schuster, Upper Saddle River, 1998b), p. 205
10. R.E. Walpole, R.H. Meyers, S.L. Meyers, *Probability and Statistics for Engineers and Scientists*, 6th edn. (Simon & Schuster, Upper Saddle River, 1998c), p. 207
11. R.E. Walpole, R.H. Meyers, S.L. Meyers, *Probability and Statistics for Engineers and Scientists*, 6th edn. (Simon & Schuster, Upper Saddle River, 1998d), p. 115
12. R.E. Walpole, R.H. Meyers, S.L. Meyers, *Probability and Statistics for Engineers and Scientists*, 6th edn. (Simon & Schuster, Upper Saddle River, 1998e), p. 145
13. R.E. Walpole, R.H. Meyers, S.L. Meyers, *Probability and Statistics for Engineers and Scientists*, 6th edn. (Simon & Schuster, Upper Saddle River, 1998f), p. 149
14. G.H. Winberg, J.A. Schumaker, D. Oltman, *Statistics: An Intuitive Approach*, 4th edn. (Brooks/Cole, Monterey, 1981a), p. 150
15. G.H. Winberg, J.A. Schumaker, D. Oltman, *Statistics: An Intuitive Approach*, 4th edn. (Brooks/Cole, Monterey, 1981b), p. 254
16. R.E. Walpole, R.H. Meyers, S.L. Meyers, *Probability and Statistics for Engineers and Scientists*, 6th edn. (Simon & Schuster, Upper Saddle River, 1998g), p. 266
17. R.E. Walpole, R.H. Meyers, S.L. Meyers, *Probability and Statistics for Engineers and Scientists*, 6th edn. (Simon & Schuster, Upper Saddle River, 1998h), p. 167
18. R.E. Walpole, R.H. Meyers, S.L. Meyers, *Probability and Statistics for Engineers and Scientists*, 6th edn. (Simon & Schuster, Upper Saddle River, 1998i), p. 233
19. R.E. Walpole, R.H. Meyers, S.L. Meyers, *Probability and Statistics for Engineers and Scientists*, 6th edn. (Simon & Schuster, Upper Saddle River, 1998j), p. 172
20. R.E. Walpole, R.H. Meyers, S.L. Meyers, *Probability and Statistics for Engineers and Scientists*, 6th edn. (Simon & Schuster, Upper Saddle River, 1998k), p. 366
21. R.E. Walpole, R.H. Meyers, S.L. Meyers, *Probability and Statistics for Engineers and Scientists*, 6th edn. (Simon & Schuster, Upper Saddle River, 1998l), p. 396
22. R.E. Walpole, R.H. Meyers, S.L. Meyers, *Probability and Statistics for Engineers and Scientists*, 6th edn. (Simon & Schuster, Upper Saddle River, 1998m), p. 398
23. R.E. Walpole, R.H. Meyers, S.L. Meyers, *Probability and Statistics for Engineers and Scientists*, 6th edn. (Simon & Schuster, Upper Saddle River, 1998n), p. 173
24. R.E. Walpole, R.H. Meyers, S.L. Meyers, *Probability and Statistics for Engineers and Scientists*, 6th edn. (Simon & Schuster, Upper Saddle River, 1998o), p. 174

25. R.E. Walpole, R.H. Meyers, S.L. Meyers, *Probability and Statistics for Engineers and Scientists*, 6th edn. (Simon & Schuster, Upper Saddle River, 1998p), p. 168
26. R.E. Walpole, R.H. Meyers, S.L. Meyers, *Probability and Statistics for Engineers and Scientists*, 6th edn. (Simon & Schuster, Upper Saddle River, 1998q), p. 601

Problems

(12–1) The following data are obtained for copper recovery using process A: 67, 69, 72, 63, 76 in contrast to those for process B: 69, 80, 76, 81, 72. At what confidence level can it be stated that process B is better than process A?

(12–2) A company plant operator needs to be sure that the average level of dissolved lead in a discharge stream is less than 30 ppb in order to ensure regulatory compliance. Consequently, a technician is sent to gather 9 random solution samples from the process stream for lead analysis. The results from the solution analysis are 27.10, 28.90, 26.60, 27.80, 29.30, 29.20, 28.40, 27.70, and 26.70 ppb lead. Can the operator have at least 99.5% confidence that the mean lead level is less than 30 ppb?

(12–3) A technician runs a series of 5 reaction experiments, each at a different temperature, to determine the activation energy associated with a process reaction. A plot of the natural logarithm of the reaction constant [ln(k)] versus the inverse of the absolute temperature ($1/T$) is constructed and a spreadsheet software package calculates the linear regression through the data yielding the equation $\ln(k) = 21{,}512(1/T) + 35.103$ with an r^2 value of 0.990. Estimate the mean activation energy for this reaction with its associated 95% confidence interval using the Arrhenius equation and appropriate statistical analysis.

Appendix A
Atomic Weights

(from data in *General Chemistry* by Linus Pauling, Dover Publications, New York 1970).

Aluminum	Al	26.98	Molybdenum	Mo	95.94
Antimony	Sb	121.75	Nickel	Ni	58.71
Argon	Ar	39.95	Niobium	Nb	92.91
Arsenic	As	74.92	Nitrogen	N	14.01
Barium	Ba	137.34	Osmium	Os	190.2
Beryllium	Be	9.01	Oxygen	O	16.00
Bismuth	Bi	208.98	Palladium	Pd	106.4
Boron	B	10.81	Phosphorus	P	30.97
Bromine	Br	79.91	Platinum	Pt	195.09
Cadmium	Cd	112.40	Plutonium	Pu	242
Calcium	Ca	40.08	Potassium	K	39.10
Carbon	C	12.01	Radium	Ra	226
Cerium	Ce	140.12	Rhenium	Re	186.2
Cesium	Cs	132.91	Rhodium	Rh	102.91
Chlorine	Cl	35.45	Rubidium	Rb	85.47
Chromium	Cr	52.00	Samarium	Sm	150.35
Cobalt	Co	58.93	Scandium	Sc	44.96
Copper	Cu	63.54	Selenium	Se	78.96
Fluorine	F	19.00	Silicon	Si	28.09
Gallium	Ga	69.72	Silver	Ag	107.87
Germanium	Ge	72.59	Sodium	Na	22.99
Gold	Au	196.97	Strontium	Sr	87.62
Helium	He	4.00	Sulfur	S	32.06
Hydrogen	H	1.01	Tantalum	Ta	180.95
Iodine	I	126.90	Tellurium	Te	127.60

(continued)

© The Minerals, Metals & Materials Society 2022
M. L. Free, *Hydrometallurgy*, The Minerals, Metals & Materials Series,
https://doi.org/10.1007/978-3-030-88087-3

(continued)

Aluminum	Al	26.98	Molybdenum	Mo	95.94
Iridium	Ir	192.20	Tin	Sn	118.69
Iron	Fe	55.85	Titanium	Ti	47.90
Lead	Pb	207.19	Tungsten	W	183.85
Lithium	Li	6.94	Uranium	U	238.03
Magnesium	Mg	24.31	Vanadium	V	50.94
Manganese	Mn	54.94	Zinc	Zn	65.37
Mercury	Hg	200.59	Zirconium	Zr	91.22

Appendix B
Miscellaneous Constants

Constant	Symbol	Value
Elementary charge	e	1.6022×10^{-19} C
Faraday constant	F	96,485 Coulombs/mole
Boltzman constant	k	1.3807×10^{-23} J/K
Gas constant	R	8.3144 J/Kelvin or 0.08206 L · atm/mole · K
Planck constant	h	6.6262×10^{-34} J · sec
Avogadro constant	N_A	6.0221×10^{23} /mole
Atomic mass unit	AMU	1.6606×10^{-27} kg
Electron mass	m_e	9.1095×10^{-31} kg
Proton mass	m_p	1.6727×10^{-27} kg
Neutron mass	m_n	1.6750×10^{-27} kg
Vacuum permittivity	ε_o	8.8542×10^{-12} C^2/J · m
Speed of light in vacuum	c	2.9979×10^{8} m/s

© The Minerals, Metals & Materials Society 2022
M. L. Free, *Hydrometallurgy*, The Minerals, Metals & Materials Series,
https://doi.org/10.1007/978-3-030-88087-3

Appendix C
Conversion Factors

Conversions	Nomenclature
1 atmosphere = 101,325 Pascals	m = meter (distance)
1 atmosphere = 760 Torr	kg = kilogram (mass)
1 atmosphere = 14.7 PSI	g = gram (mass)
1 Torr = 133.322 Pascals	C = Coulomb (charge)
1 mm Hg = 133.3224 Pascals	s = second (time)
1 electron volt = 1.6022×10^{-19} J	J = Joule (energy)
1 cal = 4.184 J	K = Kelvin (temperature)
1 J = 10^7 ergs = 10^7 dyne · cm	w = Watt (power)
1 W = 1 J/s	A = ampere (current)
1 Amp = 1 C/second	V = Volt (electrical potential)
1 V = 1 J/Coulomb	N = Newton (force)
1 N = 1 J/meter or 10^5 dynes	Pa = Pascal (pressure)
1 Pa = 1 N/m^2	atm = atmosphere (pressure)
1 foot = 12 inches	in = inch (distance)
1 inch = 2.54 cm	ft = foot (distance)
1 kg = 2.205 pounds	PSI = pounds/sq. inch (press.)
1 cubic foot = 7.48 U. S. gallons	Btu = British therm. unit (en.)
1 cubic foot = 28.32 L	cp = centipoise (viscosity)
1 Btu = 252 cal	l = liter (volume)
1 horsepower = 0.746 kilowatt	lb = pound (mass)
1 centipoise = 0.01 g/cm · second	v = kinem. visc. (visc/dens)
1 troy ounce = 31.1 g	cS = centistoke = 0.01 cm^2/sec
1 oz = 28.35 g	ton = short ton = 2000 pounds
1 pound = 453.59 g	tonne = metric ton = 1000 kg
1 °C = 1.8 °F 273 K = 0 °C = 32 °F	1 L gas = 0.0446 mol (STP)

© The Minerals, Metals & Materials Society 2022
M. L. Free, *Hydrometallurgy*, The Minerals, Metals & Materials Series,
https://doi.org/10.1007/978-3-030-88087-3

Appendix D
Free Energy Data

Compound	ΔG_r° (J/mole)	References
Ag^1	77,100	[1]
Ag^{2+}	268,200	[3]
$AgCl$	−109,720	[3]
Ag_2O	−10,820	[3]
AgO	10,880	[3]
$Ag(OH)$	−91,970	[3]
$Ag(CN)_2^-$	301,500	[3]
$Ag(NH3)_2^+$	−17,400	[3]
$AgNO_3$	−32,180	[3]
$Ag(S_2O_3)_2^{3-}$	−1,036,000	[3]
$Ag(SO_3)_2^{3-}$	−943,100	[3]
Al^{3+}	−489,400	[2]
AlO_2^-	−839,700	[3]
$Al(OH)_3$	−1,154,900	[2]
Al_2O_3 (gamma)	−1,562,702	[2]
$Al_2O_33H_2O$ (gibbsite)	−2,320,400	[3]
AsO_4^{3-}	−636,000	[3]
$AsO2-$	−349,900	[8]
AsO^+	−163,650	[8]
AsO_3^{3-}	−447,300	[8]
$AsH_3(g)$	68,840	[8]
$HAsO_4^{2-}$	−707,130	[3]
$H_2AsO_4^-$	−748,550	[3]
H_3AsO_4	−769,060	[3]
AsS	−70,320	[2]

(continued)

© The Minerals, Metals & Materials Society 2022
M. L. Free, *Hydrometallurgy*, The Minerals, Metals & Materials Series,
https://doi.org/10.1007/978-3-030-88087-3

(continued)

Compound	ΔG_r° (J/mole)	References
Au^+	163,200	[3]
Au^{3+}	433,500	[3]
$AuCl_2^-$	−47,630	[7]
$AuCl_4^-$	−235,000	[9]
$Au(CN)_2^-$	289,300	[7]
$Au(CNS)_2^-$	241,000	[9]
$Au(CNS)_4^-$	544,000	[9]
$Au(OH)_3$	−290,000	[9]
Au_2O_3	163,200	[3]
$Au(S_2O_3)_2^{3-}$	−1,065,000	[7]
Ba^{2+}	−560,700	[3]
$Ba_3(AsO_4)_2$	−3,074,000	[8]
$Ba(AsO_2)_2$	−1,284,000	[8]
$BaCO_3$	−1,138,900	[3]
$BaSO_4$	−1,353,000	[3]
$BaSeO_4$	−1,062,000	[3]
$BaWO_4$	−1,563,000	[3]
Br-	−104,010	[2]
Ca^{2+}	−553,540	[2]
$Ca_3(AsO_4)_2$	−3,058,000	[8]
$Ca(AsO_2)_2$	−1,291,000	[8]
CaF^+	−835,700	[10]
CaF_2	−1,162,000	[3]
$Ca(OH)^+$	−717,800	[3]
$Ca(OH)_2$	−898,408	[2]
$CaCO_3$ (calcite)	−1,128,842	[2]
CaO	−603,487	[2]
$CaSO_4$	−1,291,000	[2]
$CaSO_4(2H_2O)$	−1,797,197	[2]
Cd^{2+}	−77,580	[2]
$CdCO_3$	−669,440	[2]
CdO	−228,515	[2]
$Cd(OH)_2$	−470,550	[3]
CdS	−145,630	[2]
Cl^-	−131,270	[2]
$Cl_2(aq)$	6,900	[3]
HCl	−131,170	[3]
ClO^-	−37,240	[3]
HClO	−79,960	[3]
ClO_3^-	−2,590	[3]

(continued)

(continued)

Compound	ΔG_r° (J/mole)	References
$HClO_3$	−2,590	[3]
ClO_4^-	−10,750	[3]
$HClO_4$	−10,340	[3]
CN^- (cyanide)	172,400	[4]
HCN	119,700	[4]
HCN(g)	124,700	[4]
OCN^-	−97,400	[4]
HOCN	117,100	[4]
CO	−131,171	[2]
CO_2(g)	−394,375	[2]
CO_2(aq)	−385,980	[4]
CO_3^{2-}	−527,900	[2]
HCO_3^-	−586,850	[2]
H_2CO_3 (aq)	−623,170	[2]
CH_3OH(aq)	−175,200	[3]
HCO_2H(aq)	356,000	[3]
HCHO(aq)	−129,700	[3]
HCO_2(aq)	−334,700	[3]
HCOO-	−351,000	[4]
H_2CO_2	−372,300	[4]
Co^{2+}	−54,400	[2]
Co^{3+}	134,000	[2]
CoO	−214,194	[2]
$Co(OH)_2$	−456,100	[3]
$Co(OH)_3$	−596,700	[3]
Cr^{2+}	−176,155	[3]
$Cr^{3+}(Cr(6H_2O)^{3+})$	−215,490	[3]
CrO_4^{2-}	−736,800	[3]
$HCrO_4^-$	−773,700	[3]
Cr_2O_3	−1,047,000	[3]
$Cr_2O_7^{2-}$	−1,320,000	[3]
$Cr(OH)_2$	−587,900	[3]
$Cr(OH)_3$	−900,900	[3]
Cu^+	49,980	[4]
Cu^{2+}	65,520	[2]
$Cu_3(AsO_4)_2$	−1,299,600	[8]
$Cu(AsO_2)_2$	−701,000	[8]
$Cu(CN)_2^-$	257,800	[4]
CuCl	−119,860	[4]
$CuCl^+$	−68,200	[4]

(continued)

(continued)

Compound	ΔG_r° (J/mole)	References
$CuCl_3^{2-}$	−376,000	[4]
$CuCl_2$	−175,700	[4]
$CuCl_2^-$	−240,100	[4]
$CuCO_3$ (aq)	−501,700	[3]
Cu_2O	−146,030	[2]
CuO	−129,564	[2]
$Cu(OH)_2$	−356,900	[3]
$CuCO_3 \cdot Cu(OH)_2$	−893,600	[4]
$(CuCO_3)2Cu(OH)_2$	1,315,500	[4]
CuS	−49,080	[2]
Cu_2S	−86,868	[2]
$CuSO_4$	−662,310	[2]
Cu_2SO_4	−652,700	[3]
F^-	−276,500	[3]
$HF(aq)$	−294,300	[3]
Fe^{2+}	−78,870	[2]
Fe^{3+}	−4,600	[2]
$FeAsO_4 \cdot 2H_2O$	−1,263,520	[5]
$FeAsO_4$	−771,600	[6]
$FeCl_3$	−333,754	[2]
FeO	−251,156	[2]
$FeOOH(goethite)$	−488,550	[2]
$Fe(OH)^+$	−277,400	[4]
$Fe(OH)_2$	−486,500	[4]
$Fe(OH)_2^+$	−438,000	[4]
$Fe(OH)_3$	−714,000	[11]
Fe_3O_4	−1,012,566	[2]
Fe_2O_3	−742,683	[2]
FeS	−101,333	[2]
FeS_2 (pyrite)	−160,229	[2]
$FeSO_4$	−820,800	[4]
$Fe(SO_4)^+$	−772,700	[4]
$Fe(SO_4)_2^-$	−1,524,500	[4]
$Fe_2(SO_4)_3$	−2,249,555	[2]
$H+$	0	[2]
H_2 (aq)	17,600	[4]
H_2O (l)	−237,141	[2]
HO_2^- (aq)	−67,300	[4]
H_2O_2 (aq)	−134,030	[4]
Hg_2^{2+}	−153,600	[2]

(continued)

(continued)

Compound	ΔG_r° (J/mole)	References
Hg^{2+}	164,400	[2]
Hg_2CO_3	−442,700	[3]
$HgCl$	−105,415	[2]
$HgCl_2$	−185,800	[3]
Hg_2Cl_2	−210,700	[3]
$Hg(OH)_2$	−274,900	[3]
$HgSO_4$	−590,000	[3]
Hg_2SO_4	−623,900	[3]
HgS (cinnabar)	−50,645	[2]
K^+	−282,490	[2]
KCl	−408,554	[2]
KOH	−378,932	[2]
La^+	−292,620	[2]
$LiOH$	−438,941	[2]
Mg^{2+}	−454,800	[2]
$Mg_3(AsO_4)_2$	−2,773,000	[8]
$MgCO_3$	−1,029,480	[2]
MgO	−569,196	[2]
$Mg(OH)_2$	−833,506	[2]
$MgCl_2$	−591,785	[2]
Mn^{2+}	−228,000	[2]
Mn^{3+}	−82,000	[3]
$MnCO_3$	−816,700	[4]
MnO	−362,896	[2]
MnO_2	−465,140	[4]
$MnOOH$	−557,700	[3]
MnO_4^-	−447,200	[4]
$MnSO_2$	−957,326	[2]
MoO_2	−533,053	[2]
MoO_2	−668,055	[2]
NH_3 (gas)	−16,410	[2]
NH_3 (aq)	−26,600	[2]
NH_4^+	−79,457	[2]
NH_4NO_3	−183,803	[2]
NH_4OH	−263,800	[3]
HNO_3 (aq)	−26,650	[3]
NO_3^-	−111,500	[2]
NO_2	51,310	[1]
Na^+	−261,900	[2]
$NaCl$	−384,212	[2]

(continued)

(continued)

Compound	ΔG_r° (J/mole)	References
$NaCO_3^-$	−797,300	[3]
Na_2CO_3 (aq)	−1,052,000	[3]
$NaHCO_3$ (aq)	−847,600	[3]
NaOH	−379,651	[2]
Na_2S	−362,400	[3]
Na_2SiO_3	−1,427,000	[3]
Ni^{2+}	−45,600	[2]
$NiCO_3$	−615,100	[3]
NiO	−211,581	[2]
NiO_2	−198,740	[3]
$HNiO_2^-$	−349,200	[3]
$Ni(OH)_2$	−453,100	[4]
$NiSO_4$	−773,700	[4]
$O_2(g)$	0	[1]
O_2 (aq)	16,400	[5]
O_3 (aq)	174,100	[4]
O_3 (gas)	163,000	[4]
OH^-	−157,328	[2]
P_2O_5	−1,372,797	[2]
PO_4^{3-}	−1,019,000	[2]
HPO_4^{2-}	−1,094,000	[3]
$H_2PO_4^-$	−1,135,000	[3]
H_3PO_4	−1,147,000	[3]
Pb^{2+}	−24,400	[2]
PbO (litharge-red)	−189,202	[2]
PbO (massicot -yellow)	−188,573	[2]
$Pb(OH)_2$	−420,508	[4]
Pb_3O_4	−617,700	[3]
Pb_2O_3	−411,800	[3]
PbO_2	−215,314	[2]
$PbCO_3$	−626,400	[3]
PbS	−96,075	[2]
$PbSO_4$	−813,026	[2]
PbS_2O_3	−560,700	[3]
Pt^{2+}	−229,300	[3]
S^{2-}	85,800	[2]
H_2S (gas)	−33,543	[2]
H_2S (aq)	−27,830	[2]
HS^-	12,100	[2]
HSO_4^-	−755,910	[4]

(continued)

(continued)

Compound	ΔG_r^o (J/mole)	References
H_2SO_4 (l)	−689,995	[2]
H_2SO_4(aq)	−744,530	[1]
SO_2 (gas)	−300,194	[4]
SO_2 (aq)	−300,676	[4]
$S_2O_3^{2-}$ (thiosulfate)	−522,500	[4]
$HS_2O_3^-$	−541,900	[3]
$H_2S_2O_3$	−543,500	[3]
SO_3^{2-} (sulfite)	−486,600	[2]
HSO_3^-	−527,300	[3]
SO_4^{2-} (sulfate)	−744,630	[2]
SbO^+	−175,700	[3]
Sb_2S_3	−173,470	[2]
Se^{-2}	129,000	[2]
SeO_2	−173,600	[3]
SeO_3^{2-}	−373,800	[3]
SeO_4^{2-}	−441,100	[3]
$HSeO_3^-$	−411,300	[3]
$HSeO_4^-$	−452,700	[3]
H_2ScO_3	−425,900	[3]
H_2SeO_4	−441,100	[3]
SiO_2 (quartz)	−856,288	[2]
$H_3SiO_4^-$	−1,200,000	[3]
H_4SiO_4 (aq)	−1,308,000	[2]
Sn^{2+}	−26,200	[3]
Sn^{4+}	−2720	[3]
SnO	−257,300	[3]
SnO_2	−519,902	[2]
$Sn(OH)^+$	−253,600	[3]
SnS	−104,698	[2]
Sr^{2+}	−559,440	[2]
Ti^{2+}	−314,200	[3]
Ti^{3+}	−349,800	[3]
TiO	−513,312	[2]
TiO_2 (anatase)	−883,303	[2]
TiO_2 (rutile)	−889,446	[2]
$Ti(OH)_3$	−1,050,000	[3]
U^{3+}	−520,500	[2]
U^{4+}	−579,100	[2]
UCl_3	−823,820	[2]
UCl_4	−1,018,390	[2]

(continued)

(continued)

Compound	ΔG_r° (J/mole)	References
UO_2	−1,031,770	[2]
UO_3	−1,146,461	[2]
V^{2+}	226,800	[3]
V^{3+}	−251,400	[3]
VO^{+2}	−456,100	[3]
VO_2^+	−596,700	[3]
VO	−404,219	[2]
V_2O_3	−1,139,052	[2]
V_2O_4	1,318,457	[2]
V_2O_5	−1,419,435	[2]
$V(OH)_3$	−912,200	[3]
WO_2	−533,858	[2]
WO_3	−764,062	[2]
WO_4^{2-}	−920,500	[3]
WS_2	−297,945	[2]
Zn^{2+}	−147,260	[2]
$Zn_3(AsO_4)_2$	−1,903,000	[8]
$Zn(AsO_2)_2$	−917,900	[8]
$ZnCO_3$	−746,500	[3]
ZnO	−320,477	[2]
ZnO_2^{2-}	−388,870	[3]
$Zn(OH)^+$	−330,100	[4]
$Zn(OH)_2$ (γ)	−553,810	[4]
$Zn(NH_3)_4^{2+}$	−301,900	[4]
$Zn(NH_3)_2^{2+}$	−225,000	[4]
ZnS (sphalerite)	−202,496	[2]
ZnS (wurtzite)	−190,220	[2]
$ZnSO_4$	−871,530	[2]
Zr^{-4+}	−594,000	[3]
ZrO_2	−1,036,400	[3]

References

(1) P.W. Atkins, *Physical Chemistry*, 3rd edn. (W.H. Freeman and Company, New York, 1986)
(2) R.A. Robie, B.S. Hemingway, J.R. Fisher, Thermodynamic Properties of Minerals and Related Substances at 298.15 K and 1 Bar (10^5 Pascals) Pressure and at Higher Temperatures. (Geological Survey Bulletin 1452, U. S. Department of the Interior, Washington, D.C.)

(3) R.M. Garrels, C.L. Christ, *Solutions, Minerals, and Equilibria* (Jones and Bartlett Publishers, Boston, 1990)

(4) D.D. Wagman, W.H. Evans, V.B. Parker, R.H. Schumm, I. Halow, S.M. Bailey, K.L. Churney, R.L. Nuttall, The NBS tables of chemical thermodynamic properties. J. Phys. Chem. Reference Data, **11**(#2), (National Bureau of Standards, Washington, D.C.)

(5) S. Therdkiattikul, *Treatment Methods, Safe Contaminants Precipitation Disposal and Water Reclamation, and Recycling of Effluent from Bacterial Leaching of Precious Metal Refractory Ores*, Ph.D. Dissertation, University of Utah, 1996

(6) N. Papassiopi, M. Stefanakis, A. Kontopoulos, Removal of arsenic from solutions by precipitation as ferric arsenates, in *Arsenic Metallurgy Fundamentals and Applications* ed. by R.G. Reddy, J.L. Hendrixs, P.B. Queneau, (TMS, Warrendale, 1988)

(7) C.A. Fleming, Hydrometallurgy of precious metals recovery, in *Hydrometallurgy, Theory and Practice*, vol. B, ed. by W.C. Cooper, D.B. Dreisinger (Elsevier, Amsterdam, 1992)

(8) R.G. Robins, Arsenic hydrometallurgy, in *Arsenic Metallurgy Fundamentals and Applications*, ed. by R.G. Reddy, J.L. Hendrix, P.B. Queneau (TMS, Warrendale, 1988), pp. 215–247

(9) R.W. Boyle, The geochemistry of gold and its deposits, *Canadian Geological Bulletin* 280, (Minister of Supply and Services, Canada, 1979), p. 10

(10) K.P. Anathapadmanabhan, P. Somasundaran, Role of dissolved mineral species in calcite-apatite flotation. Miner. Metall. Process. **5**, 36–42 (1984)

(11) D.C. Silverman, Corrosion **38**(8), 453–455 (1982)

Appendix E
Laboratory Calculations

Background Information

Measurement accuracy is a function of the instrument accuracy. With solution vessels it is important to note the accuracy of the volume mark(s). Volumetric flasks have much greater accuracy than beakers because they are generally calibrated *to contain* (TC) the stated volume within a narrow tolerance at a specified temperature that is usually 20 °C (changing the temperature will change the volume). The tolerance level is usually stated on the flask so that the user knows the accuracy. Most new flasks have a class designation for tolerance with class A being the most accurate. Most class A flasks are accurate to approximately 1 part in 1000. Thus, regardless of how accurate a weight balance is, the most accurate solutions are usually only accurate to within 1 part in 1000. Most class B flasks are accurate to within 2 parts per 1000. Similar tolerances are found for pipettes. However, it should be noted that most pipettes are calibrated *to dispense* (TD) the specified volume. The dispensable volume is different than the containable volume due to the residual film of liquid that does not flow out of the container during a reasonable dispensing time. An easy check on the accuracy of a given vessel can be performed by dispensing the pure water into a preweighed container that is subsequently reweighed on an accurate balance. The resulting weight of water should then be compared with the stated volume using the appropriate conversion (1 ml of pure water at 22 °C = 0.9978 g). All good flasks and reusable pipettes have the letters TD or TC on them so that the user knows if the specified volume is based on dispensing or containing. As a point of caution, never use beakers for accurate volume determinations since their volume markings are generally inaccurate.

Solution Preparation Principles

Safety is the most important aspect of solution preparation. Those preparing solutions must ensure that the desired solutions are safe to prepare and that all recommended safety equipment and procedures are utilized appropriately.

© The Minerals, Metals & Materials Society 2022
M. L. Free, *Hydrometallurgy*, The Minerals, Metals & Materials Series,
https://doi.org/10.1007/978-3-030-88087-3

Accurate solution preparation depends upon the accuracy and cleanliness of the equipment and the chemicals used during the preparation. For accurate solutions with very low concentrations, the glassware must be exceptionally clean and often requires acid cleaning in sulfuric, nitric, or hydrochloric acid solution followed by rinsing and cleaning in strong base and further thorough rinsing in purified water depending upon the solutions used in the glassware. The water used in dilute solution preparation must also be of very high purity. All contents of the solutions need to be clearly labeled on their containers.

Accurate solution preparation also requires high purity chemicals. Many chemicals that come in high purity forms are not the exact compounds that are desired. For example when ferrous sulfate is purchased it often contains seven waters of hydration, which must be considered in solution preparation calculations. Some high purity compounds such as acids are sold in a diluted form due to difficulties in concentrating the acid to a pure acid form. Other compounds absorb water or react with atmospheric gases such as oxygen and carbon dioxide to form compounds other than what is stated on the label—especially if the compound is stored for long periods of time before use. When using high purity solids that have a tendency to hydrate or react with air, it is important verify the purity using accurate analytical instruments such as ICP and AA spectroscopy that are calibrated using certified standard solutions.

Solution Preparation Calculations

The key to making accurate calculations in a laboratory setting is to keep track of what is needed and what is available to produce it, making sure the units are properly used. It is useful to make the necessary unit conversions based on relative equalities that are rearranged to give a value of unity or one. As examples:

$$1 \text{ mol NaCl} = 58.43 \text{ g NaCl} \Rightarrow \quad \frac{1 \text{ mol NaCl}}{58.43 \text{ g NaCl}} = 1$$

$$1 \text{ m} = \frac{1 \text{ mol}}{1000 \text{ g solvent}} \Rightarrow \quad \frac{1 \text{ m}}{\frac{1 \text{ mol}}{1000 \text{ g solvent}}} = 1$$

$$1 \text{ mol Fe}_3O_4 = 3 \text{ mol Fe} \Rightarrow \quad \frac{1 \text{ mol Fe}_3O_4}{3 \text{ mol Fe}} = 1$$

$$1 \text{ g HCl solution} = 0.37 \text{ g HCl} \Rightarrow \quad \frac{1 \text{ g HCl solution}}{0.37 \text{ g HCl}} = 1$$

$$1 \text{ ml H}_2O = 1 \text{ g H}_2O \Rightarrow \quad \frac{1 \text{ ml H}_2O}{1 \text{ g H}_2O} = 1$$

$$1 \text{ liter} = 1000 \text{ ml} \Rightarrow \quad \frac{1 \text{ liter}}{1000 \text{ ml}} = 1$$

When using an HCl solution that is 37% pure, a relative equality can be written as:

The overall calculation is then performed by setting up the calculation in the form of an equation, then rearranging and solving the equation with the proper units using the equalities as well as what is available. The equalities simply convert the units.

Example E.1. Prepare a 0.1 molal aqueous solution of sodium using available impure sodium chloride salt (96% pure) in a 100 ml volumetric flask.

Rewriting the problem in terms of an equation leads to:

$$0.1 \text{ m Na} = \frac{X[\text{gNaCl(impure)}]}{100 \text{ ml}}$$

or

$$X[\text{g NaCl(impure)}] = (0.1 \text{ m Na})(100 \text{ ml})(\text{equalities.} \ldots)$$
$$= (0.1 \text{ m Na})(100 \text{ mlH}_2\text{O})$$

$$\frac{1 \text{ mol Na}}{1000 \text{ g H}_2\text{O}} \frac{1 \text{ mol NaCl}}{1 \text{ mol Na}} \frac{58.43 \text{ g NaCl}}{1 \text{ mol NaCl}}$$
$$\frac{1 \text{ g NaCl(impure)}}{0.96 \text{ g NaCl}} \frac{1 \text{ g H}_2\text{O}}{1 \text{ ml H}_2\text{O}}$$

so $X[\text{gNaCl(impure)}] = 0.6086$,

Note that all of the units except gNaCl(impure) cancel out. Therefore, by adding 0.6086 g of the available impure NaCl to 100 ml of water will result in the desired 0.1 m solution of sodium.

Example E.2. Calculate the volume (in ml) of certified standard copper solution that will be needed to give a final solution copper concentration of 0.0001 molal in a 100 ml flask.

$$\frac{X \text{ ml Custd}}{100 \text{ ml H}_2\text{O}} = 0.0001 \text{ m Cu}$$

$$X \text{ ml Custd} = (0.0001 \text{ m Cu})(100 \text{ ml H}_2\text{O})(\text{equalities...})$$
$$= \frac{(0.0001 \text{ m Cu})(100 \text{ mlH}_2\text{O}) \frac{1 \text{ mol Cu}}{1000 \text{ g H}_2\text{O}} \frac{63.55 \text{ g Cu}}{1 \text{ m Cu}} \frac{1 \text{ ml Custd}}{1 \text{ mol Cu}} \frac{}{1000 \text{ µg Cu}}}{}$$

$$\frac{1 \text{ µg Cu}}{10^{-6} \text{g Cu}} \frac{1 \text{ g H}_2\text{O}}{1 \text{ ml H}_2\text{O}}$$

$X \text{ ml Custd} = 0.6355.$

Thus, 0.6355 ml of the copper standard solution would be needed to make a 0.0001 molal solution in a 100 ml flask. Obviously, a more concentrated solution would need to be made followed by another dilution in order to make an accurate solution since 0.6355 ml of the standard cannot be measured with high accuracy.

Example E.3. Calculate the approximate weight of concentrated sulfuric acid (96% purity) needed to adjust 10 L of water to pH 3.5 from pH 7.0, assuming that the acid completely dissociates into H^+ and SO_4^{2-} ions, no other reactions are occurring, and the activity coefficients are equal to one. (pH $= -\log(aH +$) or for unit activity coefficients, molality $= 10^{-pH}$) (note that $10^{-3.5} - 10^{-7.0} \approx 10^{-3.5}$)

$$\frac{X \text{ g } H_2SO_4 \text{so ln}}{10 \text{ l } H_2O} = 10^{-3.5} \text{m } H^+$$

$$X \text{ g } H_2SO_4 \text{so ln} = (10^{-3.5} \text{m } H^+)(10 \text{ l } H_2O)(\text{equalities}\ldots)$$

$$(10^{-3.5} \text{m } H^+)(10 \text{ l } H_2O)\frac{1 \text{ mol } H^+}{1000 \text{ g } H_2O}\frac{1 \text{ mol } H_2SO_4}{2 \text{ mol } H^+}\frac{98.07 \text{ g } H_2SO_4}{1 \text{ mol } H_2SO_4}$$
$$\frac{1 \text{ g } H_2SO_4 \text{so ln}}{0.96 \text{ g } H_2SO_4}\frac{1000 \text{ g } H_2O}{1 \text{ l } H_2O}$$

$$X \text{ g } H_2SO_4\text{soln} = 0.1615$$

Therefore, 0.1615 g of the concentrated sulfuric acid should be added to lower the pH from 7 to 3.5. Because this value is very low, it would be necessary to make a diluted acid stock solution, then use the stock solution to make the pH adjustment.

Problem

E.1. Determine how many milliliters of the concentrated nitric acid will be needed prepare 250 ml of a 1×10^{-3} M HNO_3 solution using concentrated nitric acid (69% HNO_3, specific gravity of 1.42).

Appendix F
Selected Ionic Species Data

Specie	ΔH_{298}	S°_{298}	Partial molal cp (25–60 °C)	Partial molal cp (25 200 °C)
	(J/mole)	(J/mole · K)	(J/mole · K)	(J/mole · K)
Ag^+	105,750	73.38	125	180
Al^{3+}	−531,000	−308	301	451
Ba^{2+}	−537,640	9.60	159	230
Be^{2+}	−383,000	−130	259	385
Ca^{2+}	−542,830	−53.1	188	276
Cd^{2+}	−75,900	−73.2	188	280
Cl^-	−167,080	56.73	−213	−276
CO_3^{2-}	−677,140	−56.90	−552	−631
Cu^{2+}	64,770	99.6	205	301
F^-	−335,350	−13.18	−196	−272
Fe^{3+}	−48,500	−316.0	293	439
H^+	0	0	96	146
HCO_3^-	−691,990	91.2	−113	146
K^+	−252,170	101.04	113	163
Li^+	−278,455	11.30	150	217
Mg^{2+}	−466,850	−138	213	314
Mn^{2+}	−220,700	−73.6	196	293
Na^+	−240,300	58.41	146	188
NO_3^-	−207,400	146.94	−205	−238
OH^-	−230,025	−10.71	−196	−272
Pb^{2+}	−1700	10.0	155	226
SO_4^{2-}	−909,270	20.0	−414	−477
Sr^{2+}	−545,800	−33.0	180	263

© The Minerals, Metals & Materials Society 2022
M. L. Free, *Hydrometallurgy*, The Minerals, Metals & Materials Series,
https://doi.org/10.1007/978-3-030-88087-3

Data from C. M. Criss, J. W. Cobble, "The Thermodynamic Properties of High Temperature Aqueous Solutions. IV. Entropies of the Ions up to 200 °C and the Correspondence Principle" J. Phys. Chem., 86, 5385; and R. A. Robie, B. S. Hemingway, and J. R. Fisher, "Thermodynamic Properties of Minerals and Related Substances at 298.15 K and 1 Bar (10^5 Pascals) Pressure and at Higher Temperatures," Geological Survey Bulletin 1452, U. S. Department of the Interior, Washington, D. C.

Appendix G
Standard Half-Cell Potentials

(for all reactions involving hydrogen assume pH = 0).

Reaction	E°(V)
$F_2 + 2e^- \leftrightarrow 2F^-$	+2.65
$O_3 + 2H^+ + 2e^- \leftrightarrow O_2 + H_2O$	+2.07
$Co^{3+} + e^- \leftrightarrow Co^{2+}$	+1.84
$H_2O_2 + 2H^+ + 2e^- \leftrightarrow 2H_2O$	+1.776
$OCl^- + 2H^+ + 2e^- \leftrightarrow H_2O + Cl^-$	+1.714
$MnO_4^- + 8H^+ + 5e^- \leftrightarrow Mn^{2+} + 4H_2O$	+1.52
$Au^{+3} + 3e^- \leftrightarrow Au$	+1.498
$ClO_3^- + 6H^+ + e^- \leftrightarrow Cl^- + 3H_2O$	+1.45
$Cl_2 + 2e^- \leftrightarrow 2Cl^-$	+1.358
$Pt^{+3} + 3e^- \leftrightarrow Pt$	+1.358
$ClO_4^- + 2H^+ + 2e^- \leftrightarrow H_2O + ClO_3^-$	+1.186
$Br_2 + 2e^- \leftrightarrow 2Br^-$	+1.065
$HNO_2 + H^+ + e^- \leftrightarrow H_2O + NO$	+0.99
$NO_3^- + 3H^+ + 2e^- \leftrightarrow H_2O + HNO_2$	+0.94
$NO_3^- + 2H^+ + e^- \leftrightarrow NO_2 + H_2O$	+0.81
$Ag^+ + e^- \leftrightarrow Ag$	+0.799
$Hg_2^{+2} + 2e^- \leftrightarrow 2Hg$	+0.788
$Fe^{+3} + e^- \leftrightarrow Fe^{+2}$	+0.770
$O_2 + 2H^+ + 2e^- \leftrightarrow H_2O_2$	+0.680
$Cu^{+2} + 2e^- \leftrightarrow Cu$	+0.337
$Ge^{2+} + 2e^- \leftrightarrow Ge$	+0.230
$Cu^{+2} + e^- \leftrightarrow Cu^+$	+0.153
$2H^+ + 2e^- \leftrightarrow H_2$	0.000

(continued)

© The Minerals, Metals & Materials Society 2022
M. L. Free, *Hydrometallurgy*, The Minerals, Metals & Materials Series,
https://doi.org/10.1007/978-3-030-88087-3

(continued)

Reaction	$E^o(V)$
$Pb^{+2} + 2e^- \leftrightarrow Pb$	−0.126
$Sn^{+2} + 2e^- \leftrightarrow Sn$	−0.136
$Ni^{+2} + 2e^- \leftrightarrow Ni$	−0.250
$Co^{+2} + 2e^- \leftrightarrow Co$	−0.277
$Cd^{+2} + 2\ e^- \leftrightarrow Cd$	−0.403
$Fe^{+2} + 2e^- \leftrightarrow Fe$	−0.410
$Ga^{+3} + 3e^- \leftrightarrow Ga$	−0.56
$Cr^{+3} + 3e^- \leftrightarrow Cr$	−0.744
$Zn^{+2} + 2e^- \leftrightarrow Zn$	−0.763
$Mn^{+2} + 2e^- \leftrightarrow Mn$	−1.18
$V^{+2} + 2e^- \leftrightarrow V$	−1.18
$Al^{+3} + 3e^- \leftrightarrow Al$	−1.662
$Ti^{+3} + 3e^- \leftrightarrow Ti$	−1.80
$U^{+3} + 3e^- \leftrightarrow U$	−1.80
$Be^{+2} + 2e^- \leftrightarrow Be$	−1.85
$Mg^{+2} + 2e^- \leftrightarrow Mg$	−2.363
$Ce^{+3} + 3e^- \leftrightarrow Ce$	−2.48
$La^{+3} + 3e^- \leftrightarrow La$	−2.52
$Na^+ + e^- \leftrightarrow Na$	−2.712
$Ca^{+2} + 2e^- \leftrightarrow Ca$	−2.87
$Sr^{+2} + 2e^- \leftrightarrow Sr$	−2.89
$Ba^{+2} + 2e^- \leftrightarrow Ba$	−2.90
$K^+ + e^- \leftrightarrow K$	−2.92
$Li^+ + e^- \leftrightarrow Li$	−3.05
$Sm^{+2} + 2e^- \leftrightarrow Sm$	−3.12

(Data obtained from References 2.2–2.5, 2.8–2.10)

Appendix H
General Terminology

Acid baking	Process in which acid is mixed with ore and heated to elevated temperatures above 100 °C
Agglomeration	In leaching operations, this is the process of immobilizing fine material into clusters of fine particles using water and binders such as cement.
Anode	The positively-charged electrode at which oxidation occurs.
Anolyte	Electrolyte solution surrounding an anode that is isolated from the solution surrounding the cathode by a membrane or porous barrier.
Breakthrough capacity	The capacity of an absorbent to absorb a given species in a vessel or column without allowing leakage of the entity above a threshold level into the effluent.
Cathode	The negatively-charged electrode at which reduction, and often metal plating, occur.
Catholyte	Electrolyte solution surrounding a cathode that is isolated from the solution surrounding the cathode by a membrane or porous barrier.
Chelating agent	A chemical agent that is capable of multiple attachments to a central metal atom in a complexed form.
Comminution	Comminution is the process of particle size reduction.
Crud	A mixture of organic, aqueous, and solid matter that can become a problem during solvent extraction processing.
Diluent	A compound used for dilution. Most often it is used to describe an organic solvent that is used as a carrier for a solvent extractant.

Effluent	Solution leaving a given process.
Eh	The electrochemical potential or oxidation reduction potential of a solution relative to the standard hydrogen half-cell. The term is often inappropriately used to represent a measured potential against a variety of reference cell potentials without conversion to the hydrogen cell reference, which represents the same potential used in free energy-based thermodynamic calculations.
Electrolyte	A medium containing dissolved ions through which charge can be transported. Most often used in association with solution used for electrochemical purposes such as electrorefining, electrowinning or electroplating.
Eluant	Solution leaving the process of elution.
Elution	Process of extraction that is most often associated with stripping of ion exchange resin.
Elutriation	Process of separating solids by means of a liquid. Also, used to describe washing of solids.
Emulsion	A dispersion of tiny immiscible droplets of one liquid in another liquid.
Extractant	An ion or compound capable of dissolving otherwise insoluble matter through chemical interaction or association.
Extraction coefficient	The ratio of metal in the extractant phase relative to that in the aqueous phase in solvent extraction or ion exchange processes.
Hydrometallurgy	The study of metal chemistry in an aqueous or water-bearing environment.
Inversion (phase)	Transition from one continuous phase to another in processes involving two immiscible liquids such as solvent extraction.
Leaching	The process of solubilizing, extracting or removing a metal or mineral compound by means of a solution.
Liquor	Process liquid (generally used in association with solution that contains a desired entity).
Ligand	Species such as chloride that complexes with a desired species such as a metal ion.
Lixiviant	An ion or compound capable of dissolving otherwise insoluble matter through chemical interaction or association.
Modifier	A chemical additive used to modify the action of another species. Most often associated with additives used to enhance solvent extraction effectiveness.

ORP	Oxidation-reduction potential, which is the same as Eh if a standard hydrogen cell is the reference or if the data is converted to it. (see Eh)
Oxidant	An ion or compound capable of increasing the oxidation state of another entity by means of an electrochemical reaction.
pH	$\log(a_{H+})$, where a_{H+} is the hydrogen ion activity.
pK -	$\log(K)$, where K is the equilibrium constant.
pK_a -	$\log(K_a)$, where K_a is the equilibrium constant for acidification equilibria when the activities of all species is unity except for the hydrogen ion. Thus, pK_a represents the equilibrium pH for reactions involving hydrogen ion association/disassociation.
pH_{50}	pH at which an ion is equally partitioned between aqueous and organic phases during solvent extraction, or the pH at which the extraction coefficient is equal to one.
Pregnant solution	A solution that contains the desired entity. In other words, it is a loaded solution that is ready to release its value.
Raffinate	Solution from which a valuable entity has been removed. Most commonly used to describe the aqueous solution that leaves a solvent extraction loading operation containing a low level of the desired entity.
Reductant	An ion or compound capable of reducing the oxidation state of another entity by means of a chemical reaction.
Refractory	In gold extraction, refractory indicates the ore's resistance to traditional leaching. Ores that are not amenable to traditional extraction are referred to as refractory ores. The ore's resistance to traditional extraction is often associated with sub-micron size gold particles that are dispersed in a sulfide matrix, making traditional leaching ineffective without pretreatment of the ore by some form of oxidation, which makes these gold particles accessible to the lixiviant.
Scrubbing	Used to describe the process of selectively removing contamination from a chemical process such as solvent extraction.
Slurry	A medium consisting of liquid and suspended particles.
Solute	A dissolved entity.
TCLP	Stands for toxicity characteristic leaching procedure. This procedure is designed to extract potentially toxic ions from solid waste in a simulated landfill environment to determine if the material is a potential hazardous waste.

Appendix I
Common Sieve Sizes

MESH[a]	Aperture size (mm)
4	4.75
8	2.36
20	0.850
60	0.250
80	0.180
100	0.150
170	0.090
200	0.075
270	0.053
325	0.045
400	0.038

[a] TYLER, USA (ASTM-E-11-70), and Canadian Standard Sieve Series (8-GP-1d)

Appendix J
Metals and Minerals

Metal element	Mineral sources	Chemical formulas
Aluminum (Al)	Bauxite[a] ore consists of:	
	Diaspore	AlO(OH),
	Gibbsite	Al(OH)$_3$,
	Boehmite	AlO(OH)
	Alumina	Al$_2$O$_3$
Antimony (Sb)	Stibnite[a]	Sb$_2$S$_3$
Arsenic (As)	Arsenopyrite	FeAsS
	Realgar	AsS
	Orpiment	As$_2$S$_3$
Beryllium (Be)	Beryl[a]	Be$_3$Al$_2$Si$_6$O$_{18}$
	Bertrandite	Be$_3$Al$_2$(SiO$_3$)$_6$
Bismuth (Bi)	Bismuth metal	Bi
Cadmium (Cd)	Greenockite	CdS
Cesium (Cs)	Pollucite[a]	Cs$_4$Al$_4$Si$_9$O$_{26}$ · H$_2$O
	Lepidolite	K(Li,Al)$_3$(Si,Al)$_4$O$_{10}$(OH,F)$_2$
Chromium (Cr)	Chromite[a]	FeCr$_2$O$_4$
Cobalt (Co)	Smaltite	CoAs$_2$
	Cobaltite	CoAsS
	Carrolite	CuCo$_2$S$_4$
	Linnaeite	CoS$_4$

(continued)

© The Minerals, Metals & Materials Society 2022
M. L. Free, *Hydrometallurgy*, The Minerals, Metals & Materials Series,
https://doi.org/10.1007/978-3-030-88087-3

(continued)

Metal element	Mineral sources	Chemical formulas
Copper (Cu)	Chalcopyrite[a]	$CuFeS_2$
	Chalcocite[a]	Cu_2S
	Bornite[a]	Cu_5FeS_4
	Covellite[a]	CuS
	Cuprite[a]	Cu_2O
	Malachite[a]	$CuCO_3 \cdot Cu(OH)_2$
	Native metal	Cu
	Tennantite	$Cu_8As_2S_7$ (varying stoich
	Tetrahedrite	$4Cu_2S \cdot Sb_2S_3$
	Azurite	$2CuCO_3 \cdot Cu(OH)_2$
	Enargite	$Cu_3As_5S_4$
Gallium (Ga)	No important minerals, by-product	
Germanium (Ge)	Argyrodite	$3Ag_2S \cdot GeS_2$
Gold (Au)	Native metal[a]	Au
	Sylvanite	$(AuAg)Te_2$
	Calaverite	$AuTe_2$
Hafnium (Ha)	No important minerals, co-product with Zr	
Indium (I)	No important minerals, by-product	
Iron (Fe)	Hematite[a]	Fe_2O_3
	Magnetite[a]	Fe_3O_4
	Goethite	$Fe_2O_3 \cdot H_2O$
	Limonite	hydrous ferric oxides
	Siderite	$FeCO_3$
	Pyrrhotite	FeS (variable stoichiom.)
	Pyrite	FeS_2
Lead (Pb)	Galena[a]	PbS
	Cerussite	$PbCO_3$
	Anglesite	$PbSO_4$
	Jamesonite	$Pb_4FeSb_6S_{14}$
Lithium (Li)	Spodumene[a]	$LiAlSi_2O_6$
	Amblygonite	$2LiF \cdot Al_2O_3 \cdot P_2O_5$
	Lepidolite	$LiF \cdot KF \cdot Al_2O_3 \cdot 3SiO$
Magnesium (Mg)	Dolomite	$MgCa(CO_3)_2$
	Magnesite	$MgCO_3$
	Carnallite	$KMgCl_3 \cdot 6H_2O$
	Brucite	$Mg(OH)_2$
Manganese (Mn)	Pyrolusite[a]	MnO_2
	Manganite	Mn_2O_3
	Braunite	$3Mn_2O_3 \cdot MnSiO_3$
	Psilomelane	oxide mixture

(continued)

(continued)

Metal element	Mineral sources	Chemical formulas
Mercury (Hg)	Cinnabar[a]	HgS
Molybdenum (Mo)	Molybdenite	MoS_2
	Wulfenite	$PbMoO_4$
Nickel (Ni)	Pentlandite[a]	$(FeNi)S$
	Garnierite[a]	hydrated Ni-Mg silicate
	Niccolite	$NiAs$
	Millerite	NiS
Niobium (Nb)	Pyrochlore[a]	$(Ca,Na)_2(Nb,Ta)_2O_6(O,OH,F)$
	Columbite[a]	$(Fe,Mn)(Nb,Ta)_2O_6$
Platinum group metals (Ru, Rh, Pd, Os, Ir, Pt)		
Platinum (Pt)	Native metal[a]	Pt
	Sperrylite[a]	$PtAs_2$
Osmium (Os)	Os-Ir alloy	$Os-Ir$
Rare earth metals (Ce, Pr, Nd, Sm, Eu, Gd, Tb, Dy, Ho, Er, Tm, Yb, Lu)		
Cerium (Ce)	Bastnaesite[a]	$(Ce,La)(CO_3)F$
Lanthanum (La)	Bastnaesite[a]	$(Ce,La)(CO_3)F$
Yittrium (Y)	Xenotime[a]	YPO_4
Thorium (Th)	Monazite[a]	$(Ce,La,Th)PO_4$
	Thorianite	$ThO_2U_3O_8$
Radium (Ra)	By-product	
Rhenium (Re)	No important minerals, by-product from MoS_2	
Rubidium (Rb)	No important minerals, by-product	
Silicon (Si)	Quartz[a]	SiO_2
Selenium (Se)	Naumanite	Ag_2Se
	Clausthalite	$PbSe$
	Eurcairite	$(AgCu)_2Se$
	Berzelianite	Cu_2Se
Silver (Ag)	Argentite[a]	Ag_2S
Tantalum (Ta)	Pyrochlore[a]	$(Ca,Na)_2(Nb,Ta)_2O_6(O,OH,F)$
	Columbite[a]	$(Fe,Mn)(Nb,Ta)_2O_6$
Tellurium (Te)	Sylvanite	$(AuAg)Te_2$
	Calavernite	$AuTe_2$
Thallium (Tl)	No important minerals, by-product	
Thorium (Th)	See rare earths	
Tin (Sn)	Cassiterite[a]	SnO_2
Titanium (Ti)	Ilmenite[a]	$FeTiO_3$
	Rutile	TiO_2
	Anatase	TiO_2
Tungsten (W)	Wolfram[a]	$(Fe,Mn)WO_4$
	Scheelite[a]	$CaWO_4$

(continued)

(continued)

Metal element	Mineral sources	Chemical formulas
Uranium (U)	Pitchblende[a]	UO_2
	Uranite[a]	U_3O_8
	Carnotite	$K_2(UO_2)_2(VO_4)_2 \cdot 3H_2O$
	Autunite	$Ca(UO_2)_2(PO_4)_2 10\text{-}12H_2O$
	Tobernite	$Cu(UO_2)_2(PO_4)_2 \cdot 12H_2O$
Vanadium (V)	Patronite[a]	VS_4
	Carnotite[a]	$K_2(UO_2)_2(VO_4)_2 \cdot 3H_2O$
	Roscoelite	$H_8K(MgFe)(AlV)_4(SiO_3)_{12}$
	Vanadinite	$(PbCl)V_4(PO_4)_3$
Zinc (Zn)	Sphalerite[a]	ZnS
	Smithsonite	$ZnCO_3$
	Marmatite	$(Zn,Fe)S$
	Zincite	ZnO
	Willemite	Zn_2SiO_4
Zirconium (Zr)	Zircon	$ZrSiO_4$

[a] Indicates major commercial mineral source primary information source: *Mineral Processing Technology*, 3rd edition, B. A. Wills, Pergamon Press, 1985

Index

Printed in the United States
by Baker & Taylor Publisher Services